Biodegradable Polymers

Studying (bio)degradable polymers value chain can help one understand the importance of these to the environment and human health. This book provides an overview of the biodegradable polymer along the value chain, identifies and analyses existing practices for biodegradable plastics and assesses the relevant legal, regulatory, economic and practical reasons for the importance of proper use and proper recycling of biodegradable plastics. It covers related materials development, environmental impacts, their synthesis by traditional and biotechnological routes, policy and certification, manufacturing processes, (bio)degradable polymer properties and so forth.

Features:

- Gives a clear idea of the present state of the art and future trends in the research of the biodegradable polymers in the context of circular economy.
- Describes the entire value chain and life cycle of bioplastics, considering different types of polymers.
- Clarifies the life safety of (bio)degradable polymeric materials.
- Presents novel opportunities and ideas for developing or improving technologies.
- Determines the course of degradation during prediction study.

This book is aimed at researchers, graduate students and professionals in the polymer processing industry (petrochemical polymer industry, industry producing bio-based and (bio)degradable polymers), food packaging industry, industry involved in waste management, pharma industry, chemical engineering, product engineering and biotechnology.

Biodegradable Polymers

Value Chain in the Circular Economy

Edited by Joanna Rydz

CRC Press
Taylor & Francis Group
Boca Raton London New York

CRC Press is an imprint of the
Taylor & Francis Group, an **informa** business

Designed cover image: Maciej Rydz

First edition published 2024
by CRC Press
6000 Broken Sound Parkway NW, Suite 300, Boca Raton, FL 33487–2742

and by CRC Press
4 Park Square, Milton Park, Abingdon, Oxon, OX14 4RN

CRC Press is an imprint of Taylor & Francis Group, LLC

© 2024 Taylor & Francis Group, LLC

ISBN: 978-0-367-37067-1 (hbk)
ISBN: 978-0-367-37136-4 (pbk)
ISBN: 978-0-429-35279-9 (ebk)

DOI: 10.1201/9780429352799

Typeset in Times
by Apex CoVantage, LLC

I dedicate this book to Benjamin, to Uncle Jędrek and to the memory of my dad.

Joanna Rydz, Gliwice-Wylatkowo 2022

Contents

About the Editor

Joanna Rydz obtained a doctoral degree at the Jagiellonian University in 2005. She has been working at the Centre of Polymer and Carbon Materials, Polish Academy of Sciences since 1995, currently as an adjunct. She is a laureate of several scholarships and researcher positions: the Marie Curie Fellowship at Vienna University of Technology; Scholarship of Slovak Republic for research stay in Polymer Institute SAS in Bratislava; a position at Institute of Polymers BAS in Sofia. The area of her competence and expertise concerns environmentally friendly (bio)degradable polymers produced from renewable resources, which are currently among the most promising polymers for sustainable development with possible use in agriculture, medicine and particularly in the packaging technologies.

Contributors

Khadar Duale
Centre of Polymer and Carbon Materials,
    Polish Academy of Sciences
    M. Curie-Skłodowska 34, Zabrze,
    Poland

Marta Musioł
Centre of Polymer and Carbon
    Materials, Polish Academy of Sciences
    M. Curie-Skłodowska 34, Zabrze,
    Poland

Joanna Rydz
Centre of Polymer and Carbon Materials,
    Polish Academy of Sciences,
    M. Curie-Skłodowska 34, Zabrze,
    Poland

Wanda Sikorska
Centre of Polymer and Carbon Materials,
    Polish Academy of Sciences
    M. Curie-Skłodowska 34, Zabrze,
    Poland

Barbara Zawidlak-Węgrzyńska
Department of Chemistry, Faculty
    of Medicine in Zabrze Academy
    of Silesia in Katowice, Katowice,
    Poland

Preface

The idea of a circular economy is becoming an integral part of policy and programs at all levels of management or education by governments, non-government organisations, enterprises or institutions. The closed-loop economy implements the concept of sustainable development to reduce the negative impact of global production and consumption while ensuring economic growth and environmental sustainability. Progress in its implementation in industrial sectors, however, requires clarifying the concept from the perspective of balancing aspects covering environmental, economic and social issues, which may support the transformation process. The principles of the circular economy are based on minimising waste and pollution, streamlining the circulation of products and materials in effect resource efficiency and regeneration of nature, including primarily reducing emissions of greenhouse gases, and contributing to the preservation of biodiversity.

The use of conventional non-biodegradable plastics in various fields, especially in agricultural applications, and the packaging sector causes a significant increase in waste, which results in an increase in environmental pollution. Green polymers, i.e. polymers in line with the concept of sustainable chemistry, fit perfectly into a closed-loop economy. Green polymer materials made from (bio)degradable, renewable or recycled raw materials can help prevent and partially reduce waste and contribute to more sustainable life cycles. Furthermore, such materials could have a lower carbon footprint and, in some cases, may exhibit more favourable material properties in many applications. Products and processes that reduce or eliminate the use or production of substances that are hazardous to humans, animals, plants and the environment are in line with the concepts of pollution prevention and zero waste on both laboratory and industrial scales. In the long run, waste prevention, ecodesign and reuse can save businesses and consumers money. Environmental pressures will be reduced, security of supply of raw materials will increase, and competitiveness and innovation of companies and products will rise, resulting in economic growth. Thus, a circular economy can support the goals of reducing the overconsumption of natural resources while providing economic benefits.

Introduction

The idea of a circular economy is becoming increasingly important around the world by being an integral part of policy and programs at all levels of management or education by governments, non-government organisations, enterprises or institutions. The circular economy implements the concept of sustainable development to reduce the negative impact of global production and consumption while ensuring economic growth and environmental sustainability. Sustainable development means being aware that we should not only exploit natural resources living here and now but also leave them to future generations (Brundtland Report, 1987). It takes into account social, environmental and economic factors, and organisations that contribute to such development through social responsibility can make the greatest contribution. Environmental protection as an area of corporate social responsibility (CSR) mainly covers the issue of reducing and adapting to climate change, as well as protecting biodiversity and restoring the natural environment, which in practice means minimising the release of pollutions into the environment and the level of consumption of natural resources (ISO 26000:2010, 2010).

Conventional polymer materials, which are some of the basic engineering materials, have become ubiquitous in almost all areas of human life, effectively replacing the materials used for centuries such as wood, glass or metals. The use of these materials has many advantages, although their resistance to biological agents causes a negative impact on the environment. The use of conventional plastics in various fields, especially in agricultural applications and the packaging sector, therefore, causes a significant increase in waste, which results in an increase in environmental pollution. In addition, difficulties related to their recycling result, among other things, from contamination with oily substances from foodstuffs, cosmetics or household chemicals, which is difficult to remove, making typical utilisation options (depending on the material properties: material recycling or monomer recovery) not always possible and leaving only energy recovery.

Problems with conventional plastics are increasing, leading not only a reduction in the non-renewable raw materials, such as oil, natural gas or coal but also causing difficult-to-remove environmental pollution, such as plastic waste in the oceans and on land and their residues in the form of microplastics. The new European Commission Plastics Strategy in the circular economy obliges to accelerate the transformation towards an even more sustainable and more efficient plastics management (COM(2018) 28 final, 2018). The current focus on the efficient use of resources is forcing the plastics industry to engage more in energy savings, reducing carbon dioxide emissions, water and even consumption. Economical technologies contribute to the rational use of resources and reduce the negative environmental impact of manufactured products, which should remain in the economy as long as possible, and waste generation should be minimised, which will lead to mitigation of climate change and improve the health and safety of societies. This will limit leakage to the environment of conventional plastics but will not completely solve the problem associated with their end of life, especially plastics used as packaging.

DOI: 10.1201/9780429352799-1

In 2021 world plastic production reached around 367 million metric tons. The negative environmental effects of such a large number of polymeric material produced can be minimised by adopting good practices such as pollution avoidance procedures, e.g. through end-of-life options and proper management. Currently, end-of-life options for polymeric materials are: 34.6% recycling, 42% energy recovery and 23.4% waste storage (PlasticsEurope, 2021). As a result of the European Commission's activities, the European production of plastic (non-recycled) begins to decline noticeably after 2017 (by over nine million metric tons), which translates into a slight decrease in global production (by one million metric tons from 2019).

The European strategy for plastics in the circular economy leads to higher quality recycling of plastics, in particular, packaging recycling at 55% in 2030, by undertaking commitments to separate waste collection and improving extended producer responsibility systems (COM(2019) 190 final, 2019). Member States are also expected to reduce the use of bags to 90 per person by 2019 and to 40 by 2026; they are also expected to improve product design in such a way as to increase their durability, enable their repair and recycling and monitor and reduce marine litter. The European Union's approach to waste management is based on a "waste hierarchy" that focuses on maximum resources protection, firstly by avoiding waste generation and reuse, then recycling (including composting), other recovery through e.g. waste incineration and thus energy production (controversial in some countries) and finally disposal, e.g. through storage, which is the most harmful option for both the environment and health (Directive 2018/851, 2018; News European Parliament, 2018).

Polymers can be made from various sources. Raw materials may be of fossil origin (from non-renewable raw materials, such as oil, natural gas or coal), natural ones, such as cellulose or starch or from renewable raw materials (from non-fossil resources of natural origin, e.g. biomass or organic waste). Polymers from renewable sources differ from natural polymers in the fact that the synthesis is intentionally initiated by a chemical agent or microorganisms. Regardless of the nature of the raw materials, some polymers are also (bio)degradable. This means that, provided they are properly collected and processed together with organic waste, they can be subjected to organic recycling. However, the mere fact that polymers come from renewable raw materials does not mean that they are (bio)degradable (Vert et al., 2012).

In the ecosystem, carbon is part of a closed cycle, and its disruption leads to irreversible environmental changes; that is why it is so important that carbon-based polymer materials are sustainably adapted to its global circulation. The conversion rate of fossil resources (petroleum-based materials) by obtaining polymers from them is characterised by a total imbalance in relation to the rate of their use (consumption) and renewal, while using natural resources (biomass) as raw materials for the production of polymers (bio-based) and synthetic (biodegradable) polymers, balances the rate of carbon dioxide renewal with its consumption, which leads to environmental sustainability. Environmental issues, as well as the gradual depletion of world oil resources, lead scientists to look for alternative sources of materials, and therefore polymers from renewable raw materials play an increasingly important role in human life. The development of environmentally friendly and sustainable production of polymers both (bio)degradable and/or from renewable raw materials (with a minimised carbon footprint) is justified from an economic and, above all, from an ecological point of view and has a significant contribution to achieving the goals of the circular economy. Therefore, the use of (bio)degradable polymers with a minimal carbon footprint should be widespread due to the growing interest in organic recycling, sustainability, environmental issues and human health. From a sustainability development point of view, (bio)degradable polymers are an interesting alternative to conventional polymers. There is also a growing demand for (bio)degradable polymers that are designed as materials for a variety of applications with a predictable life span. Materials designed for specific applications not only have to perform specific functions but also have to meet acceptable safety standards during use and exhibit chemical and physical stability especially in long-term applications.

The increasing share of (bio)degradable plastics on the market creates new opportunities but also risks (COM(2018) 28 final, 2018). The rate of degradation and mechanical properties are key factors in many applications of these polymers, especially in materials for applications with a long shelf life. Therefore, it is of utmost importance to design products from (bio)degradable materials that would be safe for human health and the environment and at the same time to responsibly and sustainably indicate new areas where their unique properties could be used. Compostable polymer materials can minimise the increase in currently generated onerous waste from conventional plastics. On the other hand, (bio)degradation processes can also occur when using products from (bio)degradable polymers. The physico-chemical changes caused by degradation processes affect the thermal and mechanical properties of the material, causing its morphological and structural transformations, which entails deterioration of its quality, which in turn can be a key factor in many applications of (bio)degradable polymers. Knowledge of degradation and weakening of (bio)degradable polymer materials during use and thus predicting their behaviour determines the scope of applications, the possibilities and limitations of the final product, as well as its durability. Determining when (bio)degradable polymers can be used safely and optimisating and understanding of physico-chemical changes in their structure is crucial for their many applications.

The continuous development of new materials that are stronger, lighter or more versatile than the previous ones must not only lead to improved safety but also reduce environmental problems, as the complexity of recovering the value associated with the used product increases. Current challenges to the design of products from (bio) degradable materials lead to the development of materials that are stable in use and at the same time susceptible to microbial attack during organic recycling. For each application of polymer materials, understanding which materials are optimal for the purposefulness of their targeted applications allows accurate prediction of behaviour and performance throughout their life cycle, under real conditions. If all factors are considered at an early stage in the development of new materials, potential failures can be avoided. The main challenge is therefore to identify the conditions in which the use of (bio)degradable plastics is beneficial, as well as the criteria for their use. Proper packaging design from "plastics of the future" should lead to minimising possible failures related to the future use of such materials and significantly support the implementation of circular economy principles.

Both natural, bio-based and synthetic (bio)degradable polymers are used in many sectors. Although consumers are changing their behaviour and increasingly expect ecoproducts, the widespread introduction of (bio)degradable polymeric materials into the market must be preceded by a number of other changes, such as improvement of composting infrastructure, the development of new technology, financial capacity as well as bio-based plastics and biodegradable polymer-related policies. Nevertheless, in the coming years an increase in applications of (bio)degradable polymers can be expected, as the development of new possibilities in different areas predicts an increase in the market for bio-based materials.

By understanding the value chain of (bio)degradable polymers, one can help understand the importance of these polymers to the environment and human health. That is why it is so important to know its individual stages. The aim of this book is to provide an overview of the situation regarding the biodegradable polymer along the value chain, identify and analyse existing practices for biodegradable plastics and assess the relevant legal, regulatory, economic and practical reasons for the importance of proper use and proper recycling of biodegradable plastics. This book will be devoted to the development of (bio)degradable polymer materials development and their environmental impacts, the current and the future of (bio)degradable and bio-based polymers, green polymers, recent advances in synthesis of biodegradable polymers by traditional methodologies and by biotechnological routes, regulations, policy and certification, manufacturing processes, (bio)degradable polymer properties in view of potential applications, recycling and end-of-life disposal options.

REFERENCES

Brundtland Report. 1987. *United Nations Report of the World Commission on Environment and Development Our Common Future.* www.are.admin.ch/are/en/home/sustainable-development/international-cooperation/2030agenda/un-_-milestones-in-sustainable-development/1987--brundtland-report.html (accessed December 22, 2019).

COM(2018) 28 final. 2018. *Communication from the Commission to the European Parliament, the Council, the European Economic and Social Committee and the Committee of the Regions. A European Strategy for Plastics in a Circular Economy.* Brussels: European Commission.

COM(2019) 190 final. 2019. *Report from the Commission to the European Parliament, the Council, the European Economic and Social Committee and the Committee of the Regions on the implementation of the Circular Economy Action Plan.* European Commission, Brussels.

Directive 2018/851. 2018. Directive (EU) 2018/851 of the European Parliament and of the Council of 30 May 2018 amending Directive 2008/98/EC on waste.

ISO 26000:2010. 2010. *Guidance on social responsibility. ISO/TMBG Technical Management Board – Groups.* Geneva: International Organization for Standardization.

News European Parliament. 2018. Society. *Waste management in the EU: Infographic with facts and figures.* www.europarl.europa.eu/news/en/headlines/society/20180328STO00751/eu-waste-management-infographic-with-facts-and-figures (accessed December 23, 2019).

PlasticsEurope. 2021. Plastics – The facts 2021. *An analysis of European plastics production, demand and waste data.* https://plasticseurope.org/pl/knowledge-hub/plastics-the-facts-2021 (accessed April 19, 2022).

Vert, M., Doi, Y., Hellwich, K.-H., Hess, M., Hodge, P., Kubisa, P., Rinaudo, M., and Schué F. 2012. Terminology for biorelated polymers and applications (IUPAC Recommendations 2012). *Pure Appl. Chem.* 84(2): 377–410.

1 Recent Advances in the Synthesis of (Bio)degradable Polymers by Traditional Technologies and by Biotechnological Routes

Wanda Sikorska

CONTENTS

DOI: 10.1201/9780429352799-2

1.1 INTRODUCTION

It should be noted that, over the past several years, there has been a significant increase in the consumption of goods from conventional plastic. Plastics have become ubiquitous because they have excellent and tailor-made properties with controllable flexibility, the ability to be moulded into any shape, a cheap price point, durability, relative impermeability, serialisability and a high strength-to-mass ratio. However, the features of this material such as durability, aging resistance and high resistance to biological factors cause plastic waste to occupy an increasing landfill area at small possibilities of their utilisation. Moreover, most synthetic, non-wettable plastics are usually insoluble in typical degradation media such as water (H_2O) or buffer, which limits the activity of enzymes on their surface. Thus, the non-biodegradable nature of conventional plastics diminishes their advantages. The conventional plastics are a very useful material, but their use has become too common and irresponsible. Hence changes in the area of materials used in goods production according to consumer needs and growing ecological interest are needed. Biodegradable polymers that are stable during use and at the same time susceptible to microbial attack after usage can be one of the solutions that the world needs today (Nakajima et al., 2017; Haider et al., 2019). The European Commission in 2019 presented a directive on the reduction of the impact of conventional plastic products on the environment. The basic intention underlined in the document is a significant limitation and, for some products, even the elimination of disposable items made of plastics, such as plastic cutlery, straws or plates from the market starting in 2021. Disposable products, due to their very short life cycle, should be replaced with more balanced counterparts, e.g., ones that are reusable or made from biodegradable materials (Directive 2019/904, 2019). Reducing waste by reusing products is an important aspect of the circular economy model. The circular economy idea takes into account all stages of the product life cycle, starting from its design, through production, consumption and waste collection, to its management. It is therefore a waiver from the traditional linear economy, which often treats waste as the last stage of a product's life cycle. In a circular economy, it is important that waste, if it is generated, is treated as secondary raw materials. All pre-waste activities are intended to serve this purpose. Reducing waste production, increasing the use of biological components and advancing recycling are therefore challenges to a modern and environmentally friendly closed economy system (Kirchherr et al., 2018). A European Strategy for Plastics in a Circular Economy, published by the European Commission, draws attention to certain aspects related to the role of plastics in the circular economy, in particular the need to recycle plastics and reduce environmental pollution with plastic waste. The preparation for re-use and recycling of waste materials such as paper, metal, plastic and glass from households should account for a minimum of 65% by 2035. The Strategy postulates for all plastic packaging to be reusable or recyclable by 2030 and that at least 55% of consumer plastic packaging waste placed on the European market should be recycled. Selective collection was considered a key factor in obtaining the appropriate amount of raw material for recycling, also drawing attention to the need to increase the demand for recycled plastics. The authors of the Strategy also emphasised the importance of the proper functioning of the extended producer responsibility system and the need to develop innovations and new investments in more effective waste management and recovery of raw materials

(European Commission, 2018). The change of the paradigm from a linear economy to a circular economy will require both innovative solutions and the search for new materials. Reducing the use of materials from non-renewable resources in favour of biomaterials is an important aspect of the circular economy model, especially as non-renewable environmental resources are now overexploited. Recycling can therefore be called the core of the circular economy (EEA Report No 8, 2018).

1.2 CLASSIFICATION OF POLYMERS DUE TO THEIR BIODEGRADABILITY AND ORIGIN

Polymers can be classified according to physicochemical properties, raw material origin and susceptibility to the activity of enzymes and microorganisms. Table 1.1 presents a classification of polymers due to their biodegradability and sources of obtainment.

Due to their origin, polymers are divided into natural polymers (biopolymers) and polymers obtained from renewable resources as well as polymers from petrochemical (non-renewable) sources. The properties and degradation mechanism of polymers result from their specific structure, not their origin. Plastics are technological materials that are obtained by adding additives and fillers to the raw polymers. Plastics are defined by their plasticity (the state of a viscous liquid during processing). Bioplastics are usually semi-synthetic polymers produced both from petrochemical sources and renewable resources. Bioplastics are sustainable plastics during usage and may be able to biodegrade under appropriate conditions. The bioplastics that decomposed by microorganisms via organic recycling in this way are recycled after use (see Table 1.1, groups I, II and III; Greene, 2014).

Often, the term *biopolymers* is used improperly for both natural polymers and artificial bio-based polymers. The term should be only used for natural polymers produced by the cells of living organisms and not be confused with the term *bioplastics*. Thus,

---

TABLE 1.1
Classification of Polymers Due to Their Biodegradability and Sources of Obtainment.

	Biodegradability	
Origin	Biodegradable	Non-biodegradable
Renewable resources	cellulose, starch, PHA, PGA, PLA, PLGA (group I)	bio-PE, bio-PP, bio-PET (group III)
Petrochemical sources	PCL, PBT, PBS, PBAT, PBTS, PBSA (group II)	PE, PP, PET, PS, PVC (group IV)

PHA – polyhydroxyalkanoate, PGA – polyglycolide, PLA – polylactide, PLGA – poly(lactide-*co*-glycolide), PCL – poly(ε-caprolactone), PBT – poly(1,4-butylene terephthalate), PBS – poly(1,4-butylene succinate), PBAT – poly(1,4-butylene adipate-*co*-1,4-butylene terephthalate), PBTS – poly(1,4-butylene terephthalate-*co*-1,4-butylene succinate) (PBTS), PBSA – poly(1,4-butylene succinate-*co*-1,4-butylene adipate), PE – polyethylene, PP – polypropylene, PET – poly(ethylene terephthalate), PS – polystyrene, PVC – poly(vinyl chloride), bio-PE – bio-based PE, bio-PP – bio-based PP, bio-PET – bio-based PET

bioplastics are plastics from renewable sources biodegradable or non-biodegradable and biodegradable plastics from petrochemical (non-renewable) sources. The bioplastics available on the market at a larger scale include Natureflex cellulose films (Futamura), polylactide (NatureWorks LLC) and its blend with poly(1,4-butylene adipate-*co*-1,4-butylene terephthalate; Ecovio®, BASF), biodegradable polyester/starch blend Mater-Bi (Novamont), bio-based high-density and low-density polyethylene (Braskem). However, bioplastics are still in the early stages of development and occupy a small market niche. Their further development is related to the improvement of properties, availability and lowering the price, as well as for biodegradable plastics with the introduction of organic waste collection systems for composting. As shown by the life cycle assessments (LCA) at the production stage, materials from renewable raw materials are more favourable in terms of carbon dioxide (CO_2) emissions than petrochemical materials, because plants absorb CO_2 from the air in photosynthesis process (Álvarez-Chávez et al., 2012). The renewable resources used for bioplastics production can be divided into first generation feedstock such as sugar, sugarcane, beet, soy, cassava, rice, wheat, potato, corn and hemp; more preferred second generation feedstock (non-food crops such as waste from food crops, agricultural and wood residues, etc.) and third-generation feedstock such as biomass derived from algae or methane (CH_4) made from waste (Song et al., 2009).

1.2.1 BIOPOLYMERS AND (BIO)DEGRADABLE SYNTHETIC POLYMERS FROM RENEWABLE RESOURCES

Group I in Table 1.1 includes polysaccharides such as cellulose or starch and polymers of biological origin including polyhydroxyalkanoates (PHA)s (e.g. poly(3-hydroxybutyrate) (PHB)) as well as polymers from bio-derived monomers – aliphatic polyesters: polylactide (PLA), polyglycolide (PGA) and their copolymer poly(lactide-*co*-glycolide) (PLGA). The PHA and PLA are often called green polymers, obtained in a sustainable way. Biodegradable polymers are a sustainable material when biomass resources are used in their production. Such polymers do not contribute to global warming and save petroleum resources (Moshood et al., 2022).

1.2.1.1 Polysaccharides

Polysaccharides are polymers with high molar mass, ranging from several thousand to several million, occurring naturally in plants and extracted directly from them. Due to their structure, they are divided into two main groups: homoglycans, i.e. single-component polysaccharides – including cellulose and starch – and heteroglycans, i.e. multi-component polysaccharides, including heparin and hyaluronic acid. Cellulose ($C_6H_{10}O_5$)$_n$, whose *D*-glucose molecules are linearly linked by β-1,4-glycosidic bonds, is an important structural component of the primary cell wall of higher plants and some algae. As a non-toxic, biodegradable biopolymer with high tensile and compressive strength, cellulose is widely used in various fields of nanotechnology, pharmaceutical, food, cosmetic, textile and paper industries, as well medicine as drug delivery systems for the treatment of cancer and other diseases. The great industrial importance possesses derivatives of cellulose such as microcrystalline cellulose and bacterial cellulose. The most commonly used is microcrystalline cellulose, which is an important assistant substance due to its binding and tablet-forming properties, characterised by plasticity and cohesiveness, compared to a conventional plastic (Shaghaleh et al., 2018). Bacterial cellulose is produced by gram-negative acetic fermentation bacteria of the *Komagataeibacter species* (formerly *Gluconacetobacter*) but can also be synthesised by bacteria type *Agrobacterium, Pseudomonas* and *Rhizobium*. Opposite to cellulose of plant origin, bacterial cellulose fibres are much longer and stronger, and their supramolecular structure can be significantly modified at the stage of synthesis. Bacterial cellulose is characterised by low toxicity, biocompatibility and biodegradability as well as possessing also high purity and, unlike plant cellulose, it does not require an expensive and complicated purification process. Due to such properties, it has found a number of applications, including medicine (biomaterials for dressings), food, cosmetics and paper industries (Shigematsu et al., 2005; Esa et al., 2014; Wang et al., 2019). The production of cellulose on an industrial scale is still small, mainly due to the problems connected with the selection of highly active strains capable of cellulose biosynthesis and also due to the high cost of the substrate components. Cellulose production costs can be reduced by using as media for the bacteria cultivation substrates containing waste substances such as glycerol, whey, vegetable processing waste or molasses (Molina-Ramírez et al., 2018). A valuable raw material for the microbial synthesis of bacterial cellulose for the *Komagataeibacter xylinus* strain is xylose, a monosaccharide formed in the process of hydrolysis in the leaves of lignocellulose grasses (Yang et al., 2013). Starch, which acts as an energy store in plants, consists of two polysaccharides, i.e. amylose and amylopectin, linked by α-glycosidic bonds. Amylose forms linear chains composed of glucose molecules linked together by α-1,4 bonds. Amylopectin is composed of glucose molecules additionally linked by an α-1,6 bonds, which forms the characteristic for starch molecules branching. In its pure form, the so-called native, starch is not suitable for processing with the available processing methods. Its plasticisation is required. The preparation of thermoplastic starch takes place in several stages. In the first stage, a mixture of natural starch with a plasticiser and other additives is prepared. Then, during the extrusion process at elevated temperature, the starch is destructurised under the action of shear forces in the extruder. The crystalline phase of natural starch is destroyed, its bonds are broken and there is a physical interaction between the starch and the plasticiser. In this way, a thermoplastic material in the form of granules is obtained, which can be processed with conventional plastics processing equipment. Thermoplastic starch is a fully biodegradable and 100% natural polymer, obtained from potato starch, with no environmental toxicity. Thermoplastic starch is home compostable and fully biodegradable in less than six months (Kundu and Payal, 2021). Unfortunately, due to their hydrophilic nature, materials based only on starch and plasticisers show poor mechanical properties and degrade quickly, especially in high humidity conditions. Therefore, thermoplastic starch is mixed with other polymers with better strength parameters, which are also biodegradable and obtained based on renewable raw materials, such as PLA or poly(1,4-butylene succinate) (PBS). There are only a few producers of thermoplastic starch materials in the world that are completely biodegradable: Novamont (Mater-Bi), National Starch

(ECO-FOAM) and Capsulgel/Warner Lambert (CAPILL; Martinez Villadiego et al., 2022). In 2021, Azoty Company, Poland launched a pilot thermoplastic starch production line under the trade name of Envifill®, based on its own patented technology. Disposable packaging and bags, trays, pads, foils or pots can be produced from it, among other things (Grupa Azoty Tarnów).

1.2.1.2 Poly(3-hydroxyalkanoate)s (PHA)

PHAs seem to be a perfect alternative for conventional plastics because they possess characteristics similar to common plastics but unlike them are biodegradable. PHAs are best fit to those applications where biodegradation is necessary that is products for medical purposes or that inevitably end up in the environment. PHAs are used to encapsulate grains and fertilisers and to produce biodegradable food containers and foils. Due to their high biocompatibility, they are used as sutures, orthopaedic implants and mechanical barriers to prevent postoperative adhesions, internal drug release systems in the form of stents, bone marrow scaffolding and wound supporting healing bandages. PHAs are a family of polyhydroxyesters from 3-, 4-, 5- or more hydroxyalkanoic acids synthesised by biotechnological rout using microorganisms and plants (Mozejko-Ciesielska and Kiewisz, 2016). More than 150 types of PHAs have been found and there are many known microorganisms capable of producing it (Li et al., 2016). Already in the 1930s, the ability of various bacteria to internally store carbon and energy in the form of PHB (first time in *Bacillus megaterium* by Lemoignea) was discovered. Since then, microorganisms of many groups have been shown to be able to produce PHA, including eubacteria (e.g. *Pseudomonas, Bacillus, Citrobacter, Enterobacter, Klebsiella, Escherichia, Ralstonia*), cyanobacteria and archaea. PHB synthesis is widespread in molecular nitrogen-bonding bacteria, e.g. *Rhizobium* and *Azotobacter*. Some methylotrophs are capable of producing PHB but in small amounts (Anjum et al., 2016). The microorganisms were divided into two groups depending on the culture conditions necessary to obtain the PHA. The first group are microorganisms that require an excess of the carbon source in the medium and limitation of nutrient such as oxygen and nitrogen. In this case the microorganisms use PHAs to store energy in response to stress conditions (e.g. *Cupriavidus necator* H16 formerly *Ralstonia eutropha* H16), *Protomonas extorquens* and *Pseudomonas oleovorans*). The second group consists of microorganisms that do not require nutrient limitation, and PHAs are accumulated in growth phase (np. *Alcaligenes latus, Azotobacter vinelandii* and recombined *Escherichia coli*). Recombined *Escherichia coli* strains are especially interesting for PHA production, because it is easy to genetically modify, it grows fast and it can use different carbon sources, such as fatty acids. In general, genetic manipulation of bacteria for PHA production is based on using the genes responsible for PHA metabolism, together with other metabolic genes (Koller et al., 2017). Photosynthetic microorganisms (cyanobacteria) are capable of producing PHB with the sunlight energy participation using as a carbon source agro-industrial pre-treated wastewater from the olive mill. It is worth noting that the use of cyanobacteria allows the production of PHA together with bio-hydrogen, which is an additional benefit in addition to the use of solar energy and can significantly reduce the total cost of PHAs production (Padovani et al., 2016). Among archaeon capable of producing PHA, the most attention is paid to halophilic organisms (living in saline environments), which do not require strictly sterile conditions for cultivation, due to the high concentration of salt (required to maintain the stability of the cell wall), eliminating undesirable microflora. In low or no salt concentrations (especially in distilled water), the halophilic cell wall disintegrates, releasing the cell contents and making it easier to obtain PHA. The types of halophilic archaeon that produce PHA include but are not limited to *Haloferax, Haloarcula, Haloquadratum, Haloalkalicoccus, Natrinema, Natronobacterium, Halopiger, Halococcus* (Tekin et al., 2012). PHAs are produced intracellularly as the reserve material of carbon and energy during the fermentation process and can constitute up to 90% of the mass of a bacterial cell. Fermentation is a biochemical process in which organic compounds are transformed with the partic-ipation of microorganisms or enzymes produced from them. This process takes place under anaerobic conditions. From a technological point of view, two types of fermen-tation have been distinguished: batch fermentation (in this also fed-batch fermenta-tion) and continuous fermentation (in which the biomass concentration and substrates are on the constant level at steady-state conditions). The percentage and properties of PHA obtained can be modified through the selection of adequate substrates and microorganism use during fermentation and the controlling of the process parame-ters such as the degree of strain growth, final cell density, aeration, pH and availabil-ity of carbon, nitrogen and phosphorus sources. Product (polymer) isolation also plays an important role and is associated with cell degradation that occurs either chemically or enzymatically. Sodium chlorate is used for chemical disintegration, often in combination with chloroform. The addition of surfactants helps to reduce the degree of polymer degradation. In the case of enzymatic degradation, proteolytic enzymes (papain, pepsin, trypsin etc.) are used. Therefore, many efforts are focused on optimising the fermentation process (Mozejko-Ciesielska et al., 2019). Fed-batch fermentation occurs when the growth medium is supplemented with an additional portion of a substrate, which provides a constant regulation of nutrients, avoids car-bon limitation and enables the efficient growth of microorganisms and the accumu-lation of bioproducts. It also allows one to achieve a high density of bacterial cells and the highest possible concentration of the bioproducts. The batch fermentation process can take place as one-stage or two-stage cultivation, carried out under strictly defined conditions. In the one-stage process, the phases of microbial growth and accumulation of the produced polymers occur in parallel, while in the two-stage process they follow one another (Kaur et al., 2012). *Pseudomonas putida* KT2440 is able to accumulate even up to 75.5% of middle-chain-length PHA (mcl-PHA) using acrylic acid together with nonanoic acid and glucose as co-substrate in a batch cul-ture (Jiang et al., 2013). The lower mcl-PHA content (32%) was obtained when nona-noic acid was delivered to cells grown from glucose by the previously mentioned strain without supply of pure oxygen (O_2) or oxygen-enriched air (Davis et al., 2014). Promising results were also obtained with the use of palm kernel oil and the newly isolated TO7 strain belonging to *Pseudomonas mosseli*, which was able to produce mcl-PHA at a concentration of 47.1% (Liu et al., 2018). Thus, it can be assumed that the kind of bacteria (species) used will determine the type of PHA produced. On the other hand, also fed medium composition and growth conditions influence the type of PHA produced. The medium for the microorganisms producing PHA are low-molar-mass organic acids as well as glucose and fructose. *Alcaligenes latus* very often uses sucrose and the more complex organic compounds found in whey and molasses. *Methylobacterium organophilum* synthesises poly(3-hydroxybutyrate) in the presence of methanol. The substrates for *Pseudomonas putida* are fructose, glucose or glycerol but also sodium gluconate or oleic acid (Mozejko-Ciesielska et al., 2018). In addition to homopolymers, microorganisms are capable of accumulating copolymers, i.e. PHA containing at least two different units. To obtain PHAs other than PHB, the appropriate feeding medium must be used. Propionate and valerate acids are responsible for 3-hydroxyvalerate synthesis in poly(3-hydroxybutyrate-*co*-3-hydroxyvalerate) (PHBV). Butyrate is a precursor for hydroxyhexonate units in poly(3-hydroxybutyrate-*co*-3-hydroxyhexanoate) (PHBHHx). Unfortunately, additional precursors in fermentation media increase the cost of PHA production. However, copolymerisation of PHB with various amounts (5–20%) of valerate units allows the production PHBV copolymers with improved mechanical properties. PHBV copolymer can be produced from microbial consortia of *Cupriavidus necator*, *Bacillus subtilis* and recombinant *Cupriavidus necator* bacterium, which were used for PHBHHx copolymer synthesis. The use in PHBV fermentation process the PHA-accumulating *Ralstonia eutropha* 5119 together with sucrose-hydrolysing *Bacillus subtilis* allow reducing the cost of PHA production. This is because *Bacillus subtilis* also ferments sugars into organic acids (acetic acid and propionic acid) and no additional precursors are needed (Anis et al., 2013; Bhatia et al., 2018). Moreover, it is possible to adapt the bacteria to different carbon sources and use modified microorganism cultures to produce polymers with a given composition by direct synthesis. The use of modified PHA synthases allows the production of PHA with chemically modifiable functional groups and with special, controllable properties. Functional PHA can be prepared via biosynthetic processes by incorporating various functional monomers or other metabolites in the polymer sequence. For example, fermentation of *Escherichia coli* in a culture containing poly(ethylene glycol) (PEG) has given a block copolymer consisting of PHB with PEG. The PEG influences the properties of the synthesised copolymer, especially on its molar mass and end-group structure (Tomizawa et al., 2010). Other possibilities for PHA production are also known. An interesting example is the genetically modified plants such as *Arabidopsis thaliana*, which accumulate PHA in plastids, cotton, corn and oil plants, e.g. rape (Somleva et al., 2013).

The physical properties of PHAs vary from crystalline-brittle to soft-sticky materials depending on the length of the side aliphatic chain on β-carbon. The short-chain polymers are more crystalline, hard and brittle. The medium-chain ones are more resilient (Li et al., 2016).

Degradation of polyesters in environment occurs via enzymatic processes as a result of the action of specific microorganisms (biodegradation) or by hydrolysis of ester bonds (hydrolytic degradation). But most often both of these mechanisms occur at the same time, in the appropriate sequence. The most easily biodegradable polymers are characterised by the lack of side branches and the greatest possible linearity, which increase the susceptibility of macromolecules to the action of enzymes. Moreover, the biodegradation susceptibility is the greater; the more chemical groups in the macromolecule are sensitive to the microorganism action (e.g. ester, hydroxyl, carboxyl or ether groups). The degree of crystallinity, molar mass and no crosslink also affect this process. In addition, biodegradation also depends on factors such as the type of microorganisms and their activity, environmental conditions and the shape of the finished product.

The environmental advantage of bioplastics (bio-based and biodegradable) is connected with saving the limited sources of fossil raw materials by using renewable raw materials in their production (Emadian et al., 2017). Furthermore, the products of biodegradation such as CO_2, H_2O and biomass occur naturally in the environment and after the biodegradation process return to it, creating a natural cycle of matter and a closed carbon cycle. In this context, the biodegradability appears to be the most important and favourable property of PHAs. Bioplastics produced from renewable raw materials allow also the reduction of greenhouse gases from non-renewable petrochemical raw materials used for the conventional plastics production. In this way, PHAs perfectly fit into the latest trends of the circular economy and can be used as responsible and sustainable new products with unique properties (Dietrich et al., 2017). PHAs on an industrial scale are produced through the bacterial fermentation route in an industrial bioreactor. However, due to the high production and purification costs of the resulting polymer, the industry only uses those microorganisms that store more than 80% of PHA. They are *Cupriviadus necator*, *Alcaligenes latus*, *Azotobacter vinelandii*, *Methylobacterium organophilum* and *Pseudomonas putida* or genetically recombined *Escherichia coli*. One of the solutions to lower PHA production costs seems to be the use of waste from agro-food production as a cheap, easily accessible source of carbon and nitrogen. Much of the agro-food waste can serve as a nutrient medium for the microorganisms to produce PHA. As carbon sources are used, among others is beet molasses (production of 36 g PHA per litter by *Azotobacter vinelandii*) or sugar cane (60% of the dry mass of *Pseudomonas aeruginosa* cells; Tripathi et al., 2012). Whey and its hydrolysates are also used (96.2 g PHA per litter by recombinant *Escherichia coli* CGSC 4401 possess genes from *Alcaligenes latus*; Fonseca and Antonio, 2006). Lignocellulosic materials (e.g. hay, bran, sawdust and husks) are a promising waste, but they often require pre-treatment in order to release simpler carbon sources (e.g. sugars) that are easier to absorb by microorganisms. A fairly commonly used substrate is waste glycerol, produced, among other ways, in the production of biofuels. Microorganisms used for the industrial production of PHA more and more often use fatty acids as a carbon source, which are waste from the food industry. Fatty acids provide more energy than carbohydrates, but the biggest challenge in using them is their hydrophobic nature. A wide spectrum of fatty substrates is used, including coconut oil, palm oil, olive oil, corn oil and other vegetable and animal fats (e.g. tallow). Many bacteria, including *Pseudomonas*, *Caulobacter*, *Ralstonia*, *Acinetobacter*, *Sphingobacterium*, *Burhkholderia* and *Yorkenella*, are able to use these substrates for the production of PHA (Kaur et al., 2017). Using the coffee waste oil as well as free sugars with *Cupriviadus necator* or utilisation of animal waste or plant waste oils is also part of the strategy reducing the cost of PHA production (Ciesielski et al., 2015; Riedel et al., 2015; Bhatia et al., 2018). To obtain PHA with good quality regarding biodegradability, biocompatibility and mechanical properties, the engineered microorganism such as the engineered strain *Cupriavidus necator* Re2133 with respective coenzyme-A hydratase (phaJ) and PHA synthetase (phaC2) was also used. Recently *Ralstonia eutropha* 5119 was used for PHA production from lignocellulosic biomass hydroly-sate containing glucose, xylose and various biomass-derived by-products (furfural, hydroxymethylfurfural, vanillin and acetate). Unfortunately, these side-products act as inhibitors of microbial fermentation and affect microbial growth and productivity as they require a special and costly pre-treatment method of lignin (Bhatia et al., 2019). The use of waste frying oil and waste animal hydrolysates of chicken feather as carbon and inexpensive nitrogen sources was also described. The use of animal waste has a positive effect on the production of PHBV copolymer by improving the production of biomass and increasing the efficiency of incorporation of the sodium propionate as a precursor into the copolymer structure (Benesova et al., 2017). Meat and bone meal, silage juice or proteolytic hydrolysates of cheese whey have been also used as inexpensive sources of nitrogen (Obruca et al., 2014; Koller et al., 2017). Nowadays, a production of value-added PHA by the bioconversion of polystyrene (PS) or polypropylene (PP) solid waste in shake flask experiments with the bacterial strain *Cupriavidus necator* H16 are proposed (Johnston et al., 2018; Johnston et al., 2019). The PP waste was generated as supplementary carbon sources to tryptone soya broth for 48 h fermentations with the bacterial strain *Cupriavidus necator* H16 as it is non-pathogenic, genetically stable, robust and one of the best-known producers of PHA. The yield of PHAs varied from 22 wt% in nitrogen rich only controls to 66 wt% in nitrogen rich media where PP was used as an additional carbon source. New sources of carbon from agricultural waste for the production of bio-based poly-mers are also tested by Bio-on Company at its own demonstration plant near Bologna, Italy with a capacity of 1,000 metric tons per year. The new plant produces solely the special PHAs for advanced niche products, for Bio-on's own sales (mainly for cosmetic packages). Bio-on produces PHB from molasses and sugar beet by-products at its demonstration plant. The Bio-on strategy is not about building high-capacity plants but about licensing its technology, and a demonstration plant confirms the effectiveness and feasibility of PHB production. The plant's production is dedicated to specialist products for applications in the cosmetics and pharmaceutical sector. The Bio-on PHA, MINERV-PHA™, can also be used for the packaging of fresh fruit and vegetables or even for the production of products such as toys, car parts and furniture. The company already has a number of PHA licenses under its belt. The family-owned company SECI plans to build a plant of 5,000 metric tons per year in Italy. The Russian company TAIF has purchased a license to build a plant with a mass of 10,000 metric tons in the Republic of Tatarstan. Bio-on has also licensed its trial to companies in France, Spain and Mexico (Bio-on). Technology of Danimer Scientific's of PHA production is based on Procter & Gamble's Nodax® license. The company is already a large producer of PLA-based blends with a reactive extrusion capacity of 50,000 metric tons per year. Its new commercial manufacturing facility in Winchester, Kentucky biosynthesised PHA by fermentation rout from inexpensive oils derived from the seeds of plants such as canola and soy. Danimer Scientific's PHA is also accepted for contact with food by the U.S. Food and Drug Administration (FDA). Together with PepsiCo's Frito-Lay division, Danimer Scientific is developing a snack bag material to replace the current polypropylene foils. PepsiCo is working also on PHA water bottles with Nestlé (Danimer Scientific). The Mango Materials company uses waste CH_4 gas fed to naturally occurring microorganisms to produce a biodegradable polymer. PHAs can biodegrade in many different environments, including those where no oxygen is present, producing CH_4 and closing the loop to create more polymer from that CH_4. In a pilot facility the Mango Materials company produces PHA from waste biogas from the water treatment plant containing CH_4, CO_2 and hydrogen sulphide (Hyde, 2019).

In summary, the type of PHA obtained is determined by the choice of carbon source and the optimisation of the PHA industrial fermentation processes. Improving PHA production, together with greater ecological awareness in society, cause growth in interest in such polymers as the perfect alternative for conventional plastics. Some factors, such as the degree of strain growth, final cell density, PHA content in dry mass of cells, substrates price and purification methods decide profitability of PHAs production. Furthermore, the use of waste materials as inexpensive carbon sources for PHA fermentation together with PHAs biodegradability without toxic by-products should give a biotechnological product of great industrial interest (Nielsen et al., 2017).

Apart from valuable advantages, biodegradable polymeric materials also have disadvantages. First of all, they have limited application possibilities, which makes them inferior in many applications to their non-biodegradable counterparts. These are also materials more expensive than those currently available on the market. However, their price is constantly decreasing and it is expected that in the coming years it may be equal to the price of classic polymers of petrochemical origin. Many biodegradable polymeric materials also have unfavourable mechanical properties, i.e. they are too brittle or stiff or have too low tensile strength. Due to frequent use, biodegradable materials are also required to have favourable barrier properties due to the permeability of O_2, CO_2 and H_2O vapour, which can adversely affect the product. Moreover, due to the sensitivity of biodegradable polymers to heat, humidity and shear stresses, they are more difficult to process. The previously mentioned main disadvantages of biodegradable polymeric materials are the basis for conducting research in the field of improving their properties or reducing the occurring defects. Mechanical, thermal and processing properties of materials can be modified by various methods. The main ones include the production of composites or nanocomposites with additional ingredients, e.g. fillers or fibres. The selected natural origin PHAs can be also functionalised by chemical modification such as grafting or block copolymerisation to produce modified copolymers containing structural segments derived from PHA and other polymers that results in different and tailored properties. The other method is the preparation of blends with various polymers or additives. Methods of chemical and physical modification of PHA make it possible to obtain copolymers with an appropriately changed structure and with tailor-made properties. Different polymerisation techniques make it possible to synthesise a wide range of new functional materials based on PHA with many new properties for specific applications. One of them is poly(methyl methacrylate) graft copolymers (PHA-*g*-PMMA) that were obtained by radiation-induced radical polymerisation on both PHB and PHBV. It was found that the higher crystallinity of PHB compared to PHBV resulted in a lower degree of grafting for PHB-*g*-PMMA than for PHBV-*g*-PMMA (Li et al., 2016). Atactic or isotactic PHB macromonomers functionalised by methacrylate end groups were copolymerised by a one-step atom transfer radical polymerisation (ATRP), resulting also in PHA-*g*-PMMA copolymers. Moreover, using modification by grafting technique allowed the introduction of hydrophobic or hydrophilic segments into copolymers obtained (Neugebauer et al., 2007). PMMA was also synthesised with synthetic atactic PHB (aPHB) side chains, where monomer (β-butyrolactone) was anionically polymerised on a multifunctional PMMA macroinitiator (Rydz et al., 2015). The water-soluble brush copolymer composed of synthetic aPHB and PEG brushes was prepared by applying a three-step procedure, including ring-opening anionic polymerisation of β-butyrolactone and ATRP processes (Koseva et al., 2010). More recently, the controlled synthesis of PHB-PEG-PHB triblock copolymer is another route of β-butyrolactone polymerisation on PEG macroinitiators via a crown ether-free anionic ring-opening polymerisation with good control on molar mass, and molar mass distribution of the final triblock copolymer was reported (Liu et al., 2008). The macroinitiators of natural PHA were obtained by the partial saponification of ester linkages followed by an elimination reaction. Macroinitiators obtained were then used in diblock copolymers synthesis via anionic ring-opening polymerisation of β-butyrolactone (Adamus et al., 2021). The application of obtained block copolymer as a compatibiliser for a polyhydroxyoctanoate blend with synthetic aPHB was tested. The suitability of obtained polymeric materials for cardiovascular engineering was also confirmed. The studies on permeability show that vascular prosthesis covered with PHB-*block*-(aPHB) copolymer are more watertight as compared with pure prosthesis (Adamus et al., 2012). The amorphous, amphiphilic PHBV diblock copolymers with PEG were obtained in a one-step melt reaction and use for nanoparticle preparation (Shah et al., 2012).

The commercial PHA market includes only several PHA from few companies: PHB under brand name Biogreen® and Biocycle™ (Mitsubishi Gas Chemical Company Inc. Japan and PHB Industrial S.A., Brazil, respectively), PHB/PHBV – Biomer® (Biomer Inc., Germany), PHBV – Enmat™ (Tianan Biologic, Ningbo China), poly(4-hydroxybutyrate) (P4HB) – TephaFLEX® (Tepha Inc., United States), PHBHHx – Aonilex®, Nodax™ (Kaneka Co., Japan; Danimer Scientific, United States), poly(3-hydroxybutyrate-*co*-4-hydroxybutyrate) (P3HB4HB) – GreenBio® (Tianjin Green Bio-Science Co., China/The Netherlands), PHAs – GreenBio's PHA, CJ PHA®, TerraBio® and Mirel® PHA (Tianjin GreenBio Materials Co., Ltd., China; CJ CheilJedang Corporation, Korea; Alterra Holdings, United States; Metabolix, Inc., United States, respectively) (MarketsandMarkets, 2021). The first attempt to commercialise PHA packaging was to produce a completely biodegradable product – Wella™ (Germany) shampoo bottles using Biopol®. However, due to its high price, this product did not conquer the market. Another example is the project of the U.S. company Cove, which launched a bottle of PHB. Nestle and PepsiCo plan to use Nodax™ to construct not only bottles but also other packaging for their products. PHAs have also been used in the production of wrapping films, shopping bags, paper containers and coatings, disposable items such as razors, dishes, cutlery, kitchen utensils, diapers, feminine hygiene products, cosmetic containers and cups, as well as found staples for medical purposes: production of surgical clothing, mats, covers or packaging. The products were produced, for example, by the companies: Metabolix (United States), Biomer (Germany), BASF SE (Germany), P&G Chemicals (United States), ICI (Great Britain) and BIOPOL (Brazil). Numerous filaments for three-dimensional

(3D) printing are also produced on the basis of PHA, and they can be purchased from such manufacturers as Zortrax (Poland), colorFabb (the Netherlands) and 3D Printlife (United States; Guzik et al., 2020).

1.2.1.3 Polylactide (PLA)

PLA is the most common polymer called green because is both (bio)degradable and obtained from renewable raw materials in sustainable way. The most common way to produce PLA is ring-opening polymerisation of lactide in the presence of metal catalysts in solution or suspension. PLA is produced from lactic acid by fermentation of sugar previously obtained from vegetable starch, for example, corn cassava, sugar cane and beet pulp. In an industrial process, the lactic acid is then processed via the condensation reaction into a chiral compound, the cyclic lactide dimer. Thus, if *L,L*-lactide is used in the polymerisation process, the obtained polymer will be poly(*L*-lactide) (PLLA) and analogously *D*-lactide and *D,L*-lactide will give poly(*D*-lactide) (PDLA) and poly(*D,L*-lactide) (PDLLA), respectively. PLA is soluble in organic solvents such as benzene, tetrahydrofuran, dioxane etc. Considering physical properties, PLA can be amorphous, semi-crystalline and sometimes even highly crystalline polymer (Maharana et al., 2009). The problem that hinders the development of the PLA market is precisely the costs of the intermediate stages of dimer production. Scientists from KU Leuven (the Netherlands) have presented a new PLA production process characterised by, inter alia, no waste. A mineral catalyst with a porous surface – zeolite – was used to accelerate the biomass conversion processes taking place in the reactor. By selecting, on the basis of the porosity, the appropriate type and shape of the catalyst, the lactic acid is converted directly into PLA with omission of the intermediate stages of dimer production. This allows production of PLA without metals and reduction of the amount of waste generated (Dusselier et al., 2015). Thus, a new PLA production process is cheaper and more ecological. In addition, if PLA-based materials after use are collected and sorted properly, they can be organically recycled in industrial composting plants (Musioł et al., 2015). PLA finds various applications in our everyday life. This biodegradable material is used, among other things, for the production of cups or foil for packing vegetables and can also be used as filament in 3D printing technology. PLA water bottles, commercially available from Biota, were obtained from NatureWorks™ PLA (Blair, NE). In the medical sector PLA is used to obtain interference screws (Pl-Fix, Arthrex and Bio Screw, Phusiline) or biodegradable sutures (DePuy; Bergstrom and Hayman, 2016; James et al., 2016). The PLA is often used to prepare the physical or chemical connections with other polyesters to achieve improvement of the physical properties of blends or copolymers obtained (Kumar et al., 2019). The different blends of PLA with biodegradable or non-biodegradable polymers have been investigated in terms of their miscibility, crystallisation, morphology and mechanical properties. It was found that PLA/PHB blends are immiscible and exhibit different properties over the composition range. However, the low-molar-mass PHB shows limited or partial miscibility with PLA. During a study of PLA/PHB blends in various proportions, it was found that blending PLA with 25 wt% of PHB causes some interactions between both polymers due to the fact that PHB acts as a nucleating agent (Zhang and Thomas, 2011). Furthermore, the temperature used during blend preparation also has a significant influence on the miscibility between both polymers. During the nonisothermal crystallisation, the addition of low-molar-mass PHB yields a remarkable effect on the cold crystallisation of PLA in the blends, especially when the PHB content is relatively low (Hu et al., 2008). Thus, the PLA/PHB blends prepared at high temperature exhibited greater miscibility than those prepared by solvent casting at room temperature since PLA/PHB systems are fully miscible in the melt state. This effect could be due to the transesterification reaction between PLA and PHB chains (Bartczak et al., 2013). The PHB blends with poly(ε-caprolactone) (PCL), poly(butylene adipate) (PBA) and poly(vinyl acetate) (PVAc) were also studied. It was shown that only blend PHB with PVAc was miscible. With increase in the amount of PBA, PVAc or PCL blend the enzymatic degradation of these composition with depolymerase from *Alcaligenes feacalis* T1 decreased linearly (Tokiwa et al., 2009).

1.2.1.4   Polyglycolide (PGA)

PGA is a polymer of glycolic acid found naturally in, e.g., sugar cane. As a result of the polycondensation reaction of glycolic acid, the glycolide is formed. Then glycolide undergoes a ring-opening polymerisation process, resulting in a high-molar-mass polymer containing 1–3% unreacted glycolide used as monomer. PGA is a crystalline material (45–55%) with a softening point of 220–225°C and a glass transition temperature (T_g) of 35–40°C. PGA is also characterised by tight packing of chains, due to which it is characterised by good mechanical properties with a modulus of elasticity above 10 GPa and strength of 100 MPa, so it is the material with the highest stiffness and brittleness among biodegradable poly(α-hydroxy ester)s. Moreover, the products of PGA hydrolytic degradation are metabolised in the human body. PGA is fibre-forming and first was used to produce the synthetic resorbable surgical sutures known as Dexon. Due to the high content of the crystalline phase, PGA is insoluble in most organic solvents, except for perfluorinated alcohols (Yoon et al., 2021). In order to reduce the stiffness, PGA is copolymerised with other monomers, especially with PLA in direction of PLGA copolymers obtaining. The properties and degradation time of PLGA vary and depend on the content of its constituent comonomers. The most commonly used are systems with lactide to glycolide ratios of 90/10, 80/20, 85/15 or 75/25. However, the change in copolymer properties is not a linear function of the individual comonomers sharing in its composition. In general, the content of lactic acid units in the copolymer improves the elastic properties, while the addition of glycolic acid improves the mechanical properties and regulates the time of the copolymer degradation, which can be from several weeks to several months (Blasi, 2019; Sharma et al., 2019). The final properties of the PLGA copolymer depend on its crystallinity, molar mass and processing method. PLGA is used for different biomedical applications, often as delivery systems (Roointan et al., 2018).

1.2.2   BIODEGRADABLE SYNTHETIC POLYMERS FROM PETROCHEMICAL SOURCES

The group II in Table 1.1 includes PCL, PBS and their copolymers poly(1,4-butylene terephthalate-*co*-1,4-butylene succinate) (PBTS) or poly(1,4-butylene succinate-*co*-1,4-butylene adipate) (PBSA), poly(1,4-butylene terephthalate) (PBT) and poly(1,4-butylene adipate-*co*-1,4-butylene terephthalate) (PBAT, Ecoflex®).

1.2.2.1   Poly(ε-caprolactone) (PCL)

PCL, aliphatic linear polyester, is produced via ring-opening polymerisation of ε-caprolactone using a different types of anionic, cationic and coordination catalysts (Chen et al., 2015). As a polyester, PCL has the ability to non-enzymatic hydrolytic degradation leading to sequential fragmentation in the first step followed by decomposition and complete degradation over a period of about two years to non-toxic caproic acid as a degradation product. It is a rubber-like material with an average degree of crystallinity and, as a plasticiser, it has ability to lower the modulus of elasticity. PCL has good processing capabilities; it is also the most flexible among biodegradable polymers (with a deformation up to 700%). Given its interesting properties, this polymer is used in industry as an additive to biodegradable packaging and films and is also blended with starch as well as can be used to make plastic disposable plates or cups. PCL also forms blends with other polymers, which allows for the improvement of their properties, including the degree and rate of degradability (El-Bakary et al., 2019). PCL exhibits tissue compatibility, which means that it can be used in the pharmaceutical and medical industries as an ideal material for long-time medical devices. This polymer has been used in tissue engineering such as staples, dressings or drug delivery systems. Recently, a new method of obtaining PCL using pressure polymerisation of ε-caprolactone has been patented. Changing the conditions under which the ε-caprolactone polymerisation process is carried out allows the production of higher-purity polymers. An alternative turned out to be the use of water as the initiator of the chemical reaction and high pressure as its catalyst. The presence of water allows one to control the course of the reaction, while carrying it out under high pressure allows one to obtain a product of high purity, meaning, among other things, no metal ions and no organic and inorganic pollutants. The PCL obtained in this way can be used not only in industry but also in medicine, e.g. for the production of surgical sutures, as a drug carrier or as a skeleton in tissue engineering. Moreover, the proposed PCL producing allows the composition of the reaction mixture to be simplified, which results in lower production costs (Das et al., 2020).

1.2.2.2   Poly(1,4-butylene succinate) (PBS)

PBS is produced by polymerisation of succinic acid and butylene glycol. PBS has great impact resistance, tearing strength, ductility and malleability. The heat resistance and processing behaviour of PBS are shown to be the greatest among all biodegradable polymers. PBS can be readily decomposed into H_2O and CO_2 by microbes or enzymes existing in animals and plants, while it is highly stable when used and stored properly. PBS is compostable according to the Biodegradable Products Institute and is approved for contact with food. PBS found applications, among others, as films and packaging materials, bottles for cosmetics or drinks, disposable medical products, garbage bags and meal boxes. Go Yen Chemical Industrial Co, Ltd (GYC GROUP), Hexing Chemical (Anhui, China), Xinfu Pharmaceutical (Hangzhou, China) and IRe Chemical (South Korea) are leading manufacturers of poly(1,4-butylene succinate). GOYENCHEM-PBS102, developed by GYC GROUP, features remarkable mechanical properties, processing behaviour and formability, which makes it ideal for extrusion moulding and extrusion blow moulding in producing agricultural films, food packaging materials, zipper bags, semi-transparent containers or masterbatches. Mitsubishi Chemicals produced PBS under trade name GS Pla (Green and Sustainable Plastic) (Rafiqah et al., 2021). Recently biomass resources have been used for chemosynthetic polymer production. 1,4-butanediol and succinic acid were obtained from furfural and used as monomers for first fully biomass-based poly(1,4-butylene succinate) (Tachibana et al., 2017). Bio-based succinic acid was used not only for the production of PBS but also for PBSA. The share of PBA aliphatic blocks in copolymers improves their biodegradation. However, despite its good biodegradability in various environments, PBSA is not widely used due to the relatively high cost of production (Seggiani et al., 2019).

In general, aliphatic polyesters are more prone to degradation but have poorer mechanical properties compared to aromatic polymers. In order to improve the physicochemical properties of biodegradable aliphatic polyesters, other monomers are incorporated into their molecular chains. On the other hand, incorporation of aliphatic components into the aromatic polyester chains results in the degradability of the obtained copolymer. One of such copolymers obtained from aliphatic and aromatic monomers is poly(1,4-butylene adipate-co-1,4-butylene terephthalate) (Siegenthaler et al., 2012).

1.2.2.3 Poly(1,4-butylene adipate-co-1,4-butylene terephthalate) (PBAT)

PBAT is a terpolymer consisting of three comonomers arranged statistically in a chain. It is produced under the trade name Ecoflex® by BASF. This polymer is obtained from adipic acid, terephthalic acid and 1,4-butanediol via polycondensation reaction, and its mechanical properties resemble low-density polyethylene (PE) and PCL. PBAT can be processed at a temperature of 110 to 130°C by means of extrusion or injection. It should be emphasised that PBAT (unlike some polyesters) does not require drying before processing. It exhibits a low glass transition temperature of –22°C and its properties can be tailored using different content of terephthalate units. PBAT is used for the production of agricultural and hygienic films, for coating paper and the production of disposable packaging. PBAT is very easily biodegradable and within a few weeks disintegrated into CO_2, H_2O and biomass. After 45 days of incubation in industrial compost it reaches 80% of biodegradability (Muroi et al., 2016, Jian et al., 2020).

1.2.3 BIOSTABLE SYNTHETIC POLYMERS FROM RENEWABLE RAW MATERIALS

The bio-based poly(ethylene terephthalate) (bio-based PET), bio-based PP or bio-based PE constituting group III in Table 1.1 are examples of the bio-related non-biodegradable bioplastics friendly to nature. Non-biodegradable polymers obtained from renewable raw materials have appeared on world markets relatively recently. They are gaining more and more interest both in industry and science. The main reason for starting their production was the exhausting resources of fossil raw materials and thus the exploration of alternative sources of raw materials for the production of conventional thermoplastic polymers. It is estimated that nature produces around 170 billion tons of biomass annually. Of this huge amount of biomass, humanity uses only about 3.5% (6 billion tons), of which approximately 62% is intended for food production, 33% is for energy and paper production and 5% is converted into chemical compounds (Rebouillat and Pla, 2016).

1.2.3.1 Bio-based Polyethylene (bio-based PE)

Technologies for the production of non-biodegradable polymers from renewable raw materials (mainly of plant origin) have already been developed by several global companies. The most advanced work in this area is carried out by Braskem in Brazil, which currently produces, among other things, PE and PP from renewable raw materials. I'm green™ PE produced by Braskem is a bioplastic whose production is carried out via biotechnological and chemical processes. The corn starch, sugarcane and wheat starch are used as sugar sources for the fermentative conversion of ethanol by microorganisms. Next, according to the company's technology, ethanol is dehydrated and the ethylene obtained on this route is a substrate in the polymerisation processes. The obtained I'm green™ PE, with the same chemical structure, has also the same properties as their counterparts made from raw petrochemicals materials. Bio-based PE with a wide variety of applications and processing methods is also recyclable in the same stream as conventional PE. I'm green™ PE is used especially in the production of rigid and flexible packaging (for food, drinks, detergents and cosmetics), toys, cans and plastic bags. To identify products that contain green PE in their composition, Braskem has created a special I'm green™ logo to help customers identify this series. However, it should be remembered that although I'm green™ PE is bio-based, it is not biodegradable (Braskem, 2014).

1.2.3.2 Bio-based Poly(ethylene terephthalate) (Bio-based PET)

Bio-based PET is a high-performance engineering plastic with physical properties that are suitable enough to be applied as bottles, fibres, films and engineering applications but with poor sustainability and degradability. The bio-based PET is also produced from green ethanol from sugar cane or corn starch, which are converted to mono(ethylene glycol). Unfortunately, the aromatic part of PET – terephthalic acid – is still derived from fossil sources and obtained by the standard transesterification process. The leading manufacturer of bio-based PET is Plant PET Technology Collaborative including Coca Cola, Ford, Heinz, Nike and Proctor & Gamble, which worked on the bio-based PET production under the name "PlantBottle" which will be fully derived from biomass (Andreeßen and Steinbüchel, 2019).

Classic polymers from renewable raw materials are also produced by companies such as DuPont (production of polydioxanone), SolVin (production of poly(vinyl chloride) (PVC)) or BASF (production of certain types of polyamides). This new group of polymers is developing dynamically, especially in South America, rich in renewable resources such as sugar cane. However, currently the land used for the cultivation of renewable raw materials for the production of bioplastics amounted to approximately 0.79 million ha in 2019, which was less than 0.02% of the global 4.8 billion ha of agricultural area. These indicate great opportunities for the development of new bioplastics and indirectly also of agriculture and other sectors of the economy. In 2021, the global production capacity of bioplastics amounted to approximately 2.42 million metric tons, of which almost 64% (1.5 million metric tons) were biodegradable plastics, in this PHA, PLA, PBAT and others. As far as non-biodegradable bioplastics, bio-based PE and bio-based PET show the highest relative share in the global production of bioplastics. However, bio-based PP, which had high commercial potential, entered mass production. Its production capacity is expected to increase almost twice by 2026 (Bioplastics market data).

1.2.4 BIOSTABLE SYNTHETIC POLYMERS FROM PETROCHEMICAL SOURCES

Group IV (Table 1.1) contain the non-biodegradable polymers, the so-called "big five" with the largest share in the plastics market: PE, PP, PVC, PS and PET. In 2020, the demand for plastics was at the level of 367 million metric tons and the packaging, construction and automotive industries accounted for 70% of the final plastics market in the European Union (EU) countries (World Plastics Production, 2020).

1.2.4.1 Polyethylene (PE) and Polypropylene (PP)

Polyethylene is a white, thermoplastic, porous substance with a density of 0.92–0.97 $g \cdot cm^{-3}$ and a melting point of 110–137°C. PE is obtained by the polymerisation of ethylene and is used in the production of films, pipes, hoses, containers, electrical insulating materials, hockey sticks, skis, sails, ropes, bulletproof vests, toys and packaging. PE is an excellent dielectric, characterised by considerable flexibility, good mechanical properties and high resistance to alkalis, acids and salts (Nakajima et al., 2017).

Polypropylene has a density of 0.90–0.91 $g \cdot cm^{-3}$ and is obtained by the polymerisation of propane. It is a polymer lighter than PE. It shows a higher melting point and greater strength. It is used in the production of films, linings, gears, containers and pipes. Properties of PP, like other plastics, depend on the molar mass, preparation methods, degree of molar-mass, dispersity and crystallinity. PP is characterised by a higher melting and boiling point, lower resistance to low temperature and resistance to the oxidation process than PE. It is non-polar and shows quite good resistance to acids (excluding oxidising acids), bases, alcohols and various types of solvents. It is not resistant to aromatic hydrocarbons, and it can dissolve in warm xylene. It is characterised by high resistance to scratches and damage. PP is used in industry for the production of printing rollers, gears and dishes (Alsabri et al., 2021).

1.2.4.2 Poly(vinyl chloride) (PVC)

Poly(vinyl chloride) is a thermoplastic white substance with a small share of the crystalline phase. PVC has a powder consistency and density of 1.35–1.46 $g \cdot m^{-3}$, a molar mass of 30–150 thousand $g \cdot mol^{-1}$, high stiffness and good strength properties. It does not dissolve in many non-polar solvents and it is resistant to water, concentrated and diluted acids and bases as well as mineral oils. PVC is one of the most important and most produced thermoplastics. PVC is used in the construction industry (tiles and linings, pipes and fittings), the electrical and radio engineering industry and for the production of cellulose and paper, elastomers and fibres, clothing and footwear, as well as packaging. It is used in the power industry (electrical insulating material), in medicine (drains, probes, catheters, syringes) as well for covering volleyball, basketball and handball fields. PVC is resistant to hydrochloric acid, sulfuric acid, diluted nitric acid, NaOH, KOH, ammonia, soda solution, alcohol, gasoline. It can dissolve in acetone, cyclohexanone, esters, dioxane, toluene, pyridine, xylene, carbon disulphide, ethylene chloride and dimethylformamide. Hard PVC can be used in the production of pipes used in the chemical industry, water and sewage pipes, drawing tools, packaging, ventilation systems, pumps and tanks. Soft PVC can be used in the production of garden hoses, hoses for the chemical industry or for electrical insulation, for various types of cladding, gaskets, roof panels, foil and artificial leather (Omnexus).

1.2.4.3   Polystyrene (PS)

Polystyrene is a relatively hard, colourless and brittle material, characterised by a low melting point. PS is used for the production of Styrofoam – a very light insulating material, in the production of containers, toys, packaging, artificial jewellery, toothbrushes and CD boxes. PS is brittle, has a low thermal conductivity and is suitable for polishing. PS is a thermoplastic material; it softens at 70°C, while at a temperature of 100–110°C it can be mouldable. PS is resistant to acids, bases, alcohols, saturated hydrocarbons, mineral and vegetable oils. It can dissolve in styrene, benzene, toluene, carbon disulphide, dioxane and cyclohexane. Expanded polystyrene is a low-temperature PS emulsion with a blowing agent. PS is used for the production of electrical insulating materials, parts for refrigeration devices, dishes, containers, toys and accessories. PS is an excellent dielectric, characterised by considerable flexibility, good mechanical properties and high resistance to alkalis, acids and salts (Johnston et al., 2018).

1.2.4.4   Poly(ethylene terephthalate) (PET)

Poly(ethylene terephthalate) is a thermoplastic aliphatic polyester with good dimensional stability, resistance to impact, moisture, alcohols and solvents, oils, grease and diluted acids. Chemically, PET is very much similar to poly(butylene terephthalate). PET is obtained via polycondensation reaction of the monomers obtained by the esterification reaction between terephthalic acid and ethylene glycol or by transesterification reaction between ethylene glycol and dimethyl terephthalate. The use of 2,5-furandicarboxylic acid and mono(ethylene glycol) makes it possible to obtain poly(ethylene 2,5-furandicarboxylate) (PEF), which is a 100% bio-based polyester with barrier and thermal properties similar to PET. 2,5-Furandicarboxylic acid obtained from biosources (sugars) in the biofermentation process can replace terephthalic acid from crude oil (Eerhart et al., 2012). PET is recyclable and transparent to microwave radiation. Recycled PET can be used as fibres for fabrics, sheets for packaging and manufacturing automotive parts. Blending PET with other polymers results in new materials having improved properties at favourable cost profiles, which opens up new markets and application potential without much investment and development. The thermoplastic polymers that are used to produce blends with PET are PE, PP, polycarbonates, PS, poly(ethylene-*co*-vinyl acetate) (EVA copolymer) and acrylonitrile-butadiene-styrene terpolymer (ABS). Polyolefin-modified PET is often reinforced with glass fibre and used in injection moulded automotive and industrial applications. PET with PC blends includes those requiring a combination of excellent toughness, chemical and heat resistance along with high impact, tensile and flexural strength. Blending significantly improves thermal and mechanical elements, impact resistance and flame retardant properties (Zimmermann, 2020). PET structure, depending on the production and thermal processing conditions, consists of both amorphous and semi-crystalline phases. The more flexible amorphous regions are more susceptible to hydrolysis by polyester hydrolases such as lipases and cutinases from fungi and bacteria. It was found that PET degradation leads to bis(2-hydroxyethyl) and mono(2-hydroxyethyl) terephthalate as intermediate products. The final degradation products are the monomers: terephthalic acid and ethylene glycol (Wei et al., 2016; Austin et al., 2018).

1.3 CONCLUSIONS

In summary, the biodegradable polymers can be of natural origin (biopolymers) or synthetically obtained by biotechnological means (PHA) or from bio-derived monomers, which will later be polymerised, as in the case of PLA obtained from lactic acid or by traditional synthesis, as in the case of polymers, such as PCL, PBAT or PBS. PBS and PBAT are now the biodegradable plastics that have been used frequently. Unfortunately, biodegradable polymers are still not considered a serious alternative to conventional plastics due to high production costs. The industrial interest in the production of biodegradable polymers is depending first on the price of oil because the production of conventional plastics compared with PHAs is much cheaper and therefore seems to be more profitable. However, the application of different strategies to reduce the production cost of PHA, in particular through the utilisation of alternative carbon sources, should influence the growing interest in biodegradable polymers.

REFERENCES

Adamus, G., Dominski, A., Kowalczuk, M., Kurcok, P., and Radecka, I. 2021. From anionic ring-opening polymerization of β-butyrolactone to biodegradable poly(hydroxyalkanoate)s: Our contributions in this field. *Polymers* 13: 4365.

Adamus, G., Sikorska, W., Janeczek, H., Kwiecień, M., Sobota, M., and Kowalczuk, M. 2012. Novel block copolymers of atactic PHB with natural PHA for cardiovascular engineering: Synthesis and characterization. *Eur. Polym. J.* 48(3): 621–631.

Alsabri, A., Tahir, F., and Al-Ghamdi, S.G. 2021. Life-cycle assessment of polypropylene production in the Gulf Cooperation. *Polymers* 13(21): 3793.

Álvarez-Chávez, C.R., Edwards, S., Moure-Eraso, R., and Geiser, K. 2012. Sustainability of bio-based plastics: General comparative analysis and recommendations for improvement. *J. Clean. Prod.* 23: 47–56.

Andreeßen, Ch., and Steinbüchel, A. 2019. Recent developments in non-biodegradable biopolymers: Precursors, production processes, and future perspectives. *Appl. Microbiol. Biotechnol.* 103: 143–157.

Anis, S.N.S., Nurhezreen, M.I., Kumar, S., and Amirul, A.A. 2013. Effect of different recovery strategies of P(3HB-*co*-3HHx) copolymer from Cupriavidusnecator recombinant harboring the PHA synthase of *Chromobacterium sp.* USM2. *Sep. Purif. Technol.* 102: 111–117.

Anjum, A., Zuber, M., Zia, K.M., Noreen, A., Anjum, M.N., and Tabasum, S. 2016. Microbial production of polyhydroxyalkanoates (PHAs) and its copolymers: A review of recent advancements. *Int. J. Biol. Macromol.* 89: 161–174.

Austin, H.P., Allen, M.D., Donohoe, B.S., et al. 2018. Characterization and engineering of a plastic-degrading aromatic polyesterase. *PNAS USA* 115: E4350–E4357.

Bartczak, Z., Galeski, A., Kowalczuk, M., Sobota, M., and Malinowski, R. 2013. Tough blends of poly(lactide) and amorphous poly([*R,S*]-3-hydroxy butyrate) – Morphology and properties. *Eur. Polym. J.* 49: 3630–3641.

Benesova, P., Kucera, D., Marova, I., and Obruca, S., 2017. Chicken feather hydrolysate as an inexpensive complex nitrogen source for PHA production by Cupriavidusnecator on waste frying oils. *Lett. Appl. Microbiol.* 65(2): 182–188.

Bergstrom, J.S., and Hayman, D. 2016. An overview of mechanical properties and material modeling of polylactide (PLA) for medical applications. *Ann. Biomed. Eng.* 44(2): 330–340.

Bhatia, S.K., Gurav, R., Choi, T.R., et al. 2019. Bioconversion of plant biomass hydrolysate into bioplastic (polyhydroxyalkanoates) using *Ralstonia eutropha* 5119. *Bioresour. Technol.* 271: 306–315.

Bhatia, S.K., Kim, J.H., Kim, M.S., et al. 2018. Production of (3-hydroxybutyrate-*co*-3-hydroxyhexanoate) copolymer from coffee waste oil using engineered *Ralstonia eutropha*. *Bioprocess. Biosyst. Eng.* 41(2): 229–235.

Bhatia, S.K., Yoon, J.J., Kim, H.J., et al. 2018. Engineering of artificial microbial consortia of *Ralstonia eutropha* and *Bacillus subtilis* for poly(3-hydroxybutyrate-*co*-3-hydroxyvalerate) copolymer production from sugarcane sugar without precursor feeding. *Bioresour. Technol.* 257: 92–101.

Bio-on. *Special PHAs Production Plant* (inauguration summer 2018). www.bio-on.it/production.php#p1 (accessed February 4, 2022).

Bioplastics Market Data. www.european-bioplastics.org/market (accessed April 29, 2022).

Blasi, P. 2019. Poly(lactic acid)/poly(lactic-*co*-glycolic acid)-based microparticles: An overview. *J. Pharm. Investig.* 49: 337–346.

Braskem. 2014. *I'm Green™ Polyethylene. Innovation and Differentiation for Your Product.* www.braskem.com.br/Portal/Principal/Arquivos/ModuloHTML/Documentos/846/AF_Catalogo_PE%20Verde_2014_ING_site.pdf (accessed May 16, 2022).

Chen, T., Cai, T., Jin, Q., and Ji, J. 2015. Design and fabrication of functional polycaprolactone. *e-Polymers* 15(1): 3–13.

Ciesielski, S., Mozejko, J., and Pisutpaisal, N. 2015. Plant oils as promising substrates for polyhydroxyalkanoates production. *J. Clean. Prod.* 106: 408–421.

Danimer Scientific. *Danimer Scientific Opens World's First Commercial PHA Production Facility in Winchester, KY.* https://danimerscientific.com/commercial-manufacturing-new-plant (accessed February 4, 2022).

Das, M., Mandal, B., and Katiyar, V. 2020. Sustainable routes for synthesis of poly(ε-caprolactone): Prospects in chemical industries. In *Advances in Sustainable Polymers. Synthesis, Fabrication and Characterization*, eds. Katiyar, V., Kumar, A., and Mulchandani, N., pp. 21–33. Cham: Springer International Publishing.

Davis, R., Duane, G., Kenny S.T., et al. 2014. High cell density cultivation of *Pseudomonas putida* KT2440 using glucose without the need for oxygen enriched air supply. *Biotechnol Bioeng.* 112: 725–733.

Dietrich, K., Durmont, M., Del Rio, F., and Orsat, V. 2017. Producing PHAs in the bioeconomy – Towards a sustainable bioplastic. *Sustain. Prod. Consum.* 9: 58–70.

Directive 2019/904. 2019. Directive (EU) 2019/904 of the European Parliament and of the Council of 5 June 2019 on the reduction of the impact of certain plastic products on the environment.

Dusselier, M., Van Woude, P., Dewaele, A., Jacobs, P., and Sels, B. 2015. Shape-selective zeolite catalysis for bioplastics production. *Science* 349: 6243.

EEA Report No 8/2018. *The Circular Economy and the Bioeconomy – Partners in Sustainability.* www.eea.europa.eu/publications/circular-economy-and-bioeconomy (accessed December 15, 2021).

Eerhart, A.J.J.E., Faaij, A.P.C., Patel, M.K. 2012. Replacing fossil-based PET with biobased PEF; Process analysis, energy and GHG balance. *Energy Environ. Sci.* 5: 6407.

El-Bakary, M.A., El-Farahaty, K.A., and El-Sayed, N.M. 2019. In vitro degradation characteristics of polyglycolic/polycaprolactone (PGA/PCL) copolymer material using Mach-Zehnder interferometer. *Mater. Res. Express.* 6: 105374.

Emadian, S.M., Onay, T.T., and Demirel, B. 2017. Biodegradation of bioplastics in natural environments. *Waste Manage.* 59: 526–536.

Esa, F., Tasirin, S.M., and Rahma, N. 2014. Overview of bacterial cellulose production and application. *Agric. Agric. Sci. Procedia.* 2: 113–119.

European Commission. 2018. *The Commission's Plan for Plastics.* https://ec.europa.eu/info/research-and-innovation/research-area/environment/circular-economy/plastics-circular-economy_en (accessed December 21, 2021).

Fonseca, G.G., and Antonio, R.V. 2006. Polyhydroxyalkanoates production by recombinant *Escherichia coli* harboring the structural genes of the polyhydroxyalkanoate synthases of *Ralstonia eutropha* and *Pseudomonas aeruginosa* using low cost substrate. *J. Appl. Sci.* 6: 1745–1750.

Greene, J.P. 2014. *Sustainable Plastics: Environmental Assessments of Biobased, Biodegradable, and Recycled Plastics.* Hoboken: John Wiley & Sons, Inc.

Grupa Azoty Tarnów. *Envifill® skrobia termoplastyczna.* https://tarnow.grupaazoty.com/nasza-oferta/polimery-biodegradowalne/envifill-skrobia-termoplastyczna (accessed February 4, 2022) (in Polish).

Guzik, M., Witko, T., Steinbüchel, A., Wojnarowska, M., Sołtysik, M., and Wawak, S. 2020. What has been trending in the research of polyhydroxyalkanoates? A systematic review. *Front. Bioeng. Biotechnol.* 8: 959.

Haider, T.P., Völker, C., Kramm, J., Landfester, K., and Wurm, F.R. 2019. Plastics of the future? The impact of biodegradable polymers on the environment and on society. *Angew. Chem. Int. Ed. Engl.* 58(1): 50–62.

Hu, Y., Sato, H., Zhang J., Noda, I., and Ozaki, Y. 2008. Crystallization behavior of poly(*L*-lactic acid) affected by the addition of a small amount of poly(3-hydroxybutrate). *Polymer* 49: 4204–4210.

Hyde, E. 2019. *A Two-for-one Punch: How Mango Materials Turns Waste Methane into Biodegradable Materials.* www.builtwithbiology.com/read/a-two-for-one-punch-how-mango-materials-turns-waste-methane-into-biodegradable-materials (accessed February 4, 2022).

James, R., Manoukian, O.S., and Kumbar, S.G. 2016. Poly(lactic acid) for delivery of bioactive macromolecules. *Adv. Drug Deliv. Rev.* 107: 277–288.

Jian, J., Xiangbin, Z., and Xianbo, H. 2020. An overview on synthesis, properties and applications of poly(butylene-adipate-*co*-terephthalate) – PBAT. *Adv. Ind. Eng. Polym. Res.* 3(1): 19–26.

Jiang, X.J., Sun, Z., Ramsay, J.A., and Ramsay, B.A. 2013. Fed batch production of MCL-PHA with elevated 3-hydroxynonanoate content. *AMB Express* 3: 50.

Johnston, B., Radecka, I., Chiellini, E., et al. 2019. Mass spectrometry reveals molecular structure of polyhydroxyalkanoatesa attained by bioconversion of oxidized polypropylene waste fragments. *Polymers* 11(10): 1580.

Johnston, B., Radecka, I., Hill, D., et al. 2018. The microbial production of polyhydroxyalkanoates from waste polystyrene fragments attained using oxidative degradation. *Polymers* 10: 957.

Kaur, G., Srivastava, A., and Chand, S. 2012. Advances in biotechnological production of 1,3-propanediol. *Biochem. Eng. J.* 64: 106–118.

Kaur, L., Khajuria, R., Parihar, L., and Dimpal Singh, G. 2017. Polyhydroxyalkanoates: Biosynthesis to commercial production – A review. *J. Microbiol. Biotechnol. Food Sci.* 6: 1098–1106.

Kirchherr, J., Piscicelli, L., Boura, R., et al. 2018. Barriers to the circular economy: Evidence from the European Union (EU). *Ecol Econ.* 150: 264–272.

Koller, M., Marsalek, L., Sousa Dias, M.M., and Braunegg, G. 2017. Producing microbial polyhydroxyalkanoate (PHA) biopolyesters in a sustainable manner. *New Biotechnol.* 37: 24–38.

Koseva, N.S., Novakov, Ch.P., Rydz, J., Kurcok, P., and Kowalczuk, M. 2010. Synthesis of aPHB-PEG brush co-polymers through ATRP in a macroinitiator-macromonomer feed system and their characterization. *Des. Monomers Polym.* 13: 579–595.

Kumar, R., Jha, D., and Panda, A.K. 2019. Antimicrobial therapeutics delivery systems based on biodegradable polylactide/polylactide-*co*-glycolide particles. *Environ. Chem. Lett.* 17: 1237–1249.

Kundu, R., and Payal, P. 2021. Biodegradation study of potato starch-based bioplastic. *Curr. Chin. Chem.* e190421192895.

Li, Z., Yang, J., and Loh, X.J. 2016. Polyhydroxyalkanoates: Opening doors for a sustainable future. *NPG Asia Mater.* 8: e265.

Liu, K.L., Goh, S.H., and Li, J. 2008. Controlled synthesis and characterizations of amphiphilic poly[(R,S)-3-hydroxybutyrate]-poly(ethylene glycol)-poly[(R,S)-3-hydroxybutyrate] triblock copolymers. *Polymer* 49: 732–741.

Liu, M.H., Chen, Y.J., and Lee, C.Y. 2018. Characterization of medium chain-length polyhydroxyalkanoate biosynthesis by *Pseudomonas mosselii* TO7 using crude glycerol. *Biosci. Biotechnol. Biochem.* 82(3): 532–539.

Maharana, T., Mohanty, B., and Negi, Y.S. 2009. Melt-solid polycondensation of lactic acid and its biodegradability. *Prog. Polym. Sci.* 34: 99–124.

MarketsandMarkets. 2021. *Polyhydroxyalkanoate (PHA) Market.* www.marketsandmarkets. com/Market-Reports/pha-market-395.html (accessed December 15, 2021).

Martinez Villadiego, M., Arias Tapia, M.J., Useche, J., and Escobar Macías, D. 2022. Thermoplastic starch (TPS)/polylactic acid (PLA) blending methodologies: A review. *J. Polym. Environ.* 30: 75–91.

Molina-Ramírez, C., Enciso, C., Torres-Taborda, M., et al. 2018. Effects of alternative energy sources on bacterial cellulose characteristics produced by *Komagataeibactermedellinensis*. *Int. J. Biol. Macromol.* 1(117): 735–741.

Moshood, T.D., Nawanir, G., Mahmud, F., Mohamad, F., Ahmad, M.H., and AbdulGhani, A. 2022. Sustainability of biodegradable plastics: New problem or solution to solve the global plastic pollution? *Curr. Res. Green Sustain. Chem.* 5: 100273.

Mozejko-Ciesielska, J., and Kiewisz, R. 2016. Bacterial polyhydroxyalkanoates: Still fabulous? *Microbiol. Res.* 192: 271–282.

Mozejko-Ciesielska, J., Pokoj, T., and Ciesielski, S. 2018. Transcriptome remodeling of *Pseudomonas putida* KT2440 during mcl-PHAs synthesis: Effect of different carbon sources and response to nitrogen stress. *J. Ind. Microbiol. Biotechnol.* 45: 433–446.

Mozejko-Ciesielska, J., Szacherska, K., and Marciniak, P. 2019. *Pseudomonas species* as producers of eco-friendly polyhydroxyalkanoates. *J. Polym. Environ.* 27: 1151–1166.

Muroi, F., Tachibana, Y., Kobayashi, Y., Sakurai, T., and Kasuya, K. 2016. Influences of poly(butylene adipate-*co*-terephthalate) on soil microbiota and plant growth. *Polym. Degrad. Stab.* 129: 338–346.

Musioł, M., Sikorska, W., Adamus G., et al. 2015. (Bio)degradable polymers as a potential material for food packaging: Studies on the (bio)degradation process of PLA/(R,S)-PHB rigid foils under industrial composting conditions. *Eur. Food Res. Technol.* 242: 815–823.

Nakajima, H., Dijkstra, P., and Loos K. 2017. The recent developments in biobased polymers toward general and engineering applications: Polymers that are upgraded from biodegradable polymers, analogous to petroleum-derived polymers, and newly developed. *Polymers* 9: 523.

Neugebauer, D., Rydz, J., Goebel, I., Dacko, P., and Kowalczuk, M. 2007. Synthesis of graft copolymers containing biodegradable poly(3-hydroxybutyrate) chains. *Macromolecules* 40: 1767–1773.

Nielsen, C., Rahman, A., Rehman, A.U., Walsh, M.K., and Miller, C.D. 2017. Food waste conversion to microbial polyhydroxyalkanoates. *Microb. Biotechnol.* 10: 1338–1352.

Obruca, S., Benesova, P., Oborna, J., and Marova, I. 2014. Application of protease-hydrolyzed whey as a complexnitrogen source to increase poly(3-hydroxybutyrate)production from oils by *Cupriavidus necator. Biotechnol. Lett.* 36: 775–781.

Omnexus. *Comprehensive Guide on Polyvinyl Chloride (PVC).* https://omnexus.specialchem. com/selection-guide/polyvinyl-chloride-pvc-plastic?gclid=EAIaIQobChMIytWh-KC-59wIVHkiRBR3vAQiMEAMYASAAEgJn8fD_BwE (accessed April 29, 2022).

Padovani, G., Carlozzi, P., Seggiani, M., Cinelli, P., Vitolo, S., and Lazzeri, A. 2016. PHB-rich biomass and BioH$_2$ by means of photosynthetic microorganisms. *Chem. Eng. Trans.* 49: 55–60.

Rafiqah, S.A., Khalina, A., Harmaen, A.S., et al. 2021. A review on properties and application of bio-based poly(butylene succinate). *Polymers* 13: 1436.

Rebouillat, S., and Pla, F. 2016. Recent strategies for the development of biosourced-monomers, oligomers and polymers-based materials: A review with an innovation and a bigger data focus. *J. Biomater. Nanobiotechnol.* 7(4): 167–213.

Riedel, S.L., Jahns, S., Koenig, S., et al. 2015. Polyhydroxyalkanoates production with *Ralstoniaeutropha* from low quality waste animal fats. *J. Biotechnol.* 214: 119–127.

Roointan, A., Kianpour, S., Memari, F., Gandomani, M., Gheibi Hayat, S.M., and Mohammadi-Samani, S. 2018. Poly(lactic-*co*-glycolic acid): The most ardent and flexible candidate in biomedicine! *Int. J. Polym. Mater. Polym. Biomater.* 67: 1028.

Rydz, J., Sikorska, W., Kyulavska, M., and Christova, D. 2015. Polyester-based (bio)degradable polymers as environmentally friendly materials for sustainable development. *Int. J. Mol. Sci.* 16(1): 564–596.

Seggiani, M., Gigante, V., Cinelli, P., et al. 2019. Processing and mechanical performances of poly(butylene succinate-*co*-adipate) (PBSA) and raw hydrolyzed collagen (HC) thermoplastic blend. *Polym. Test.* 77: 105900.

Shaghaleh, H., Xu, X., and Wang, S. 2018. Current progress in production of biopolymeric materials based on cellulose, cellulose nanofibers, and cellulose derivatives. *RSC Adv.* 8: 825–842.

Shah, M., Ullah, N., Choi, M.H., Kim, M.O., and Yoon, S.C. 2012. Amorphous amphiphilic P(3HV-*co*-4HB)-*b*-mPEG block copolymer synthesized from bacterial copolyester via melttransesterification: Nanoparticle preparation, cisplatin-loading for cancer therapy and *in vitro* evaluation. *Eur. J. Pharm. Sci.* 80: 518–527.

Sharma, D., Lipp, L., Arora, S., and Singh, J. 2019. Diblock and triblock copolymers of polylactide and polyglycolide. In *Materials for Biomedical Engineering: Thermoset and Thermoplastic Polymers*, eds. Grumezescu, V., and Grumezescu, A., pp. 449–477. Amsterdam: Elsevier.

Shigematsu, T., Takamine, K., Kitazato M., et al. 2005. Cellulose production from glucose using glucose dehydrogenase gene (*gdh*)-deficient mutant of *Gluconacetobacterxylinus* and its use for bioconversion of sweet potato pulp. *J. Biosci. Bioeng.* 99: 415–422.

Siegenthaler, K., Kunkel, A., Skupin, G., and Yamamoto, M. 2012. Ecoflex® and Ecovio®: Biodegradable, performance-enabling plastics. *Adv. Polym. Sci.* 245: 91–136.

Somleva, M.N., Peoples, O.P., and Snell, K.D. 2013. PHA bioplastics, biochemicals, and energy from crops. *Plant Biotechnol.* 11: 233–252.

Song, J.H., Murphy, R.J., Narayan R., and Davies, G.B.H. 2009. Biodegradable and compostable alternatives to conventional plastics. *Philos. Trans. R. Soc. Lond. B Biol. Sci.* 364: 2127–2139.

Tachibana, Y., Yamahata, M., Ichihara, H., and Kasuya, K. 2017. Biodegradability of polyesters comprising a bio-based monomer derived from furfural. *Polym. Degrad. Stab.* 146: 121–125.

Tekin, E., Ates, M., and Kahraman, O. 2012. Poly-3-hydroxybutyrate-producing extreme halophilic archaeon, *Haloferax sp.* MA10 isolated from Çamaltı Saltern, İzmir. *Turk. J. Biol.* 36: 303–312.

Tokiwa, Y., Calabia, B.P., Ugwu, Ch.U., and Aiba, S. 2009. Biodegradability of plastics. *Int. J. Mol. Sci.* 10: 3722–3742.

Tomizawa, S., Saito, Y., Hyakutake, M., Nakamura J., Abe H., and Tsuge T. 2010. Chain transfer reaction catalyzed by various polyhydroxyalkanoate synthases with poly(ethylene glycol) as an exogenous chain transfer agent. *Appl. Microbiol. Biotechnol.* 87: 1427–1435.

Tripathi, A.D., Yadav, A., Jha, A., and Srivastava, S.K. 2012. Utilizing of sugar refinery waste (cane molasses) for production of bioplastic under submerged fermentation process. *J. Polm. Environ.* 20: 446–453.

Wang, J., Tavakoli, J., and Tang, Y. 2019. Bacterial cellulose production, properties and applications with different culture methods – A review. *Carbohydr. Polym.* 219: 63–76.

Wei, R., Oeser, T., Schmidt, J., et al. 2016. Engineered bacterial polyester hydrolases efficiently degrade polyethylene terephthalate due to relieved product inhibition. *Biotechnol. Bioeng.* 113: 1658–1665.

World Plastics Production. 2020. *Plastics Europe 2021.* https://plasticseurope.org/knowledge-hub/plastics-the-facts-2021 (accessed April 29, 2022).

Yang, X.Y., Huang, C., Guo, H.J., et al. 2013. Bioconversion of elephant grass (Pennisetum purpureum) acid hydrolysate to bacterial cellulose by *Gluconacetobacterxylinus. J. Appl. Microbiol.* 115: 995–1002.

Yoon, S.K., Yang, J.H., Lim, H.T., et al. 2021. *In vitro* and *in vivo* biosafety analysis of resorbable polyglycolic acid-polylactic acid block copolymer composites for spinal fixation. *Polymers* 13: 29.

Zhang, M., and Thomas, N.L. 2011. Blending polylactic acid with polyhydroxybutyrate: The effect on thermal, mechanical, and biodegradation properties. *Adv. Polym. Technol.* 30(2): 67–79.

Zimmermann, W. 2020. Biocatalytic recycling of polyethylene terephthalate plastic. *Philos. Trans. R. Soc.* A378: 20190273.

2 Degradation Principle and Regulations

Wanda Sikorska and Marta Musioł

CONTENTS

2.1 REGULATIONS – POLICY AND CERTIFICATION INCLUDING MARINE BIODEGRADABILITY STANDARDS

Plastic waste is becoming a growing problem. This is because most plastics are not recycled but are thrown away and pollute the environment. The increasing amount of plastic waste, especially that ends up in the oceans and other water reservoirs, makes it necessary to take action to eliminate the polymer litter harmful to all ecosystems. Much of the plastic waste entering in the aquatic environment comes from inappropriate waste management on land. Single-use plastic and fishing nets are some of the more problematic types of waste. Hundreds of thousands of sea turtles, seals, whales and seabirds die as a result of ingesting or becoming entangled in them. Waste of lightweight plastic bags that blow in the wind or float on the surface of the water is also harmful and dangerous materials for the environment. It has been found that plastic waste accounts for more than 60% of all marine litter that moves from land to water (Jambeck et al., 2015).

In the context of the aquatic environment litter by plastic waste, a European Union (EU) Single-use Plastics Directive has been announced (Directive 2019/904, 2019). The directive aims to limit the impact of certain plastic products on the environment. The Single-use Plastics Directive bans the use of certain single-use plastic products (e.g. straws, plates and cutlery) and for other products (e.g. cups or food packaging) requires Member States to introduce programs to reduce the consumption of these products. The danger is also the release of toxic chemicals from plastics into the marine environment that cumulate in water (Directive 2019/904, 2019). Therefore, it is necessary to understand and investigate the mechanisms of biodegradation in the marine environment. There are only a few standards for assessing the biodegradability of plastics in seawater, due to the varying conditions in this environment.

DOI: 10.1201/9780429352799-3

Two U.S. standards mainly used for determining aerobic biodegradation of plastic materials in marine habitats are ASTM D6691 and ASTM D7991 (ASTM D6691-17; ASTM D7991-15). The ASTM D6691 standard can be used for tests carried out under controlled laboratory conditions for polymer materials with potential biodegradability in a marine environment. According to the ASTM D6691 materials should achieve at least of 30% biodegradability, which is measured by the released carbon dioxide (CO_2, biogas) in a respirometer experiment run between 10 and 90 days (ASTM D6691-17). According to ASTM D7991 film of the tested materials should be incubated for up to 24 months in sandy marine sediment and seawater, and biodegradability is determined by measurements of CO_2 evolution (ASTM D7991-15; Harrison et al., 2018). The phytotoxicological aspect of the biodegradation process of commercially used biodegradable plastics is also important, plant ecotoxicity tests in accordance with the requirements of the Organization for Economic Co-operation and Development (OECD) should be performed. During this test, visible damages, such as necrosis and chlorosis or other inhibitory effects, inflicted on the selected plants (radish, cress and monocotyledonous oat) are studied (Rychter et al., 2010). In Europe, the ISO 14852 and ISO 16221 standards, which are similar to the ASTM standards, are used to assess biodegradability in the marine environment. The ISO 16221 standard is a guideline for determining biodegradability in the marine environment and for measuring water quality (ISO 14852:2021, 2021; ISO 16221:2001, 2001). The degradation tests in various aquatic laboratory conditions were conducted for particular polymers. Poly(3-hydroxybutyrate) (PHB) and thermoplastic starch showed that they were completely degraded under all tested conditions, i.e. in seawater (30°C), aerobic fresh water (21°C) and anaerobic sludge (35°C). The blend of PHB and polyhydroxyoctanoate (PHO) (85/15) under marine conditions after 56 days of incubation achieved the same level of biodegradation as PHB. Thus, the addition of PHO did not accelerate the biodegradation rate of PHB/PHO blends studied. Poly(ε-caprolactone) (PCL) also showed good biodegradation in the seawater, but again mixing PHB with PCL had no effect on the biodegradation efficiency in the tested marine environment. However, the global problem of plastic waste in the real aquatic environment still remains unsolved, despite numerous actions by all relevant bodies, including industry (Narancic et al., 2018). The need to reduce waste of the marine environment has been included in EU legal regulations as one of the components of a circular economy. At the same time, the plastics industry points out that plastics are a material of high value as a secondary raw material in recycling processes and, inter alia, for this reason, they should not end up in the environment at all. To reduce environmental pollution, selective waste collection is advocated as the basis for increased recycling and recovery of secondary raw materials (Allison and Bassett, 2015). The global plastics industry implements the voluntary Operation Clean Sweep® (OCS), which aims to prevent plastic pellets from being lost and released into the environment. The loss of plastic pellets can take place at any time during the manufacturing process of plastic products, including at the stage of production, transport and processing. Plastics Europe, which supervises the implementation of the OCS program in Europe, encourages producers to handle the pellets more consciously and publishes an annual report on this subject as well on the results of such actions (PlasticsEurope AISBL, 2021). Moreover, as plastics are becoming more prevalent in ship waste, their end-of-life disposal must be considered avoiding the dumping of plastics at sea. Recently the Global Plastics Alliance has developed many programs aimed at reliably estimating the extent of water pollution by plastic waste and its real impact on the ecosystem as well as to promote appropriate pro-environmental attitudes such as "Recycling Cruise – recover plastics" (Marine Litter Solutions). Particularly dangerous are oxo-degradable plastics made from conventional plastics and additives designed to mimic the biodegradation process. Unfortunately, the main effect of adding oxo-degradable additives is only the fragmentation of plastics into microplastics that remain in the environment, and thus these materials do not meet the compostability requirements in accordance with EN13432 standard (BS EN 13432, 2000). Therefore, such plastics with "oxo" additives should not be treated as biodegradable. What's more, they should be considered as hazardous to the environment and human health. Limiting the negative impact of waste requires a change in the approach to its production and disposal in line with the principles of green chemistry (Carbery et al., 2018). Compostability of plastics in industrial composting plants for packaging is determined by the EU 13432 standard. To assess the compostability of non-packaging plastics, the EN 14995 European standard is used (BS EN 14995, 2006). The biodegradability of mulch films in soil for use in agriculture is described by the EN 17033 standard. According to the requirements of the EN 17033 standard, at least 90% biodegradability at 25°C should be achieved within two years. Materials that meet the aforementioned standards can be certified and appropriately labelled. In addition to the aforementioned, the compostability of plastics is defined by numerous other national and international standards, the already mentioned European standard EU 13432 and the U.S. standard ASTM D-6400 (ASTM D6400-21). In order to meet these standards, the disintegration within 12 weeks must be over 90% for the 2 mm sieve fraction and the biodegradability within 180 days must be over 90%. According to these standards, the lack of phytotoxicity is proved by the fact that the values of the dry and fresh mass of seedlings as well as the number of germinated plants were greater than 90% for two selected plants. There are also various standards for measuring the content of renewable organic materials, including bioplastics, such as EN 16640, which describes how to measure the ^{14}C isotope (radiocarbon method). EN 16785-1 covers other bioelements in the polymer by elemental analysis and EN 16785-2 describes the material balance method to determine the renewable content of a bio-product and requires a minimum 50% of organic compounds as well minimum of 20% renewable carbon and non-toxicity in materials (BS EN 16640, 2017; BS EN 16785-1, 2015; BS EN 16785-2, 2018). The other standards for packaging and compostable plastics are ISO 18606 and ISO 17088 which standards contain similar requirements as EU 13432 (ISO 18606:2013, 2013; ISO 17088:2021, 2021).

A significant area of application of biodegradable polymer materials is medicine, where their use results from the in vivo susceptibility of materials to degradation in the environment such as body fluids to non-toxic components. It is most often associated with the release into the environment of the organism of degradation products resulting from the chemical and/or enzymatic degradation process of the materials. Chemical degradation is induced by chemical factors such as bases, acids, solvents or reactive gases. Aquatic environment promotes the hydrolytic degradation of polymers (hydrolysis) and leads to a decrease in the molar mass of the test sample as a result of the statistical cutting of the polymer chain. This causes deterioration of the mechanical properties of the material, leading to its disintegration. During hydrolytic degradation, the crystallinity, degree of crosslinking and other material properties related to the morphology of the polymers affect the penetration of the chemical agent into the polymer. The rate of hydrolysis also depends on the reaction temperature, types of degradation media, pH value, buffer capacity, ionic strength, hydrophilicity or sample thickness (see Table 2.1).

TABLE 2.1

Factors Affecting the Rate of (Bio)degradation Processes (Rydz et al., 2015; Rydz et al., 2017).

Population of organisms	Type of polymer/chemical structure/properties	Environment conditions
• Composition and abundance of microbial populations (bacteria, actinomycetes or fungi) and of other organisms (mites, nematodes or insects). • Type of enzymes produced.	• Functional group position in the polymer chain: end, in-chain, side (end-functional, telechelic, site-specific functional, multifunctional polymers or macromonomers). • Hydrolytically enzymatically susceptible chemical linkages: ester (-C(O)-O-), anhydride (-C(O)-O-C(O)-), orthoester (-C(OR)$_3$-), carbonate (-O-C(O)-O-), amide (-C(O)-NH-), urea (-NH-C(O)-NH-) or urethane (-NH-C(O)-O). • Hydrophilicity/hydrophobicity (hydrocarbon moieties). • Degree of crystallinity. • Polymer topology (linear, crosslinked or branched polymers in particular starlike, hyperbranched polymers or dendrimers). • Polymer composition (random, periodic, block, graft, homo- or copolymers). • Monomeric units (homopolymer or type of copolymer). • Molar mass/chain length. • Pre-treatment of material (annealing or stretching). • Surface to volume ratio: thickness, porosity or texture/surface development. • Orientation, additives and tension.	• Temperature. • Presence/absence of oxygen (5–7% content of O$_2$ determines the proper growth of microorganisms). • Humidity (optimum 50–60% or high humidity limits O$_2$ availability, which promotes an increase in the activity of anaerobic microorganisms). • pH (high facilitates the growth of bacteria, low fungi). • Light (UV radiation). • Availability of nutrients (salts or metals accelerate decomposition). • C/N ratio (20–30 allowing for the active development of microorganisms). • Hydrostatic pressure and ionic strength.

Hydrolytic degradation tests of biodegradable polymers, especially for medical applications inside the body, are carried out at 37°C. However, according to ISO 13781, accelerated degradation tests could be also carried out at 70°C (ISO 13781:2017, 2017). When analysing the hydrolysis processes, many others factors must also be taken into account, including the chemical composition and the influence of the resulting degradation products. The biological effect of medical devices on tissues is evaluated according to a number of standards. ISO 10993 applies to the biocompatibility assessment of all medical devices that come into direct or indirect contact with the body, but there are also standards for the biological assessment of specific types of medical devices, in this ISO 7405 or ISO 18562 (ISO 7405:2018, 2018; ISO 10993-1:2018, 2018; ISO 18562-1:2017, 2017).

2.2 BIODEGRADATION

2.2.1 BIODEGRADABILITY AND COMPOSTABILITY

Biodegradability is the ability of organic materials to decompose into basic substances by the action of enzymes produced by microorganisms under aerobic or anaerobic conditions. The end products of the biodegradation process under aerobic conditions are CO_2, water and biomass, whereas under anaerobic conditions, methane is the major product. Biodegradation is part of the natural carbon-based life cycle. Figure 2.1 illustrates the sustainable carbon life cycle of biodegradable products. The combination of biodegradation with the use of renewable resources for the production of biodegradable polymers is a unique opportunity to adapt the entire life cycle of plastics to the natural cycle of the matter: plastics products are sourced

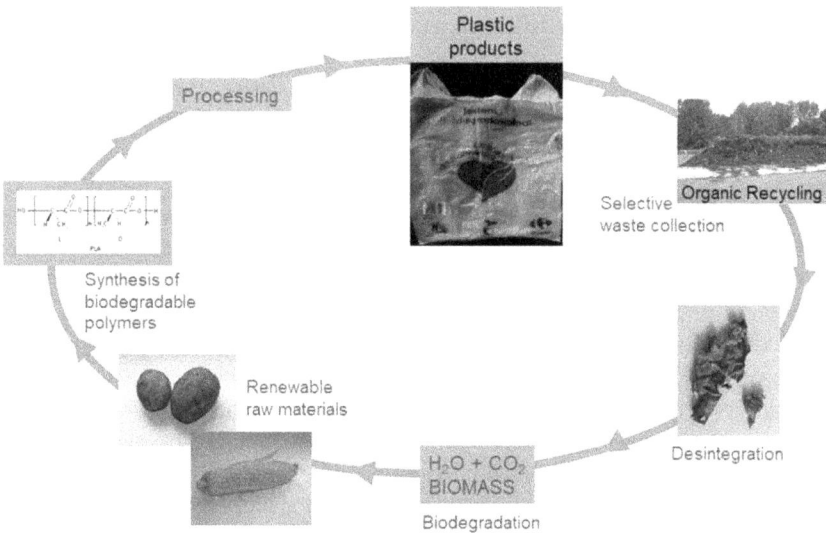

FIGURE 2.1 Sustainable carbon life cycle of biodegradable products.

from renewable natural raw materials and return to nature. This sustainable carbon life cycle cannot be achieved with conventional plastics. Most of the conventional plastics are not degraded by environmental factors such as water, air, sunlight or the action of microorganisms.

The biodegradable polymers such as polyhydroxyalkanoates (PHA)s are both compostable and biodegradable in marine environments according to EN 13432, ASTM D7991 and ASTM D6691 standards and fit perfectly into the carbon life cycle presented earlier (Figure 2.1; ASTM D6691-17; ASTM D7991-15; BS EN 13432:2000, 2000; Greene, 2012; Sikorska et al., 2021).

Composting is an aerobic organic waste treatment process in which organic materials are processed by naturally occurring microorganisms. Composting ensures an optimal environment for the growth of bacteria, fungi, etc. During industrial composting, the temperature in the pile may reach up to 70°C. Composting takes place in humid conditions. This process requires carbon (C), nitrogen (N), oxygen (O_2), water and heat. Microorganisms that break down organic waste use carbon as an energy source and build nitrogen into the cell structure. Therefore, maintaining the appropriate proportions of these two elements (C/N – 30/1) has a positive effect on the development of microorganisms suitable for this process. The most commonly used criterion for the biodegradation of plastics during composting is that the period of their fragmentation must keep up with the cycle of the composting process, but the mineralisation may take longer, e.g. in the case of biodegradable plastics used in agriculture. It is assumed that compostable plastics are always biodegradable, while biodegradable plastics are not necessarily compostable (as biodegradation may take longer than required for composting). Consequently, compostable plastics are a subgroup of biodegradable plastics (Sikorska et al., 2018).

A European Strategy for Plastics in a Circular Economy, published by the European Commission, presents the concept of a circular economy in terms of bioeconomy and outlines the challenges and benefits of transforming the traditional economic model into a circular economy. This strategy aims to bring benefits for the protection of the environment and increase the innovation and competitiveness of European enterprises. All this is due to the increased efficiency of the recycling system, the reduction of waste and the emission of harmful pollutants and the reduction of energy consumption. At the same time, the circular economy approach is also implemented in the context of product design or production processes. The transformation of the traditional economy model into a circular economy aims to bring benefits in terms of environmental protection and pollution reduction (Kirchherr et al., 2017).

One of the most advanced ways to calculate the environmental footprint, i.e. the impact of a product, service or organisation on the environment, is to use the life cycle assessment (LCA) methodology. According to the ISO 14040 standard, the life cycle is defined as "successive and interrelated stages of a product, from obtaining or producing a raw material from natural resources to its final disposal". The basic stage of the LCA test is carrying out a detailed balance of materials and energy use. The LCA method is based on quantitative data. Performing such a test may contribute not only to reducing the negative impact on the environment but also to measurable financial savings, e.g. by identifying the least effective stages of the production cycle (increasing the efficiency of energy and raw materials use). In addition, LCA analysis is necessary for obtaining certain environmental labels – especially those that are used in Western European countries. The EU Strategy for Plastics sets out the most important actions that national and regional authorities and industry should take. The list of future actions for implementing the strategy includes, inter alia, activities in the field of biodegradable plastics that are compostable. They relate in particular to preparing harmonised provisions on the definition and labelling of compostable and biodegradable plastics; carrying out an LCA to determine the conditions under which the use of such plastics is beneficial as well as the criteria for their use. In addition, these activities are intended to commence the process of reducing the use of materials that undergo oxidative degradation (Environmental Footprint).

Consumers and environmentalists have high hopes for bioplastics, but the experts' opinion on these materials is much more balanced. It is worth noting that most of the technologies for producing plastics from renewable raw materials known today lead to materials that are not biodegradable – e.g. bio-based polyethylene synthesised from ethylene obtained from sugar cane processing does not differ in properties from polyethylene obtained from classic ethylene from crude oil or natural gas and still is not biodegradable. Currently, bioplastics, both bio-based and biodegradable, have a very small share of the plastics market (less than 1%) (Dietrich et al., 2017; PlasticsEurope, 2021; Statista, 2022). And although the growth rates of the bioplastics market are quite high, it is not possible for these materials to replace traditional plastics in most applications, even in the long term. One of the reasons is the low durability of biodegradable plastics. But the biodegradable plastics can be an excellent material products with a rather short service life, where easy disposal after using is a desirable feature. Thus, industrial compostable polymer materials are often used as disposable food packaging with appropriate physical, mechanical and barrier properties, the material recycling of which is economically unjustified. Biodegradable plastics are also used as mulching films, which after harvesting can be left in the field, where they will biodegrade, enriching the soil. Similar advantages are bags for composting rubbish (green or kitchen waste) made of biodegradable plastic foil that does not need to be separated from organic waste. Organically recyclable materials should therefore be deposited in organic waste collection containers and sent to a composting plant together with this waste into the same waste stream. However, it should be emphasised that the use of biodegradable plastics for the production of various products will not solve the problem of environmental pollution. Waste from biodegradable plastics, abandoned as garbage in the environment, will not biodegrade in a short time. The process of decomposition of biodegradable materials must be carried out in specialised composting installations, under conditions of increased temperature and humidity. The purpose of composting is the utilisation of municipal waste suitable for this type of process to produce a compost that is stabilised enough to be put into the ground or stored, not only safe for human health and life but also environmentally friendly. One of the concepts of the production of eco-friendly products is the use of polymeric materials that will be subjected to organic recycling after use, which would reduce the amount of traditional plastic waste deposited in landfills and thus reduce environmental pollution. After biodegradation process, the initial organic materials are converted into simple chemicals such as water, CO_2 and methane (Ghanbarzadeh and Almasi, 2013). Biodegradable materials appear on our market more and more often. This is due not only to the need to reduce the mass of waste but also to the growing awareness of society in this area. It is precisely the growing expectations of consumers regarding environmentally friendly polymers that is one of the reasons for replacing traditional polymer packaging with compostable materials. Of particular interest in this area are aliphatic polyesters such as polylactide (PLA) and PHA (Musioł et al., 2011).

2.2.2 BIODEGRADATION STUDIES UNDER LABORATORY AND REAL CONDITIONS

Certified biodegradable material bearing a special mark can be separated from the municipal waste stream together with organic waste generated in households and directed for organic recycling. However, the widespread introduction of biodegradable polymer materials to the market must be preceded by a number of changes, such as the development of new technologies, improvement of the composting infrastructure, as well as financial possibilities and appropriate legal adjustments. Thus, research, in particular research in the field of forecasting and predicting potential applications and related risks, is still needed at every stage of the chain of production, use and disposal of biodegradable polymer materials, from making polymers through to the consumer to waste management (Sikorska et al., 2015; Swift, 2015).

The degradation process is an irreversible process, often multi-stage, leading to clear changes in the functional properties of the material. The physico-chemical properties of the material change, such as the chemical structure, molar mass, mechanical properties, brittleness, surface structure, transparency, colour and gloss. There are a number of methods to determine the degree of biodegradation of plastics, which mainly consist of determining the time after which polymer materials decompose and analysing the type and quantity of products formed during this process. The biodegradation processes are tested under laboratory conditions or in a suitable environment (inoculum), e.g. in mineral substrate, soil or compost, in aquatic environment or in the presence of isolated microorganisms or enzymes. The course of biodegradation depends on many factors, such as humidity, temperature, pH, as well as the amount and type of present microorganisms. These factors affect different polymers in different ways, and the intensity of their impact on a given material may vary throughout the process. Therefore, laboratory methods of testing polymer degradation and biodegradation give only approximate results of this process. The most reliable results come from tests carried out in a natural environment. The assessment of the degree of biodegradation is based on the estimation of changes in the properties of the tested material as well as determination of the impact of degradation products on the environment. Nevertheless, real-life conditions of home, industrial compost, soil and water can differ from laboratory conditions and interfere with the biodegradation process. Compostable plastic products are intended for biodegradation under specific, controlled conditions in industrial composting plants (Figure 2.2).

Compostable polymeric materials are generally intended to biodegradation under certain controlled conditions in industrial composting plants. In domestic composters under partially controlled conditions or under uncontrolled environmental conditions, such materials do not necessarily compost fully. Certifications guarantee

FIGURE 2.2 Composting sites: cage with the samples, pile and container system, respectively from the left.

Source: reprinted with permission from Musioł et al., "A preliminary study of the degradation of selected commercial packaging materials in compost and aqueous environments", *Pol. J. Chem. Technol.* no 13(1) (2011):55–57. For more details, see the CC BY 4.0

that a product complies with established standards or a set of conditions defined by the certifier (Hilton et al., 2020), whereas the integrated analytic scheme can help resolve the potential complications during and after the use of biodegradable products. The prediction studies of biodegradable polymeric materials revealed that a blend of poly(1,4-butylene adipate-*co*-1,4-butylene terephthalate) (PBAT) with PLA (produced under the trade mark of Ecovio®) may be a promising candidate for manufacturing of long-shelf-life compostable packages of cosmetics, especially with oily ingredients or disposable bags (Sikorska et al., 2017; Musioł et al., 2018). Additionally, the proposed approach was used in a study of three-dimensional (3D) prototyping of cosmetic compostable packages. An important application conclusion from the presented results is that the tested final products qualify for the proposed uses while maintaining biodegradability. The biodegradation test under laboratory conditions simulating the intensive composting process was also carried out for this material (Rydz et al., 2019). Moreover, the mentioned works present the results of biodegradation under industrial composting conditions, conducted for different materials: rigid PLA foils, 3D-printed jars and final products (thermoformed trays; Musioł et al., 2016). Based on macroscopic observations, it was found that during the degradation in the compost, the examined materials lost their transparency, which may indicate an increase in their crystallinity. During the biodegradation of the final products in the container and in the composting pile, the influence of the type of the composting system on the degradation process was noticed. The observed reduction in molar masses was greater in a closed system than in a pile. The specimens from composites of blend poly(3-hydroxybutyrate)-*co*-(4-hydroxybutyrate) and PLA with jute were also incubated in the Micro-Oxymax respirometer to measure of CO_2 released during incubation of materials tested under laboratory conditions (Musioł et al., 2015). In conclusion, the research conducted revealed that the industrial composting process depends less on the method of PLA processing (rigid foils or final product) and more on the conditions prevailing in the environment during the incubation of the tested materials. Studies on biodegradation under the industrial composting conditions can therefore be referred to as in vivo biological tests, as they reflect the real conditions of organic recycling. Biodegradation carried out under laboratory composting conditions using a Micro-Oxymax respirometer is in this case a reference to in vitro tests. The biodegradation of commercial bottles from NatureWorks™ PLA was also studied under real composting conditions (Kale et al., 2006, 2007). Comprehensive degradation studies in various environments, including laboratory composting conditions, were performed for several biodegradable polymers. All materials studied achieved at least 90% biodegradation within 180 days according to ISO 14855 norm (in relation to cellulose biodegradation; ISO 14855-1:2012, 2012; ISO 14855-2:2018, 2018). PCL, PHB and their blend PHB/PCL (60/40) were characterised by the highest degradation rate, and the decomposition of the blend took place within 40 days. Individual components of PLA, PHO and poly(1,4-butylene succinate) (PBS) showed slower biodegradation. PLA alone achieved 90% biodegradation after 70 days of incubation, but after blending with PCL the total degradation time of the material was reduced to 60 days under industrial composting conditions. The slow degradation was observed also for both PHO and its blend with 85% PLA, and their decomposition lasted longer than 120 days. The slowest degradation was observed for PBS, more than 200 days, but mixing with other polymers such as PHB reduced degradation time of blends obtained (Narancic et al., 2018). The behaviour in the environment of other PLA-based materials, including poly (lactide-*co*-ethylene-*co*-succinate) copolymers was also investigated. The presence of the succinic comonomer in the copolymer improved their biodegradation due to the lack of methyl groups in the succinate molecule (Cadar et al., 2012). Poly(malic acid-*co*-L-lactide) was found to accelerate PLA degradation in physiological conditions (Iyyappan, 2019). The PLA composting process is generally a two stage process combining hydrolytic and microbial degradation. These processes are effected by the temperature present in industrial composting. A higher temperature may accelerate the PLA hydrolysis process due to the susceptibility of PLA to thermophilic microorganisms (Höglund et al., 2012; Gorrasi et al., 2013; Arrieta et al., 2014). In the process of degradation caused by microorganisms, enzymes, which are biocatalysts of the reactions, play a key role. Investigating the mechanisms of polymer degradation in the presence of enzymes produced by microorganisms in the environment may lead to the development of new materials with a defined lifetime. Such solutions will lead to the reduction of environmental pollution (Kaushal et al., 2021). Proteinase K, which was selected from *Tritirachium album*, showed the ability to enzymatically degrade PLA (Williams, 1981). The evaluation of the degradability for the PLA, PBAT and poly(3-hydroxybutyrate-*co*-4-hydroxybutyrate) (P3HB4HB) blends was done using an enzymatic degradation experiment. For this purpose, the prepared materials were placed in buffer solutions of pH 8 with the addition of proteinase K. The presence of PBAT and P3HB4HB in the blends did not limit the degradation of PLA by the enzyme; these materials retained the advantage of biodegradability under these conditions (Wang et al., 2020). Due to the limited amount of microorganisms producing proteinase K in the environment and thus the reduction of the possibility of enzymatic degradation of PLA, the enzyme-containing materials were prepared. The samples for degradation experiment were received by two different methods: solution casting and extrusion. For comparison, PLA samples without enzyme were obtained by the same methods. The PLA samples with addition of proteinase K degraded via enzymatic degradation mechanisms during incubation in water. The division of the samples into smaller elements increased the degradation rate, demonstrating the applicability of that approach regarded to microplastics degradation (Huang et al., 2020).

Research on degradation in enzymes is of great importance when the tested materials are used in medicine. Specifying how the polymer will behave during degradation in various enzymes is aimed at determining the possibility of drug release in a specific place in the human body. Six different enzymes, inter alia gastrointestinal enzymes, elastase and esterase were used in the degradation study of polyglycolide (PGA) and modified PGA. The investigated materials were prepared in the form of nanoparticles. In order to simulate drug release, a fluorescent dye was coupled to PGA and its release was observed during degradation tests. Both the type of enzyme and the modification of the polymer played a significant role in the progress of degradation. The aforementioned result indicates the possibility of making an appropriate selection of materials for specific pharmaceutical applications as a form of drug dosage optimisation (Swainson et al., 2019).

2.3 CONCLUSIONS

Solution to the problem of excess landfill waste in recent years could be the introduction of goods made of natural and/or biodegradable polymers into the market. The term *biodegradable* is used to describe those materials that are enzymatically degraded by microorganisms (bacteria or fungi). Products made from biodegradable polymers undergo organic recycling processes after use. This significantly reduces environmental pollution and allows it to be used rationally in the industrial composting process. Organic recycling is the aerobic or anaerobic treatment of that part of the waste that is susceptible to the action of microorganisms. These processes must take place under controlled conditions and require appropriate facilities, industrial composting plants. Most plastic waste is resistant to this type of recycling and therefore cannot be used for this type of waste management. In this approach, depositing in a landfill is not a form of organic recycling. Thus, the percentage of non-biodegradable materials in landfill waste is constantly increasing. The introduction of more biodegradable products to the market may change this. However, introducing new materials, including biodegradable plastics, to the market requires determining their behaviour in open ecosystems and assessing their impact on the environment. So, work is underway on standards regulating the biodegradation of bioplastics in different environments describing compliance of the biodegradation processes with the criteria. Thus, the key to the success and development of the bioeconomy is to understand both the benefits and limitations of biodegradable materials.

REFERENCES

Allison, E.H., and Bassett, H.R. 2015. Climate change in the oceans: Human impacts and responses. *Science* 350: 778–782.

Arrieta, M.P., López, J., Rayón, E., and Jiménez, A. 2014. Disintegrability under composting conditions of plasticized PLA/PHB blends. *Polym. Degrad. Stab.* 108: 307–318.

ASTM D6400-21. 2021. *Standard Specification for Labeling of Plastics Designed to be Aerobically Composted in Municipal or Industrial Facilities*. West Conshohocken: ASTM International.

ASTM D6691-17. 2017. *Standard Test Method for Determining Aerobic Biodegradation of Plastic Materials in the Marine Environment by a Defined Microbial Consortium or Natural Sea Water Inoculum*. West Conshohocken: ASTM International.

ASTM D7991-15. 2015. *Standard Test Method for Determining Aerobic Biodegradation of Plastics Buried in Sandy Marine Sediment Under Controlled Laboratory Conditions*. West Conshohocken: ASTM International.

BS EN 13432:2000. 2000. *Packaging. Requirements for packaging recoverable through composting and biodegradation. Test scheme and evaluation criteria for the final acceptance of packaging*. London: British Standards Institution (BSI).

BS EN 14995:2006. 2006. *Plastics – Evaluation of compostability – Test scheme and specifications*. London: The British Standards Institution (BSI).

BS EN 16640:2017. 2017. *Products of biological origin – Determination of the carbon content of biological origin in products by radiocarbon method*. London: The British Standards Institution (BSI).

BS EN 16785-1:2015. 2015. *Bio-based products. Bio-based content. Determination of the bio-based content using the radiocarbon analysis and elemental analysis*. London: The British Standards Institution (BSI).

BS EN 16785-2:2018. 2018. *Bio-based products. Bio-based content. Determination of the bio-based content using the material balance method*. London: The British Standards Institution (BSI).

Cadar, O., Paul, M., Roman, C., Miclean, M., and Majdik, C. 2012. Biodegradation behaviour of poly(lactic acid) and (lactic acid-ethylene glycol-malonic or succinic acid) copolymers under controlled composting conditions in a laboratory test system. *Polym. Degrad. Stab*. 97: 354–357.

Carbery, M., O'Connor, W., and Palanisami, T. 2018. Trophic transfer of microplastics and mixed contaminants in the marine food web and implications for human health. *Environ. Int*. 115: 400–409.

Dietrich, K., Durmont, M., Del Rio, F., and Orsat, V. 2017. Producing PHAs in the bioeconomy – Towards a sustainable bioplastic. *Sustain. Prod. Consump*. 9: 58–70.

Directive 2019/904. 2019. Directive (EU) 2019/904 of the European Parliament and of the Council of 5 June 2019 on the reduction of the impact of certain plastic products on the environment. https://eur-lex.europa.eu (accessed December 21, 2021).

Environmental Footprint. https://eplca.jrc.ec.europa.eu/EnvironmentalFootprint.html (accessed December 21, 2021).

Ghanbarzadeh, B., and Almasi, H. 2013. Biodegradable polymers. In *Biodegradation – Life of Science*, eds. Chamy, R., and Rosenkranz, F., p. 141. Rijeka: InTech.

Gorrasi, G., and Pantani, R. 2013. Effect of PLA grades and morphologies on hydrolytic degradation at composting temperature: Assessment of structural modification and kinetic parameters. *Polym. Degrad. Stab*. 98: 1006–1014.

Greene, J. 2012. *Marine biodegradation of PLA, PHA, and bio-additive polyethylene based on ASTM D7081*. ACADEMIA: San Francisco, CA, USA. file:///C:/Users/Dell/Downloads/Marine_Biodegradation_of_PLA_PHA_and_Bio.pdf (accessed December 21, 2021).

Harrison, J.P, Boardman, C., O'Callaghan, K., Delort, A.-M., and Song, J. 2018. Biodegradability standards for carrier bags and plastic films in aquatic environments: A critical review. *R. Soc. Open Sci*. 5: 171792.

Hilton, M., Geest Jakobsen, L., Hann, S., et al. 2020. *Relevance of Biodegradable and Compostable Consumer Plastic Products and Packaging in a Circular Economy*. Luxembourg: Publications Office of the European Union.

Höglund, A., Odelius, K., and Albertsson, A.-C. 2012. Crucial differences in the hydrolytic degradation between industrial polylactide and laboratory-scale poly(*L*-lactide). *ACS Appl. Mater. Interf.* 4: 2788–2793.

Huang, Q.Y., Hiyama, M., Kabe, T., Kimura, S., and Iwata, T. 2020. Enzymatic self-biodegradation of poly(*L*-lactic acid) films by embedded heat-treated and immobilized Proteinase K. *Biomacromolecules* 21: 3301–3307.

ISO 10993-1:2018. 2018. *Biological Evaluation of Medical Devices – Part 1: Evaluation and Testing Within a Risk Management Process.* Geneva: International Organization for Standardization. Technical Committee ISO/TC 194 Biological and Clinical Evaluation of Medical Devices.

ISO 13781:2017. 2017. *Implants for Surgery – Homopolymers, Copolymers and Blends on Poly(lactide) – In vitro Degradation Testing.* Geneva: International Organization for Standardization. Technical Committee ISO/TC 150/SC 1 Materials.

ISO 14040:2006. 2006. *Environmental Management – Life Cycle Assessment – Principles and Framework.* Geneva: International Organization for Standardization. Technical Committee ISO/TC 207/SC 5 Life Cycle Assessment.

ISO 14852:2021. 2021. *Determination of the Ultimate Aerobic Biodegradability of Plastic Materials in an Aqueous Medium – Method by Analysis of Evolved Carbon dioxide.* Geneva: International Organization for Standardization.Technical Committee ISO/TC 61/SC 14 Environmental Aspects.

ISO 14855-1:2012. 2012. *Determination of the Ultimate Aerobic Biodegradability of Plastic Materials Under Controlled Composting Conditions – Method by Analysis of Evolved Carbon dioxide – Part 1: General Method.* Geneva: International Organization for Standardization. Technical Committee ISO/TC 61/SC 14 Environmental Aspects.

ISO 14855-2:2018. 2018. *Determination of the Ultimate Aerobic Biodegradability of Plastic Materials Under Controlled Composting Conditions – Method by Analysis of Evolved Carbon dioxide – Part 2: Gravimetric Measurement of Carbon dioxide Evolved in a Laboratory-scale Test.* Geneva: International Organization for Standardization.Technical Committee: ISO/TC 61/SC 14 Environmental Aspects.

ISO 16221:2001. 2001. *Water Quality – Guidance for Determination of Biodegradability in the Marine Environment.* Geneva: International Organization for Standardization. Technical Committee ISO/TC 147/SC 5 Biological Methods.

ISO 17088:2021. 2021. *Plastics – Organic Recycling – Specifications for Compostable Plastics.* Geneva: International Organization for Standardization. Technical Committee ISO/TC 61/SC 14 Environmental Aspects.

ISO 18562-1:2017. 2017. *Biocompatibility Evaluation of Breathing Gas Pathways in Healthcare Applications – Part 1: Evaluation and Testing Within a Risk Management Process.* Geneva: International Organization for Standardization. Technical Committee ISO/TC 121/SC 3 Respiratory Devices and Related Equipment Used for Patient Care.

ISO 18606:2013. 2013. *Packaging and the Environment – Organic Recycling.* Geneva: International Organization for Standardization. Technical Committee ISO/TC 122/SC 4 Packaging and the Environment.

ISO 7405:2018. 2018. *Dentistry – Evaluation of Biocompatibility of Medical Devices used in Dentistry.* Geneva: International Organization for Standardization. Technical Committee ISO/TC 106 Dentistry.

Iyyappan, J., Bharathiraja, B., Baskar, G., and Kamalanaban, E. 2019. Process optimization and kinetic analysis of malic acid production from crude glycerol using *Aspergillus niger*. *Bioresour. Technol.* 281: 18–25.

Jambeck, J.R., Geyer, R., Wilcox, C., et al. 2015. Marine pollution. Plastic waste inputs from land into the ocean. *Science* 347(6223): 768–71.

Kale, G., Auras, R., and Singh, S. 2006. Degradation of commercial biodegradable packages under real composting and ambient exposure conditions. *J. Polym. Environ.* 14: 317–334.

Kale, G., Auras, R., Singh, S., and Narayan, R. 2007. Biodegradability of polylactide bottles in real and stimulated composting conditions. *Polym. Test.* 26: 1049–1061.

Kaushal, J., Khatri, M., and Arya, S.K. 2021. Recent insight into enzymatic degradation of plastics prevalent in the environment: A mini-review. *Cleaner Eng. Technol.* 2: 100083.

Kirchherr, J., Reike, D., and Hekkert M. 2017. Conceptualizing the circular economy: An analysis of 114 definitions. *Resour. Conserv. Recyl. Adv.* 127: 221–232.

Marine Litter Solutions. *Declaration of the Global Plastics Associations for Solutions on Marine Litter.* www.marinelittersolutions.com (accessed December 21, 2021).

Musioł, M, Janeczek, H, Jurczyk, S., et al. 2015. (Bio)degradation studies of degradable polymer composites with jute in different environments. *Fib. Polym.* 16(6): 1362–1369.

Musioł, M., Sikorska, W., Adamus, G., and Kowalczuk, M. 2016. Forensic engineering of advanced polymeric materials. Part III – Biodegradation of thermoformed rigid PLA packaging under industrial composting conditions. *Waste Manage.* 52: 69–76.

Musioł, M.T., Rydz, J., Sikorska, W.J., Rychter, P.R., and Kowalczuk, M.M. 2011. A preliminary study of the degradation of selected commercial packaging materials in compost and aqueous environments. *Pol. J. Chem. Technol.* 13(1): 55–57.

Musioł, M.T., Sikorska, W., Janeczek, H., et al. 2018. (Bio)degradable polymeric materials for a sustainable future – part 1. Organic recycling of PLA/PBAT blends in the form of prototype packages with long shelf-life. *Waste Manage.* 77: 447–454.

Narancic, T., Verstichel, S., Chaganti, S.R., et al. 2018. A biodegradable plastic blends create new possibilities for end-of-life management of plastics but they are not a panacea for plastic pollution. *Environ. Sci. Technol.* 52: 10441–10452.

PlasticsEurope 2021. Plastics – The Facts 2021. *An Analysis of European Plastics Production, Demand and Waste Data.* https://plasticseurope.org/pl/knowledge-hub/plastics-the-facts-2021 (accessed April 19, 2022).

PlasticsEurope AISBL. Port of Antwerp Activity Report 2021 Operation Clean Sweep®. 2021. www.opcleansweep.eu/news/operation-clean-sweepr-port-antwerp-activity-report-2021 (accessed December 21, 2021).

Rychter, P., Kawalec, M., Sobota, M., Kurcok, P., and Kowalczuk, M. 2010. Study of aliphatic-aromatic copolyester degradation in sandy soil and its ecotoxicological impact. *Biomacromolecules* 11: 839–847.

Rydz, J., Musioł, M., and Kowalczuk, M. 2017. Polymers tailored for controlled (bio)degradation through end-group and in-chain functionalization. *Curr. Org. Synth.* 14(8): 768–777.

Rydz, J., Sikorska, W., Kyulavska, M., and Christova, D. 2015. Polyester-based (bio)degradable polymers as environmentally friendly materials for sustainable development. *Int. J. Mol. Sci.* 16(1): 564–596.

Rydz, J., Sikorska, W., Musioł, M., et al. 2019. 3D-Printed polyester-based prototypes for cosmetic applications – Future directions at the forensic engineering of advanced polymeric materials. *Materials* 12(6): 994.

Sikorska, W., Musiol, M., Nowak, B., et al. 2015. Degradability of polylactide and its blend with poly[(R,S)-3-hydroxybutyrate] in industrial composting and compost extract. *Int. Biodeterior. Biodegrad.* 101: 32–41.

Sikorska, W., Musioł, M., Zawidlak-Węgrzyńska, B., and Rydz, J. 2018. Compostable polymeric ecomaterials: environment-friendly waste management alternative to landfills. In *Handbook of Ecomaterials*, eds. Martínez, L.M.T., Kharissova, O.V., and Kharisov, B.I., pp. 2733–2764. Cham: Springer International Publishing.

Sikorska, W., Musioł, M., Zawidlak-Węgrzyńska, B., and Rydz, J. 2021. End-of-life options for (bio)degradable polymers in the circular economy. *Adv. Polym. Technol.* 2021: 6695140.

Sikorska, W., Rydz, J., Wolna-Stypka, K., et al. 2017. Forensic engineering of advanced polymeric materials–Part V: Prediction studies of aliphatic–aromatic copolyester and polylactide commercial blends in view of potential applications as compostable cosmetic packages. *Polymers* 9(7): 257.

Statista. 2022. www.statista.com/statistics/678684/global-production-capacity-of-bioplastics-by-type (accessed May 12, 2022).

Swainson, S.M.E., Taresco, V., Pearce, A.K., et al. 2019. Exploring the enzymatic degradation of poly(glycerol adipate). *Eur. J. Pharm. Biopharm.* 142: 377–386.

Swift, G. 2015. Degradable polymers and plastics in landfill sites. In *Encyclopedia of Polymer Science and Technology*, ed. Mark, H.F., pp. 40–50. Hoboken: John Wiley & Sons, Inc.

Wang, X.Y., Pan, H.W., Jia S.L., et al. 2020. Mechanical properties, crystallization and biodegradation behavior of the polylactide/poly(3-hydroxybutyrate-*co*-4-hydroxybutyrate)/poly(butylene adipate-*co*-terephthalate) blown films. *Chinese J. Polym. Sci.* 38: 1072–1081.

Williams, D.F. 1981. Enzymic hydrolysis of polylactic acid. *Eng. Med.* 10: 5–7.

3 Manufacturing Processes of (Bio)degradable Polymers

Khadar Duale

CONTENTS

3.1 INTRODUCTION

The use of environmentally friendly and renewable biodegradable polymers has received much attention lately in a concerted effort to create sustainable plastics in a circular economy, thereby reducing the "take, make and dispose" linear economy model, which is an unsustainable manufacturing practice. Plastics are woven into the fabric of our modern lifestyle and are increasingly used in our daily activities

DOI: 10.1201/9780429352799-4

(Andrady and Neal, 2009). It would therefore be highly unreasonable to claim that synthetic polymers are non-essential for our modern society. Likewise, it is imprudent to claim that the pervasive use of fossil-fuel plastics from non-degradable synthetic polymers is not the root cause of the massive global plastic pollution that is impacting the environment, leading many countries to ban single-use plastic in global efforts to tackle this problem (Okan et al., 2019). Hence, exploring alternatives to current fossil fuel-based synthetic polymers is a burgeoning field of polymer research, in addition to the fact that environmentally friendly polymers have lately gained much attention as a substitute for non-degradable polymer material (Bahramian et al., 2016; Kabir et al., 2020). Environmentally friendly polymers are becoming increasingly important in the search for biodegradable and renewable materials for the production of everyday plastics. Polyhydroxyalkanoates (PHA) and polylactide (PLA), together with aliphatic biodegradable fossil-based polymers such as poly(ε-caprolactone) (PCL), have been known for several decades and are some of the most important and most studied groups of biodegradable polymers used as materials for various applications (Domb and Kumar, 2011; Din et al., 2020).

This chapter provides an overview of the work carried out for processing and studying the properties of biodegradable polymers in different research areas. It provides information concerning the method used in undertaking this research, viz extrusion, injection moulding and different additive manufacturing (AM) methods. It is divided into two sections. In the first section of the chapter, the idea is to give an overview of some of the most common process methods for the manufacture of polymer-based components as well as some of the drawbacks and advantages of each method. The introduction of each of these techniques will be followed by several examples from some of the published research on these biodegradable polymers and their composites.

3.2 EXTRUSION

Extrusion is a technique in which a raw plastic material (e.g. beads, granules, pellets, powders, flakes or even combinations of these forms of material) is used to create the desired shape of a polymer mould (Figure 3.1).

Extrusion plays a very important role in the processing of polymers and consists of a compressor and a screw conveyor. Either a single or twin-screw extruder is typically used in the polymer industry. The two different screw types have different advantages, and the choice depends on the type of polymer being processed and the product to be manufactured (Vlachopoulos and Strutt, 2003, Abe et al., 2021). The extruder also houses a hopper for feeding the resin from above, a motor to rotate the screw, an electrically heated metal tunnel and a nozzle for the polymer melt to exit. The process starts with the plastic granules being fed into the hopper, which is connected from above to facilitate a further transfer of the granules to the extruder and also to allow the possible inclusion of various different masterbatches (such as stabilisers, colourants and plasticisers) that are mixed with the polymer. The granules are then transported forward by the screw drive and melted as the mechanical energy generated by the turning of the screws in the heating chamber is converted into heat energy and this heat is applied directly to the material. The thick, viscous melt of

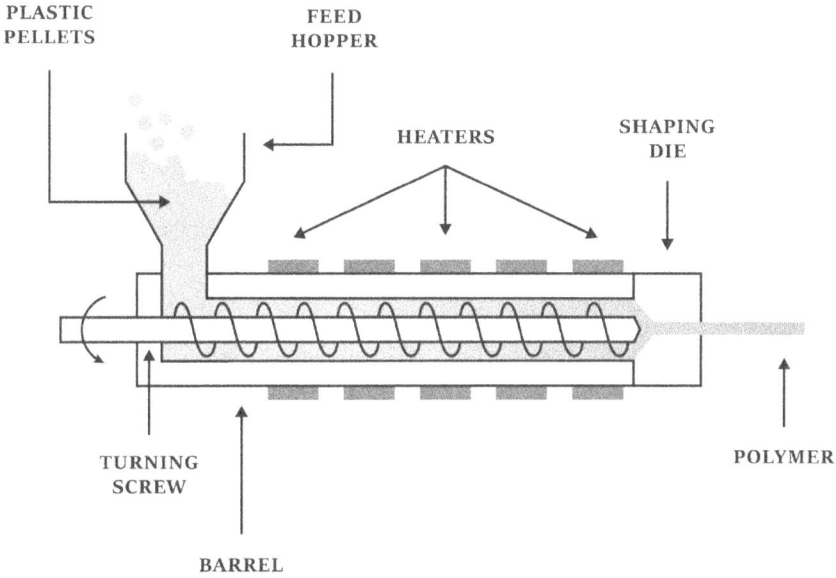

FIGURE 3.1 Screw extruder.

the polymer is finally injected into a mould and hardens under cooling, defining the shape of the moulded part after it has solidified (Vlachopoulos and Strutt, 2003).

The processing of many different biodegradable polymers using extrusion techniques has been reported and although the extrusion process is the same for both biodegradable and non-biodegradable polymers, there are significant differences in the parameters used (Sikora et al., 2020; Sikora et al., 2021; Mysiukiewicz et al., 2020; Borkar et al., 2021). It is therefore important to know how to handle biodegradable polymers properly to eliminate any potential issues while following other basic processing instructions.

3.2.1 EXTRUSION OF POLYLACTIDE (PLA)

PLA is one of the most common (bio)degradable polymers used for the extrusion process and has been investigated extensively (Taubner and Shishoo, 2001; Lim et al., 2008; Gálvez et al., 2020; Mysiukiewicz et al., 2020). PLA has a glass transition temperature (T_g) of 55 and a melting temperature (T_m) of 160–180°C, although this depends on the composition (*D*- or *L*-enantiomers content) and molar masses. PLA has several known problems such as a low T_g which can lead to loss of molar mass, mechanical properties and degradation during processing (Signori et al., 2009). Consequently, it is important to choose the right processing temperature as elevated temperatures can lead to the thermal degradation of PLA, while poor homogeneity is observed at lower temperature limits. This thermal degradation can take the form of random chain scission of the backbone and is mainly due to ester bonds (Gupta and Deshmukh, 1982; Kopinke et al., 1996; Carrasco et al., 2010). The mechanism of such random degradation reactions has been suggested to proceed in the form of intra- and intermolecular transesterification, oxidative degradation, hydrolysis and *cis*-elimination (Farah et al., 2016). One of these undesirable reactions is dominant, depending on the processing conditions (Al-Itry et al., 2012). For temperatures above 200°C, transesterification resulting in the formation of cyclic oligomers of lactic acid, carbon monoxide and carbon dioxide (CO_2), methyl ketone and acetaldehyde is the dominant degradation mechanism for PLA (Kopinke et al., 1996; Carrasco et al., 2010; Castro-Aguirre et al., 2016). Pyrolysis-mass spectrometry analysis of PLA shows that lactide is released at a temperature of 295°C, while higher temperatures of 350°C result in the removal of higher cyclic oligomers (Kopinke et al., 1996). Selecting the correct screw speed is also quite crucial as the screw speed determines the residence time of the polymer in the plastifying unit and is another parameter that can impact the degradation rate of the PLA. The impact of the extrusion processing parameters of PLA has been thoroughly investigated (Gamon et al., 2013; Taubner and Shishoo, 2001; Gálvez et al., 2020; Mysiukiewicz et al., 2020). The effect of residence time, process temperature and the intrinsic moisture content on the properties of PLA has been studied (Taubner and Shishoo, 2001). Two different processing temperatures for dry and conditioned PLA samples were taken into account, whereby two different rotational speeds of the screw were tested at each temperature to evaluate the effect of residence time. At the lower processing temperature of 210°C, the change in molar mass of the dry granules was found to be dependent on the rotational speed of the screw, as rotation at low speed caused a slight decrease in molar mass. However, increasing the temperature to 240°C causes a decrease in both residence times. On the other hand, for the higher temperature with the long residence time, it was observed that the molar mass was independent of the moisture content as both speeds showed a loss of molar mass.

Moisture content is another major concern in PLA processing as it is hygroscopic and very sensitive to high humidity (Chen et al., 2022; Castro-Aguirre et al., 2016). This implies that PLA resins supplied in containers, boxes or bags are already dried to a certain degree and should therefore be kept sealed in their original packaging until ready for use. To avoid hydrolysis, it is suggested that PLA is dried to a moisture content of less than 100 ppm (0.01 wt%) (Castro-Aguirre et al., 2016). PLA also undergoes chemical structural changes during polymer extrusion as it has different mechanical, thermal and rheological properties compared to non-biodegradable plastics. It is well documented that PLA brittleness and poor temperature resistance are some of the other drawbacks of PLA and a more comprehensive list of both advantages and shortcomings of PLA has been summarised elsewhere (Rasal et al., 2010). Several procedures proposed to improve this material property deficiency of PLA have been presented in the literature (Anderson et al., 2008). In this regard, blending with other biodegradable polymers, copolymerisation and plasticisation by adding chemical modifiers that can lead to chain extension, as well as branching and cross-linking, are some of the steps taken to overcome the shortcomings of PLA (Rasal et al., 2010; Corneillie and Smet, 2015). It is also worth mentioning that degradation is not only limited to the polymer, but additives can also suffer degradation during processing due to their nature, and some of them can even migrate from the bulk matrix to the surface, leading to the blends becoming brittle (Hahladakis et al.,

2018; Beltrán et al., 2019; Schyns and Shaver, 2021). PLA plasticisation is an area that has attracted considerable interest over the last decade for various applications, such as flexible films for specific drug delivery systems and food packaging (Ruellan et al., 2014; Ljungberg and Wesslen, 2002). There are several different plasticisers used for PLA, including citrate esters such as triethyl citrate and acetyl tributyl citrate, glyceryl triacetate, malonate oligomers, poly(ethylene glycol) (PEG), lactic acid oligomers, glucose monoesters and fatty acid esters, to name a few. A more comprehensive list can be found elsewhere (Ruellan et al., 2014).

Typically, the impact strength is a good indicator of toughness, while the elongation at break reflects the ductility of the material, which is able to sustain a large permanent deformation. The addition of these agents can improve PLA toughness and flexibility but can also have detrimental effects and lead to a reduction of some of the other intrinsic properties of PLA. For instance, it was reported that PLA was plasticised to improve ductility, although that can reduce both tensile modulus and tensile strength (Jacobsen and Fritz, 1999). The effect of extrusion screw speed and the amount of acetyl tributyl citrate as plasticiser on the mechanical properties of PLA has been also reported (Gálvez et al., 2020). The mixing was performed using a twin-screw extruder employing two different mixing zones and two screw rotation speeds of 60 and 150 rpm while the plasticiser content varied between 10–30 wt%. Thermogravimetric analysis (TGA) showed a decrease in thermal stability at both speeds with increasing plasticiser content. The blends with 20% acetyl tributyl citrate which were processed at 60 rpm resulted in PLA/acetyl tributyl citrate materials with the best mechanical properties. Although the speed changes did not generate any significant changes in tensile strength and modulus of elasticity, a significant reduction was seen when acetyl tributyl citrate was added. The elongation at break of PLA with a low loading (20 wt%) of PCL was improved by almost 60 times compared to pure PLA. However, tensile strength decreases of 20% were reported (Yang et al., 2021c).

3.2.2 EXTRUSION OF POLY(ε-CAPROLACTONE) (PCL)

Another biocompatible, degradable polymer that has received much attention and has also been processed by the excursion method is PCL, which is a hydrophobic semi-crystalline linear aliphatic polyester with a T_g of about −60°C and a melting point of 60°C (Zhang, 2016; Matzinos et al., 2002; Borkar et al., 2021). Nevertheless, despite its other advantages the low tensile strength (16–24 MPa), Young's Modulus (240–420 MPa) and a melting point of PCL limit its use as biodegradable material for food packaging. To overcome some of these limitations it has been suggested the processability of PCL is improved by mixing it with starch. However, improving one area comes at the cost of reducing other innate properties. For example, it was reported that the incorporation of starch into PCL resulted in a material with increased modulus but that both the strength and elongation were significantly reduced (Matzinos et al., 2002). The low thermal stability of PCL limits its use, although it has been used as a packaging material for perishable and chilled foods, e.g. cheese, fresh salads and meat (Correa et al., 2017; Lyu et al., 2019). These PCL-based packaging materials may also have other interesting properties such as antimicrobial activity and, due to the low processing temperature of PCL, do not lead to degradation of the antibacterial agent. In addition, the low processing temperature and the ease of application with conventional processing tools allow it to be mixed with other temperature-sensitive biodegradable polymers. This means that thermal degradation is limited in drug delivery systems as the encapsulated drug is not exposed to high temperatures and can therefore provide a better drug dispersion in the PCL matrix (Salmoria et al., 2017a; Salmoria et al., 2017b). Antimicrobial PCL composite films were obtained by incorporating different concentrations of grapefruit seed extract into PCL using a twin-screw extruder (Lyu et al., 2019). PCL/ibuprofen rods prepared by the melt extrusion technique were also reported (Salmoria et al., 2016). Here, a temperature-dependent drug release profile was observed, whereby the processing temperature affects the distribution and release rate of the drug.

3.2.3 EXTRUSION OF POLYHYDROXYALKANOATES (PHAs)

PHAs are thermoplastic polymers. The physical and thermal properties of these PHAs greatly depend on the hydroxy acid monomer composition used: they vary greatly, exhibiting a wide range of mechanical properties from hard crystalline to elastic, with medium gas permeability. For instance, PHAs can typically have melting temperatures in the 40–180°C range (Biron, 2016). For pure poly(3-hydroxybutyrate) (PHB), a T_m of around 170–180°C and a T_g of 5°C have been reported and that presents a narrow window for processing conditions (Bugnicourt et al., 2014; Arrieta et al., 2015). For poly(3-hydroxybutyrate-co-3-hydroxyvalerate) (PHBV) the T_m is reduced due to the presence of 3-hydroxyvalerate (HV) units. Extrusion is a process that is also utilised for the fabrication of PHAs based devices in the biomedical and packaging industries (Martin and Williams, 2003; Arrieta et al., 2017; Panaitescu et al., 2017). The mechanical property of PHA can be influenced by molar mass and chemical composition. PHAs with different main chain lengths as well as side chain length result in different performance during extrusion. For example – short-chain length PHA is very brittle with poor elastic properties whilst, on the other hand, medium-chain length PHA is more ductile and easier to mould (Muthuraj et al., 2021).

There are several advantages of using PHAs over their petroleum-based counterparts. PHAs are biodegradable, biocompatible, piezoelectric and truly green polymers. This is because PHAs are produced by bacteria from agricultural raw materials, industry by-products or urban waste and their thermal properties are similar to petroleum-based polymers, like polypropylene (PP; Muthuraj et al., 2021). There are also some drawbacks of PHA that limit its wider applications in packaging and biomedical devices (Li et al., 2016). Some of these issues include the high cost, brittleness, high crystallinity, low crystallisation rate and low elongation at break. For example, the secondary crystallisation of the amorphous phase creates new lamellar crystallites. Several studies point out that this phenomenon, which occurs at ambient temperature during storage of the product or through ageing, leads to an increase in the proportion of crystalline phase compared to amorphous phases of the polymer. This in turn increases the strength and brittleness of the PHB, making it less ductile due to an increase in both the stress and strain modulus (de Koning et al., 1992; Vroman and Tighzert, 2009; Burniol-Figols et al., 2020; Naser et al.,

2021). Additionally, PHAs suffer from poor thermal stability under melt processing conditions (Montano-Herrera et al., 2014). The thermal degradation can affect processing properties and therefore the viscosity and molar mass decrease. A study that has assessed the effect of treatment temperature on the crystallisation behaviour and mechanical properties of PHBV with 3 and 18% HV units has been proposed (Bossu et al., 2021). An increased temperature promoted the crystallisation of PHB-rich domains while isolating the PHV-rich domains. Based on these results, thermal treatment windows of 175–200°C and 130–145°C were proposed for PHBV3% and PHBV18%, respectively. PHAs have a T_g close to ambient temperature which limits the service temperature as it causes deterioration through secondary crystallisation that takes place in the amorphous phase after ageing at room temperature (Bugnicourt et al., 2014). The degradation temperature of PHB is just a few degrees above its T_m, which results in molecular degradation via hydrolysis and loss of molar mass at elevated processing temperatures. The nucleation density is low, and consequently it forms large spherulites with cracks and fissures with a negative impact on the mechanical properties (Zhang et al., 2016). Therefore, efforts have been made to improve the crystallisation process and enhance flexibility, involving several different approaches. The introduction of HV units into the 3-hydroxybutyrate (HB) polymer backbone during the fermentation process to create PHBV can improve the mechanical properties and can reduce the T_m (Reddy et al., 2003). It has been suggested that PHB with PHV contents in the range of 5–20% gives properties broadly similar to those of the polyolefin such as polyethylene and PP which are some of the most widely used and cost-effective thermoplastics in melt processing (Biron, 2016). The incorporation of other monomeric units can also lead to copolymers with improved properties. The addition of plasticisers and nucleating agents are some of the steps to resolve some of these PLA issues (Parulekar and Mohanty, 2007; Zhu et al., 2012; Bugnicourt et al., 2014; Ten et al., 2015). PHB has also been blended with several (bio)degradable polymers such as PLA, poly(butylene succinate) (PBS) and PCL (Lovera et al., 2007; Zhang and Thomas, 2011; Ma et al., 2014a; Ma et al., 2014b; Garcia-Garcia et al., 2016; Arrieta et al., 2017; Przybysz et al., 2018). The influence of natural polymers, such as cellulose, jute and abaca fibres has also been explored (Bledzki et al., 2009; Tábi et al., 2021). The poor affinity between the PHB/fibres interface of the nonpolar matrix and the polar fibres as well as the filler dispersion state are some of the main challenges. However, several steps are being taken to overcome this problem, involving ways to improve the wettability and to initiate close contact between the PHA and the natural polymers (Zhou et al., 2016). Some of these steps include the introduction of surface modifications to improve hydrophobicity and increase hydrophilicity or the use of an amphiphilic compatibiliser and are also used even for biomedical applications (Rasal et al., 2010; Koller, 2020). In addition, the dispersion mechanism leading to homogenised fibre dispersion provides higher mechanical performance. Melt extruded PHA devices are both compostable and biodegradable in marine environments. Using single-screw extrusion techniques, PHB and PHBV with varying structures were melt-processed into cast films by optimising the processing parameters such as barrel temperature (165–173°C) and chill rolls (60°C) (Thellen et al., 2008). The study evaluated the properties of these films paying particular attention to their oxygen and water vapour barrier properties as these play a crucial role in determining durability. Although the PHBV samples exhibited higher barrier properties against both water vapour and oxygen compared to PHB, both films were highly biodegradable in the marine environment. They exhibited 88–99% biodegradation under optimal static laboratory conditions after 49 days and a reduced degradation rate of 30–73% after 90 days under dynamic aquarium conditions. PHA films, when tested in the same way as the mechanical test, indicated tensile strength comparable to common thermoplastics such as polyethylene and poly(ethylene-*co*-vinyl acetate) films. Likewise, Young's modular values were also comparable to film-grade polyethylene and nylon-6.

The effects of the processing parameters, such as the temperature, the screw profile, speed and compatibilising route, are also important to the structure and thermal properties. The impact of the melt processing conditions on the mechanical properties and microstructure of dumbbell-shaped poly(3-hydroxybutyrate-*co*-3-hydroxyhexanoate) (PHBHHx) samples prepared by a lab-scale rotary twin-screw extruder was studied (Vanheusden et al., 2021). Several parameters viz processing temperature, mould temperature, screw speed and cooling time were evaluated. The results showed that increased values of tensile strength and elongation are obtained when processing PHBHHx in relatively low extrusion profiles between 140–145°C, with reasonably high mould temperatures of 80°C, as well as low screw speeds of 50 rpm and short mould cooling times of 60 s.

To improve the material properties of PLAs, other polymers are also blended with PHB. PLA/PHB blends were prepared in several different mass ratios (100/0, 75/25, 50/50, 25/75.0/100) by melt blending to determine whether the addition of PLA affects the properties of PHB by studying the morphology, thermal properties, mechanical properties and biodegradation behaviour of these compounds. The results indicate that PLA is able to improve the mechanical properties of PHB and, even though the two polymers are not miscible, there is still interaction between them (Zhang and Thomas, 2011).

3.3   INJECTION MOULDING

Injection moulding is a fast and efficient method of producing a wide variety of plastic components of the same shape and size in a short time, including those that can have very complex shapes and require high dimensional accuracy. Injection moulding is nowadays the most common processing method for injecting molten material into a mould and then allowing it to cool and solidify. This is a standard manufacturing method for many thermoplastic polymers due to its ease of processing and recyclability and is both a cost-effective and environmentally friendly solution. Choosing the right injection moulding machine that is suitable for your sample is very important in the first place, but it is also crucial to optimise the parameters specific to each type of polymer, particularly in the case of a biodegradable polymer. This plays a vital role not only in the quality of the moulded parts produced but also in some of the other aspects of the process, e.g. cycle times and process energy consumption (Dang, 2014).

Injection moulding machines (Figure 3.2) consist of injection units that feed the molten material into the mould and moulding units that form the moulded part.

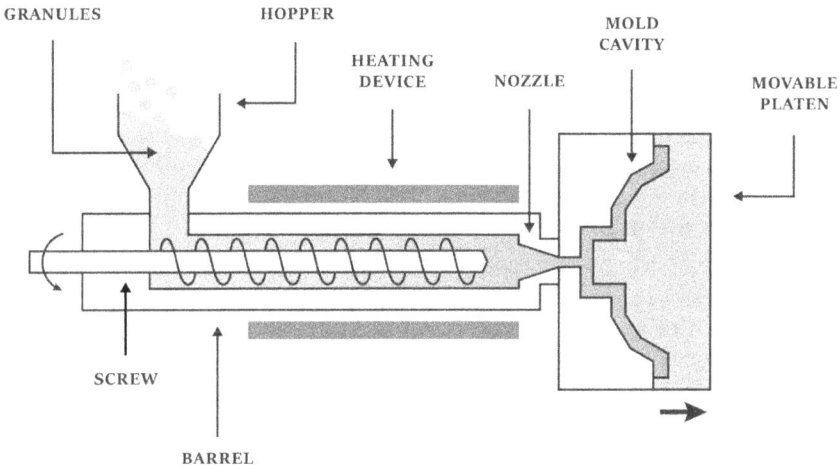

FIGURE 3.2 Injection moulding machine.

These two parts are connected to a clamping unit, which holds them together during injection and cooling. The injection unit contains a heated barrel equipped with a reciprocating screw that is driven either by a hydraulic or electric motor, which heats and plasticises the polymer melt. The screws technique is used due to its ability to generate a homogeneous melt in a very short period of time. As with the extrusion process, the polymer injection moulding process begins with plastic pellets, granules or flakes that are fed through a funnel into the injection moulding machine screw that leads to a special heated injection barrel cylinder. In the cylinder, the thermoplastic polymer melts when it is heated above its melting point. The molten plastic is then pushed forward by the screw and evenly distributed on the tool of the form clamping unit. The molten plastic formed in the mould is then cooled and solidified to the desired shape. At this stage, it is likely for some shrinkage of the material to occur as the plastic solidifies on the internal surfaces of the mould. Removing the mould may require some force but a release agent is also used to facilitate this, which is sprayed onto the surface of the mould cavity before the polymer material is injected. Once the part has been ejected, the mould can be clamped again and is ready for the next injection. In terms of processing, this method is not only an excellent alternative for producing large volumes of the same product quickly, but it is also an excellent option for plastics production as it gives the product an excellent surface finish. There are several (bio)degradable polymers suitable for injection moulding including PLA, PCL, PHA and poly(1,4-butylene adipate-*co*-1,4-butylene terephthalate) (PBAT).

3.3.1 INJECTION MOULDING OF PLA

PLA is suitable for processing with injection moulding in the manufacture of thermoplastic compounds and PLA plastics can be made into parts for a wide range of applications (Jamshidian et al., 2010; Saini et al., 2016). PLA has some of the other properties mentioned earlier that make it a very promising polymer, such as (bio)degradability, renewability and good mechanical properties to replace the fossil fuel-based thermoplastics used in injection moulding. These properties, therefore, contribute to a reduction in the societal problem of solid waste disposal. PLA is also an ideal polymer for the injection moulding process due to its low melting point, high flow rate and transparent properties. However, there are some drawbacks associated with injection moulding of PLA and the final products produced. Some of these issues are due to the crystallisation rate, the low heat deflection temperature but also the inherent brittleness of PLA, which is caused by the relatively short repeating units and the methyl side group (Ding et al., 2016; Quiles-Carrillo et al., 2018; Rosenboom et al., 2022). Both the heat tolerance and the impact strength of PLA are quite low, which means that the PLA material is not suited for more demanding applications. PLA-based materials, for example, are not intended for heavy-duty products, even in medical applications, where it is otherwise attracting a lot of attention (Hench and Jones, 2005). The process temperature and cycle time in the extruder and the hot runner are also some of the other issues since PLA is susceptible to thermal degradation in the molten state (Taubner and Shishoo, 2001). The thermal degradation reactions predominantly take place by random chain sessions along the PLA backbone. However, it was also argued that transesterification reactions are important contributors to PLA pyrolysis (Kopinke et al., 1996). For this reason, plain PLA is unsuited to compete with fossil-based polymers and requires modification to improve its processing and/or material properties (Lay et al., 2019). More durable goods can be obtained from PLA if the right modification is adopted (Tripathi et al., 2021). By blending different polymers and possibly by adding some sustainable fillers or fibres, the intrinsic properties of PLAs can be altered and fine-tuned. Some studies have reported that the PLA/PHB 75/25 blend exhibits significantly improved tensile properties compared with pure PLA (Zhang and Thomas, 2011).

There are a number of commercially available injection moulding grade PLA resins pellets that are compatible with most of the existing injection moulding machines, where the exact processing parameters are specified for each type of resin. The effects of various additives can be stressed here, just as mentioned for the extrusion-based method, where PLA is compounded with things such as initiators, chain extenders, nucleating agents, impact modifiers, compatibilisers and curing agents to produce such mixtures and composites. Many of these are also commercially available. PLA is blended with additives such as these to achieve a wide range of physical properties, including improved tensile strength and impact strength or flexural modulus, as with any other conventional plastic. Commercially available OnCap™, BIO impact T (Avient) is a transparent effect modifier for PLA, while others – such as acrylic-based additives like Arkema's Biostrength® and Dow's Paraloid™ or chain extenders from BASF Joncryl® or ethylene copolymers from DuPont's Biomax® – provide better toughness for injection moulding. These modifiers are available in either liquid or solid (granular) form and can be added during processing, as a masterbatch. Carbonates such as EMForce®Bio, UltraTalc®609 from Specialty Minerals and Terraloy™ 90,000 series masterbatches (Teknor Apex) can also be added to improve impact strength. A detailed list of many agents for modifying a variety of PLA properties can be found elsewhere (Niaounakis, 2015). In general, these additives aim to improve the various physical properties of the PLA materials: not only the strength and thermal properties but also heat deflection temperature. It is also worth highlighting that an improvement in some parameters very often leads to a reduction in other intrinsic parameters of PLA (Farah et al., 2016). The effects of epoxy-based chain extenders on the properties of PLA in injection moulding were evaluated (Pilla et al., 2009). It was reported that the addition of a chain extender leads to an increase in the molar mass of pure PLA and improved elongation at break and impact strength, but a decrease in the specific modulus and degree of crystallinity was observed.

Numerous studies have investigated the effects of talc as a nucleating agent that improves the crystallinity of neat PLA (Harris and Lee, 2008; Helanto et al., 2021; Liu et al., 2014). The addition of talc and N,N'-ethylenebis(stearamide) as nucleating agents resulted in a dramatic increase in the crystallisation rate and the final crystalline content of PLA (Helanto et al., 2021). The results also show that the processing time was reduced, while strength and modulus were improved by up to 25% due to these material and process changes. This suggests that the addition of physical nucleating agents shows a way to control the crystallisation rate and optimise the processing ratio for the injection moulding of PLA. Likewise, injection moulded PLA composites containing various natural polymers have also been investigated. The composite specimen was prepared using a twin-screw extruder before being injection moulded to test samples. The crystallisation behaviour of PLA cellulose reinforcements with microcrystalline cellulose, cellulose fibres and wood flour composites has been investigated to measure how the size, chemical composition and surface topography of cellulose materials affect the crystallisation of PLA (Mathew et al., 2006). The differential scanning calorimetry and optical microscopy analysis showed that microcrystalline cellulose and wood flour have a better nucleation ability compared to cellulose fibres.

Injection moulded PLA and hydroxyapatite composites were reported and the study showed improved thermal properties, as well as an approximately 25, 20 and 42% increase in tensile, modulus and impact properties of modified PLA/hydroxyapatite composite respectively (Akindoyo et al., 2017). The durability of injection moulded short jute fibre/PLA composites in distilled water at different temperatures has been studied (Jiang et al., 2018b). The jute fibres and PLA were mixed using injection moulding and the samples were prepared using a co-rotating twin screw extruder. The result showed the influence of the hydrothermal temperature on the stability of the jute/PLA composites since the mechanical properties of the composites and the molar mass did not change significantly at low temperatures (at 23 and 37.8°C), while a considerable decrease was seen at 60°C.

The use of surface modifiers (such as the phosphate-based Fabulase® 361 modifier) to promote a good interaction between PLA and natural polymers is common. What is also common is the use of anti-blocking additives or slipping to reduce the surface stickiness of polymer films when using additives to prevent the blocking of the machinery and to allow easier processing and handling (Fiori, 2014). This is because PLA has an inherent coefficient of friction that is mainly caused by the high density of polar ester bonds and require additives. Antiblocking agents include resins (POLYBATCH® PLA 203 AB 3 from LyondellBasell) or fatty acid amides (OnCap™ from Evient, Clariant's Cese-block or Sukanos SUKANO® Slip/Antiblock), are among some of the most frequently used slip additives that can be added to the PLA-based formulation, The additives can intercalate between the adjacent layers as they tend to migrate to the surface of the material. Nevertheless, concerns should be raised about their use as these may affect the (bio)degradability of PLA (Hahladakis et al., 2018).

3.3.2   INJECTION MOULDING OF PCL

Another biodegradable polymer that has been subjected to injection moulding is PCL, although many of the studies deal with its composites in an attempt to improve the mechanical properties (Wahit et al., 2012; Haq et al., 2019b; Herrera et al., 2020). The influence of injection moulding parameters on the mechanical, thermal and thermomechanical properties of PCL itself was also investigated (Luna et al., 2021). The study highlighted a detailed analysis of injection moulded samples, showing how changing processing parameters such as temperature and injection flow rate can lead to changes in various properties such as impact strength, tensile strength, Shore D hardness and heat deflection temperature of PCL. This result indicates several different aspects of the PCL properties and can either improve or deteriorate, depending on the required application of the polymer. For example, increasing the moulding temperature deteriorated the elongation at break, while a higher injection flow rate and lower processing temperature led to improved impact strength responses. The study also showed that the injection flow rate has a decisive effect on the heat deflection temperature of PCL as an increase indicates a negative impact on the thermomechanical resistance. This phenomenon is believed to be related to the degree of crystallinity as well as the modulus of elasticity. It is also feasible to use injection moulding technology to produce PCL scaffolds for tissue engineering (Cui et al., 2012). PCL composites with poly(ethylene oxide), NaCl and hydroxyapatite were mixed using an extruder with a twin-screw extruder to fabricate pallets that were fed into an injection moulding machine to make tensile test bars. The obtained specimens were leached with deionised water after the injection moulding process; the water-soluble and sacrificial polymer and NaCl particulates were removed to produce porous and interconnected microstructures. The study investigated not only the effect of leaching time on porosity and residual NaCl and NaCl/hydroxyapatite contents but also the effect of the addition of hydroxyapatite on the mechanical properties. The study demonstrated an 18% improvement in the modulus of elasticity and a 37% improvement in the loss modulus of the PCL composites with hydroxyapatite compared to the composites without hydroxyapatite, suggesting that the hydroxyapatite addition improved the mechanical properties of PCL scaffolds.

3.3.3   INJECTION MOULDING OF PHA

PHAs could be processed by conventional injection moulding for the fabrication of bottles and packaging films (Ashter, 2016). Despite the well-documented biodegradable and biocompatibility properties of PHAs, there are obviously the same issues as those raised for the extrusion of PLA method. In particular, the polymer characteristics in injection moulding are an issue (Zhu et al., 2012). For instance, the slow crystallisation rate of PHAs in injection moulding influences the cycle time and consequently the production speed and cost of plastic parts. The effects of reinforcing injection moulded PHA samples with natural polymers such as cellulose, jute and abaca fibres have also been investigated (Bledzki and Jaszkiewicz, 2010). The results suggested that fibre reinforcement significantly enhances the tensile stiffness and strength of all fibres, but cellulose fibres gave the best results.

The possibility of using PHB in food packaging was also investigated. Injection moulding was used to fabricate cans and capsules (Bucci et al., 2005). The results of physical, dimensional, mechanical and sensory test analyses showed that PHB can replace PP in packaging. For instance, adding PLA to PHB improves the mechanical properties of PHB. In this way the PHA compounds that are created are, for example, both high in strength and tough or low crystalline but stiff. Incorporation of these highly crystalline PHAs into the PLA matrix by melt blending also presents a way of increasing the PLA properties such as crystallinity. PHB is also used to reduce the inherently high transparency of PLA and has therefore been suggested as more suitable for food packaging materials. Transparency is another issue that is really important in food packaging. When comparing PLA and PHB, the latter acts as a better light barrier in the ultraviolet (UV) and visible light regions. As has been observed for PLA, however, the improvement of an intrinsic property can often lead to the deterioration of another, and it is therefore particularly important to maintain the biodegradable nature of PHA while improving the chemical, thermal and mechanical properties. PHB was studied as a food packaging material by manufacturing 500 mL packages (can-capsule set) through the injection moulding technique, while the degradation of these packaging materials in different environments has also been reported (Bucci et al., 2005, 2007). One thing to bear in mind regarding the use of biodegradable polymers in food packaging is to ensure that the shelf life of the packaging material is commensurate with the shelf life of the product that it is intended to preserve. This means that the environmental conditions leading to biodegradation must occur only after the packaging has been discarded and not during the storage of the food product (Petersen et al., 1999).

PHB blended with different PCL mass percentages in the injection moulding method has also been investigated to evaluate the physicochemical properties of PHB-PCL blends (Garcia-Garcia et al., 2016). PHB-PCL pellets were mixed in a co-rotating twin-screw extruder at 160°C (hopper) to 175°C (nozzle) temperature and a screw speed of 40 rpm. The extruded material was further processed by injection moulding at 175°C to obtain a standard specimen to analyse the intrinsic properties of the composites by different analytical techniques. The DSC results showed a noticeable increase in the degree of crystallinity for both PHB and PCL for the PHB blend containing 25 wt% PCL. The analysis of cracked surfaces by field emission scanning electron microscopy (FESEM) shows that PHB and PCL are immiscible.

The addition of PCL acts as an impact modifier, implying higher flexibility and ductility and accompanied by a reduction in strength and modulus. This is due to the soft and flexible nature of PCL that modifies the rigid PHB, which leads to an increase in both flexibility and ductility with the increased content of PCL. For a PCL content higher than 50 wt%, the properties of the specimens have more or less been defined by PCL, indicating the point at which matrix inversion takes place.

3.4   ADDITIVE MANUFACTURING

Additive manufacturing is a type of manufacturing process for creating a three-dimensional (3D) physical object from a digital design that has been around since the mid-1980s. The term AM is also in a non-technical context synonymous with 3D printing. The actual 3D printing process as the technique involves creating parts or prototypes in a layer-by-layer fashion directly from a computer-aided design (CAD) file that containing detailed instructions which describes the shape of an object in a triangulated form. Currently, there are many different types of geometry defining 3D print preparation programs that have specialised functions for storing information about 3D models but can also help to design, customise and prepare models for 3D printing files. There are various types of geometry designing file formats such as stereolithography (STL), wavefront OBJect (OBJ), virtual reality modelling language (WRML), COLLADA, filmbox (FBX), initial graphics exchange specification (IGES), 3D models (3DS) and AM file format for which there is support in several 3D printing and processing programs. However, due to historical and practical reasons, each industry tends to have its own 3D file format preferences. STL (standard triangulation language) file format is quite popular in many different industries and has been the standard file format since the inception of AM. Initially, the STL file format was developed to be part of the CAD package for the early SL machine. It is a so-called neutral file format as it generally accepts most 3D printers (Saptarshi and Zhou, 2019). Usually, digital 3D models can be created from a variety of different CAD software, from the more sophisticated and highly technical programs like SolidWorks or even computer graphics programs like Blender. However, they can also be created from common medical imaging techniques, like computerised tomography (CT), micro-CT, magnetic resonance imaging (MRI) scan data and 3D scanning tools as well as data created from 3D digitising systems to create 3D models of objects (Berry et al., 1997; Rengier et al., 2010; Dodziuk, 2016; Haleem and Javaid, 2019). These 3D image files are then converted into an STL file and become triangulated. Using slicing software that converts a digital model of a 3D object into printing instructions, the so-called G-code that tells the printer to build a specific 3D object is created. As such, the slicer acts as a link between a digital model and the physical model being printed. It works by virtually slicing the 3D models into many horizontal 2D layers that are later printed one by one. Besides the model, the slicer also contains other 3D printing parameters defined by the user, such as speed, layer height and temperature. In some cases, however, some of these files may contain errors that need to be rectified before being uploaded onto the 3D printer, and some of the common errors include holes, misaligned surfaces, overlapping triangles and non-manifold mess which may be especially common when scanning biological objects (Lemu and Kurtovic, 2011; Idram et al., 2019; Rimini et al., 2022). Likewise, there may be a need to place and rotate one or more 3D files in the most favourable 3D printing model. These inaccuracies are usually inherited from the processing procedure of the original file or possibly from the way in which it was created and the way in which the 3D objects to be created were processed (Oropallo and Piegl, 2016; Behm et al., 2018). Therefore, in most cases, it is common that these 3D files need to be modified before printing so that factors such as position, size and orientation are in the correct place (Naftulin et al., 2015). This is usually accomplished with a pre-processing slicer program (that is often included in AM machines) to specify printing settings and slice the digital model into layers for printing.

According to International Organization for Standardization (IOS), there are seven categories of AM processes, such as material extrusion, vat photopolymerisation, sheet laminating, directional energy deposition, powder bed fusion, binder jetting and material jetting. It is therefore not uncommon for 3D printing to be perceived and described with the various industrial abbreviations and manufacturing methods. In principle, no matter the technology involved, most AM methods build the object in a layer-by-layer fashion and these processes chiefly vary in their methods of layer manufacturing to create models. As expected, each method has its own advantages and disadvantages and it is, therefore, quite common to operate several different 3D printers in one manufacturing environment. In general, AM technology offers mass customisations, widely available solutions and fewer design restrictions, thus allowing the creation of a variety of shapes, cost-effective functional prototypes, on-demand production and distributed manufacturing. Several printing technologies are utilised in the biodegradable polymers sector of which the most commonly used processes are material extrusion (colloquially known by the registered name of fused deposition modelling (FDM) or fused filament fabrication (FFF)), powder bed fusion (inter alia colloquially selective laser sintering (SLS)) and vat photopolymerisation (inter alia colloquially stereolithography (SL)) (Puppi and Chiellini, 2020; ASTM 52900-21, 2021; Amrita et al., 2022).

There are many advantages associated with utilising 3D printing in production, for example, it has made it possible to separate the product design from the manufacturing process, where these two aspects of manufacturing can now be handled separately. This separation facilitates the outsourcing of the design part and provides flexibility in the design and adaptation of complex structures (Berman, 2012; Ngo et al., 2018). One major benefit of 3D printing as opposed to subtractive manufacturing methods is that only the amount of material needed to produce the part is used, which in return can lower the manufacturing costs (Berman, 2012; Oropallo and Piegl, 2016; de Schutter et al., 2018; Ngo et al., 2018). This is unlike, for example, other processing methods such as a computer numerical control (CNC) milling cutter which can produce a large amount of waste (Berman, 2012). The disadvantages include the fact that it is a time-consuming, stepwise layer-by-layer approach to object fabrication which in general, is not suitable for mass production compared to other mass manufacturing methods, like injection moulding that can produce a plastic part in just a few seconds, while the same part, might take a few hours to print from a 3D printer (Ngo et al., 2018). This still limits suitability for the series production of products. Nevertheless, prototypes are a good application as injection moulding tools are very expensive to manufacture. Some 3D printers may need support materials depending on the technology used (Barnett and Gosselin, 2015; Jiang et al., 2018a). A support structure consists of wires or a pole of plastic to hold an overhanging complex geometry in place during printing, and its design is crucial for 3D printing. It significantly increases both production time and costs as it has to be removed manually (Thomas and Gilbert, 2014). The support material is not placed on cavities and fragile areas which could make removal difficult (Ngo et al., 2018).

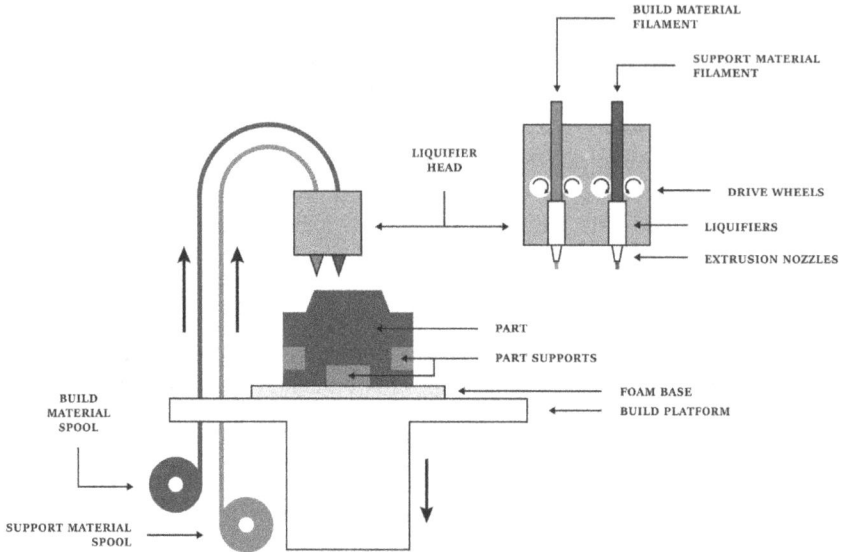

FIGURE 3.3 Schematic representation of fused deposition modelling.

3.4.1 MATERIAL EXTRUSION (FUSED DEPOSITION MODELLING)

FDM is the most widely used 3D printing technology at both industry and a consumer level, representing, therefore, the largest installed base of 3D printers worldwide (Dizon et al., 2018). The technique has been developed by Scott Crump and was trademarked in 1989 (Stratasys Inc., 1991). The technique is also known as FFF or molten polymer deposition (MPD). FDM uses material extrusion technology to print objects, whereby usually a feedstock thermoplastic polymer material in the form of either a filament or powder is pressed through a nozzle. As depicted in Figure 3.3. FDM machines have an extrusion head that consists of a motor for driving the plastic filament and an extruder nozzle for extruding the plastic.

The extruder head consists of two parts with a cold section at the top and a hot section at the bottom. The cold section pulls material from the spool and pushes the feedstock into the hot area, where the extrusion nozzle and the liquefier are located. The liquefier heats the material to just above its melting point and converts it into a thin semiliquid molten material that easily flows through the extrusion nozzle to form a thin, sticky, plastic bead. There are several different types of nozzles with different diameters usually ranging between 0.3 and 1.0 mm. Different materials require different nozzles as well as different heating methods. The position of the nozzle is directly controlled by a G-code that is generated in the slicer and moves in the x-y planes by a multi-speed numerical controlled mechanism to deposit a layer that bonds to the layer below and hardens immediately after extrusion from the nozzle. It then moves in the z-plane, to begin the deposition of a new layer according to the desired pattern defined in the 3D print preparation program. In this way, a 3D object is built by selectively depositing melted material in a pre-determined layer-by-layer fashion. FDM sometimes requires support material that helps the object during printing of bridges, overhang angles and convex surfaces to avoid deformation and enhance stability as these are parts of a print that the printer must print partially or completely over air, but these parts can also be printed without support (Fazzini et al., 2019).

Just like all other 3D printing techniques, it is worth considering the possibilities and limitations when manufacturing a model with a 3D printer. The selection of appropriate processing parameters and their optimisation as well as proper content preparation is vital to achieve the best results as this will help tailor the technology best to your desired needs (Rayegani and Onwubolu, 2014). For instance, since each 3D print material can have a different melting point, it is, therefore, important to know the correct temperature of the nozzle on the FDM printer to ensure that the extruder can correctly handle the material you wish to print (Akhoundi et al., 2020). The rheology and shearing effects of the filaments used are also important factors that should be considered, with stiffness and brittleness as the most important parameters that determine the suitability for FDM processing (Azad et al., 2020). What seems to be the main advantage of FDM machines is their simplicity, low cost and high speed, along with the ability to carry out several extrusions with different materials, as well as the relatively low technical knowledge required to run these machines (Ngo et al., 2018). Additionally, as this is a solvent-free method, it therefore avoids toxic solvents in the manufacturing process, and this is something that is crucial for products used for biomedical applications. The exclusive availability of this technique for thermoplastics in filament format is a limitation, while some of the other disadvantages of the method include a limited number of variants of usable thermoplastic materials, clogging of the nozzle, polymer drop, longer printing times, problems with layer adhesion and lack of strength (Rahim et al., 2019; Dey et al., 2021).

FDM is frequently researched for biodegradable materials and there are already myriad published results regarding various aspects of the technology (Korpela et al., 2013; Mazzanti et al., 2019; Mazurchevici et al., 2021). This is presumably due to the FDM 3D printing apparatus and the low cost of materials offered, which makes it possible for many people to become involved in the practice of 3D printing.

3.4.2 VAT PHOTOPOLYMERISATION (STEREOLITHOGRAPHY)

The SL is an AM technology that uses a focused UV laser beam (Figure 3.4) and is regarded as the first commercially available 3D printing technology (Pollack et al., 2019).

The technique was patented by Chuck Hull in 1984, although the idea of a prototyping system that uses photopolymers to create a 3D object was first published by Hideo Kodama (Hull, 1986; Kodama, 1981). The SL method was also patented by Olivier de Witte, Jean Claude André and Alain Le Mehaute, through their employer French General the Electric Company (now Alcatel-Alsthom) and CILAS (The Laser Consortium). Nevertheless, that patent application was abandoned by the company (Pollack et al., 2019). The method is based on the laser sweeping over a liquid photopolymer resin via a computer-controlled mirror and illuminating the surface of the resin according to the cross-sectional patterns of the model (Figure 3.4). Here, the UV laser beam initiates crosslinking of monomers and oligomers to form 3D solid

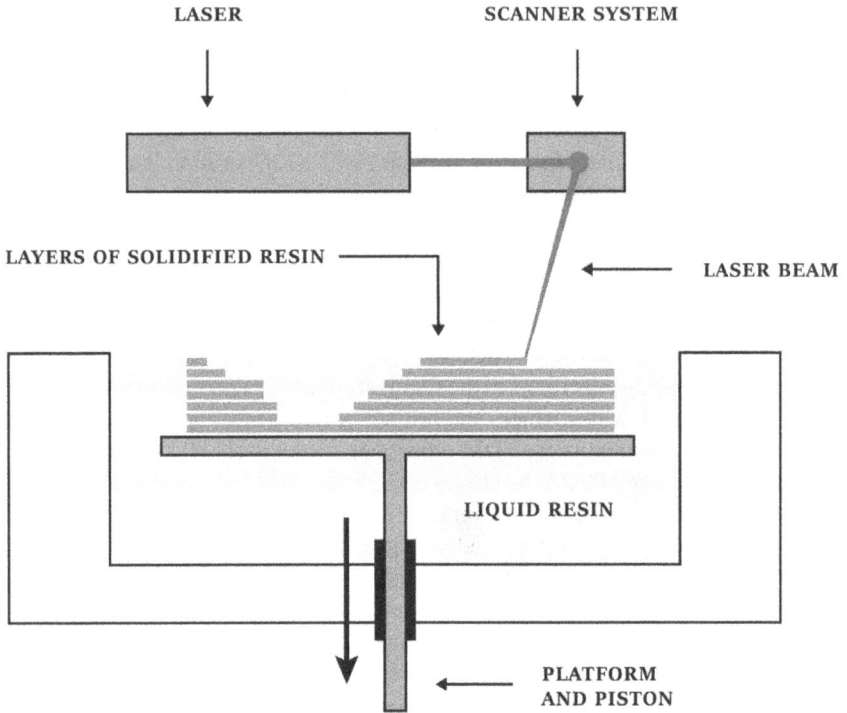

FIGURE 3.4   Schematic depiction of the SL process.

polymer objects. The laser only traces the pattern; therefore, the rest of the polymer remains liquid when the cured polymer solidifies. When the first layer photochemically solidifies, the platform is lowered by a layer thickness. This process is repeated layer by layer during the construction of the whole structure and continues until the entire object is completed. After the object is completed, it is taken out and all excess uncured polymers are washed away to clean wet resin from the surfaces, but in most cases the product is dried and placed in a UV oven to complete the crosslinking process.

One of the key advantages of the SL technology is its high accuracy among all commonly used 3D printing technologies and the ability to produce objects with a high surface finish when compared to other methods. SL machines have a good resolution to cover most needs in the manufacture of plastic parts and have a highly versatile material selection (Melchels et al., 2009). Desired flexibility, transparency and rigidity can be achieved through several possibilities of material selection. SL is a layer-based methodology just like anatomical scanning equipment such as MRI or CT and this has greatly facilitated the transfer of saturated data generated through these imaging methods. This is one of the reasons why SL technology is so useful for many applications in biomedical fields, as it allows the manufacture of highly specific custom anatomical models or implants. SL has now been widely used for the fabrication of various dental models that can be used in orthodontics and allow for direct printing of bridges/crowns using ceramic-filled special resins. However, the drawbacks of the SL include the higher cost for industrial applications.

3.4.3 POWDER BED FUSION (SELECTIVE LASER SINTERING)

SLS is an AM technique whereby a high-power laser is used to manufacture a 3D object (Figure 3.5).

The technology was developed by Carl Deckard and Joe Beaman at the University of Texas, Austin in the 1980s (Deckard, 1989). The method is usually called sintering because small solid powder beds of thermoplastic polymers (even ceramics and metals) are melted together by using a high-power laser beam. The laser beam can heat the powder either to just below its boiling point – which is then called sintering – or above its boiling point, which is then called melting. The most common type of laser used for sintering polymers and ceramics is a CO_2 laser, and depending on the capacity of the printer different lasers can be used for metals because the wavelength (10.6 µm wavelength, since polymers display high absorption rates at long wavelengths) of the laser beam interacts differently with different types of materials (Beal et al., 2009; Lupone et al., 2022). The laser sweeps over the powder material, in accordance with the cross-section of the shape defined by a 3D model. The building material is cured to a solid form by the heat of the laser beam and the model is built up through repeated iterations with the laser beam. The principal manufacturing method is the same as all other AM methods and is based on dividing a 3D geometry into horizontal layers. The laser power melts the current powder layer and about 0.02 mm of the previous layer. After each molten layer, a new powder layer is laid on top of the previous one and the powder bed is lowered with one layer. This process is

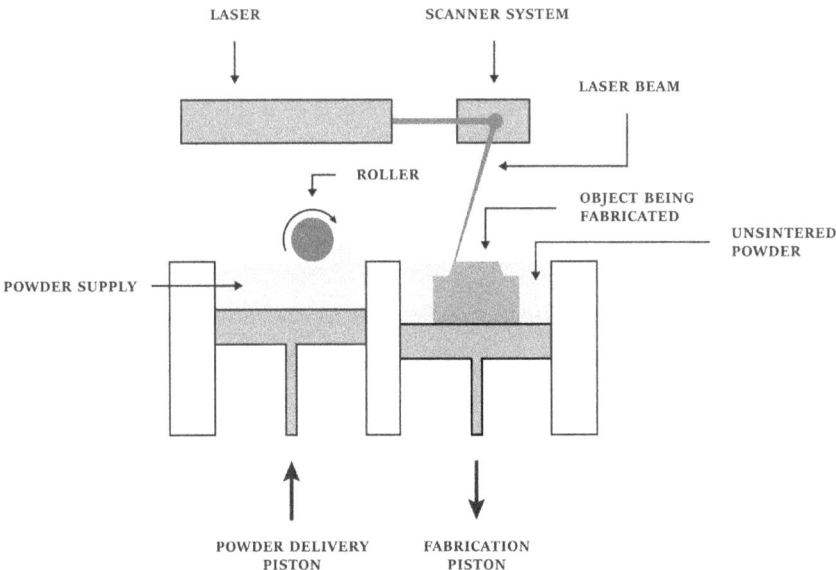

FIGURE 3.5 Schematic diagram of SLS machine.

then repeated until the sintering process is completed and the object is formed. After completion, the object must be cooled inside the machine for several hours, cleaned of excess powder by brushing or applying compressed air and finally measured for inspection.

One of the advantages of the SLS is the lack of support structure since the sintered object is held in place by adjacent powder that is not sintered (Berry et al., 1997). SLS also has a wide range of materials both synthetic as well as bio-based polymers and their composites, but even metals can be sintered (Li et al., 2017). On the other hand, a major disadvantage of the SLS technology is that the surface of the produced object can have a grainy finish requiring post-processing, such as polishing, to eliminate any residual powder trapped inside the printed object. This is because the object was created by fusing spherical particles causing a certain degree of surface roughness that can be seen after the product has been manufactured. Also, there is a significant health risk caused by the very fine SL powders. For this reason, removing parts or cleaning the pressure chamber requires a full body suit and particle filter masks. Other drawbacks include the cost, as both the machine and the material can be expensive, and a longer cooling time of the SLS technique can lead to increased production time.

3.5 BIODEGRADABLE POLYMERS USED FOR ADDITIVE MANUFACTURING

A limited number of biodegradable polymers and their composites are available for 3D printing (Amrita et al., 2022; Puppi and Chiellini, 2020). This is because printers must heat the material in order for it to be additively deposited on the construction platform so that the polymers are malleable or pliable when they are heated above a certain temperature threshold and then solidify after cooling. At the moment 3D printing materials can be produced from a range of polymers, such as proteins (e.g. keratin, gelatine or collagen) or polysaccharides (e.g. chitosan, cellulose, fibrin and hyaluronic acid or alginate) or synthetic ones such as PLA, polyurethane (PU), poly(lactide-*co*-glycolide) (PLGA) and both PCL and poly(δ-valerolactone) (PVL). However, PLA is one of the very few bio-based polymers to fulfil these criteria and that has therefore been extensively investigated (Rayna and Striukova, 2016; Ausejo et al., 2018; Rodríguez-Panes et al., 2018; Chen et al., 2019; Yao et al., 2020).

3.5.1 Additive Manufacturing of PLA through FDM

PLA is one of the most frequently studied 3D materials and forms the most popular thermoplastics filaments utilised in domestic 3D printing (Petersen and Pearce, 2017). It is, along with acrylonitrile-butadiene-styrene terpolymer (ABS), the material most widely used in filaments for FDM 3D printers due to its low cost, easy printing and (bio)degradability, and it has also a wide range of PLA filament suppliers (Rayna and Striukova, 2016; Rodríguez-Panes et al., 2018). PLA has similar properties to petroleum-based ABS since both are amorphous polymers and they create parts that can be used for a wide variety of applications (Lay et al., 2019).

The optimal printing PLA parameters for FDM technology have been explored extensively and many articles deal with issues such as temperature, filament thickness, speed of printing, the height of the layer, printer work platform, orientation, pattern infill percentage, external heating, cross-sectional area, resolution, air pressure, feed rate, nozzle distance to the printing platform, finishing methods and time (Cuiffo et al., 2017; Rodríguez-Panes et al., 2018; Valerga et al., 2018; Lee et al., 2021). Although the setting of process parameters and their range varies in different FDM machines, taking all these parameters into consideration is deemed to increase the success of the 3D printing device. It is particularly important to carefully control and select the appropriate process parameters and conditions during printing and material deposition to obtain the desired structures and to minimise potential defects such as void, gaps between lines, poor fibre-matrix bonding and surface roughness, all of which can impair mechanical properties (Ferretti et al., 2021).

PLA is considered a semi-crystalline polymer and has a T_g usually in the range of 50 to 65°C and a T_m of about 170–180°C. The proposed temperature for PLA printing differs and can range from 180°C to about 240°C. The effect of the printing temperature as one of the influential parameters on mechanical properties has been extensive investigated (Valerga et al., 2018; Zekavat et al., 2019; Hsueh et al., 2021; Syrlybayev et al., 2021). The printing temperature of 220°C was found to provide the best mechanical properties (Valerga et al., 2018). The printing temperature variations from 180 to 220°C were also investigated and it was found that the mechanical properties of PLA increase at higher printing temperatures (Hsueh et al., 2021). It was explained that the low processing temperature causes low bond strength and high porosity between different layers due to the high viscosity and low fluidity of the PLA melt, whereas, in contrast to this poor bonding, increasing the temperature causes a decrease in the viscosity of the polymer. This leads to increased bonding between the polymer melt and the layers but also a decrease in the porosity. A comprehensive study evaluating the effect of temperature on mechanical properties suggested that the different mechanical properties exhibited for different temperatures were not due to porosity but rather due to the internal geometry of the specimens (Zekavat et al., 2019). The study examined the mechanical properties of different PLA specimens fabricated between 180 and 260°C using a material extrusion printer (Ultimaker 2+, Ultimaker, Utrecht, Netherlands). The results established a correlation between the mechanical properties and the manufacturing temperature as a 6-fold decrease in porosity content was seen when the manufacturing temperature was increased from 180 to 260°C. However, the study suggested that the different mechanical properties exhibited for different temperatures were not solely due to porosity but also to the internal geometry of the specimen and the structure and orientation of porosity/air gaps with respect to the load. The internal geometry of the specimens and in particular the air gap measurement and the bonding between extruded filaments were analysed using X-ray computed tomography. This non-destructive investigation technique revealed variations in the local density of all parts regardless of the manufacturing temperature, which implies that even FDM parts printed at the recommended temperature range with 100% infill cannot achieve a homogeneous internal structure Therefore, when evaluating the mechanical strength, the internal geometry properties of the FDM printed parts, e.g. the minimum cross-sectional area, provide better information as this is affected by the printing temperature.

The specimens fabricated at the two lower temperatures of 180 and 190°C showed less than 25 mm^2, while at the next two temperatures of 200 and 230°C 27.17 and 30.28 mm^2 were shown. Finally, the same values of 35 mm^2 are observed when the temperature is raised to 250 and 260 °C (Zekavat et al., 2019).

Since the temperature has such an influence on the mechanical properties and selecting the correct printing temperature improves the quality of FDM printed parts, it is therefore good practice to measure the temperature of the extruder nozzle independently in order to accurately apply the extrusion heat. The use of cooling air velocity as an additional parameter in 3D printing by fans during the printing is also advocated in some cases, particularly if the aim is to print a very detailed object or bridge, as this allows an additional mechanism to control the PLA printing temperature and prevent the printed object from sagging down or losing details due to melting style effects. Cooling air velocity may have different effects on the dimensional quality and mechanical strength of PLA specimens printed with FDM. It was noted that the higher cooling velocities give better geometric accuracy but lower mechanical strength and, overall, a 4-fold difference in the tensile strength of the printed models between the highest cooling air velocities of 5 m/s and the lowest of 0 m/s (Lee and Liu, 2019).

Factors such as layer thickness and nozzle diameter are the two most influential parameters on tensile strength and surface roughness of FDM PLA printed parts (Yang et al., 2019). An interesting study combining experimental studies of PLA samples printed by an extrusion-based method using Intamsys Funmat HT 3D printer (INTAMSYS Technology Co. Ltd, Shanghai China) and molecular modelling simulations has looked at how residual anisotropy varies with printing conditions (Costanzo et al., 2020). A relationship among welding strength, print speed and nozzle temperature has been established. The alignment of extruded PLA layers is quantified for different nozzle temperatures and print speeds, and the residual adjustment is localised only at the welding areas between deposited filament interfaces. Additionally, residual adjustment can be directly correlated with welding hardness, and thus it is suggested that reduced welding strength is due to the reduced microstructure orientation of the polymer molecules. This materialises during the flow through the nozzle and the deposition on the building board and as such is not a result of poor inter-diffusion. The effect of layer thickness, pressure and applied heat of PLA filaments of objects obtained using a 3D printer (3DISON Plus, ROCKIT Inc., Rep. Korea) was investigated (Jo et al., 2018). For objects with improved mechanical properties, fewer voids in the internal structure and a better finish were obtained when the layer thickness, external heat and pressure were controlled. Improved mechanical properties were due to better bonding between both raster to raster as well as layer to layer of the 3D printed objects.

One of the drawbacks associated with PLA filaments is that they have a higher viscosity which can lead to the extruder head clogging up, causing a blockage that prevented it from freely extruding compared to other polymers (Tlegenov et al., 2018). It is also well documented that PLA lacks strength and resistance to heat and chemicals. Objects that are 3D printed from PLA have typically poor properties since they have very low heat resistance and often begin to soften at temperatures around 50 to 60°C. Complete drying of filaments before printing is often a requirement and normally results in an improvement in both the adhesion and mechanical properties of the printed parts. This is because PLA is highly hygroscopic and readily absorbs water from the air, resulting in poor bonding adhesion, bubble formation and degradation at 3D printing temperature, which in turn leads to a poorly finished product and its short shelf life. Likewise, the wet filament can even cause cracking sounds and splatters as the moisture in the filament is heated up in the extruder printer head. Several different methods can be applied as a desiccant to prevent PLA filaments from absorbing water, including silica gel and fumed silica due to their excellent moisture prevention properties (Gkartzou et al., 2017; Seng et al., 2020). Moisture uptake capacity of extruded samples of PLA nanocomposites with nano-silica ranging from 0, 1 to 3% by mass was examined and the research indicated that incorporation of 1 wt% nano-silica gives a 40% reduction in hygroscopicity (Seng et al., 2020). One way to prevent water uptake is to place the filament in the oven for approximately 4 to 6 h and set the temperature to just below the T_g of PLA. Some commercial filaments have an unknown or other inorganic ingredient which could react and complicate their usage in medical applications, thus changing their macromolecular structure, morphology and biocompatibility (Cuiffo et al., 2017). For that reason, extensive biocompatibility tests are usually required prior to any biomedical use.

Polymers with a semi-crystalline structure exhibit both amorphous and crystalline regions; they will therefore exhibit properties of both crystalline and amorphous structures. Due to this semi-crystalline nature, PLA tends to shrink at a higher temperature rate compared to amorphous polymers. Unlike the completely random orientation of amorphous polymers, semi-crystalline polymers are mostly structurally oriented in a relaxed state. However, the orientation is never uniform across the cross-section due to shear stress and cooling rates throughout the element. For that reason, defects arise from an unbalanced orientation with shrinkage and the shrinkage that is due to the orientation (Rydz et al., 2020).

The extrusion temperature plays an important role in the adhesion between the layers and the mechanical properties of the printed parts. The fusion between the new layers deposited on top of the previously printed layers should occur before the extruded filament has cooled below its T_g, and this bonding improves the longer the filament maintains a temperature higher than its T_g, otherwise poor bonding between the printed layers can occur. This unwanted phenomenon is also sometimes created during the 3D printing process. Low temperatures can lead to poor adhesion between the printed layers and promote flow marks, weld lines, poor surfaces, lamination and short shots. On the other hand, excessive temperatures are known to cause problems with shrinkage phenomena, such as sinking, twisting, contraction and voiding (Rydz et al., 2020). The temperature also depends on the thickness of the filament because, when using the 1.75 mm thick filament as opposed to the 3.00 mm filament, the optimum printing temperature will be closer to the lower end of this PLA filament temperature range. Likewise, when the 2.85 mm filament is used, it is advisable to be nearer to the higher end of the temperature range in order to compensate for the increased thickness of the material. It is also worth keeping in mind that with any material size the PLA extruder temperature can differ (± 10%) depending on your machine. FDM manufactured parts generally have layer lines, sometimes making a post processing step essential to achieve a smooth surface for many industrial products. There is a wide range of post-processing methods available for FDM printed parts, and some of these methods can even provide increased strength and help to mitigate the anisotropic behaviour of FDM parts (Tronvoll et al., 2018; Valerga et al., 2019). The use of low-power CO_2 lasers for post-treatment of PLA sheets was investigated, as well as the impact of different chemical post-processing methods (Nguyen and Lee, 2018; Valerga et al., 2019; Moradi et al., 2021). The use of thermal annealing as a means of improving the mechanical properties of PLA samples processed by FDM techniques was also investigated and showed an increased strength of the material. The post-treatment annealing process was performed above the T_g of PLA and below T_m to increase the degree of crystallinity and fraction of ordered regions of the polymer since crystallinity is an important factor contributing to the strength of a semi-crystalline material.

In 3D printing, due to its favourable mechanical properties, such as (bio) degradability, biocompatibility, cytocompatibility and ease of processing, PLA has been considered a potential material for a variety of uses in biotechnology and medical applications ranging from scaffolding materials in tissue engineering and regenerative medicine to cardiology, drug delivery and dental applications (Long et al., 2017; Ritz et al., 2017; Jia et al., 2018; Liu et al., 2020c; Park et al., 2021).

A great deal of research has investigated the use of FDM for the fabrication of PLA-based tissue scaffolds. The potential of solid sheets of PLA scaffolds as well as porous cage-like 3D prints of PLA has been demonstrated by material extrusion (Ultimaker 2+, Ultimaker, Utrecht, Netherlands) (Ritz et al., 2017). Coated PLA or PLA filled with collagen was used and the biocompatibility and endotoxin contamination levels were tested. Endotoxin contamination levels of 0.1–0.25 endotoxin units (EU)/ml were observed, which is clearly below the U.S. Food and Drug Administration (FDA) level of 0.5 EU/ml. The study also revealed that different cell types such as fibroblasts and both osteoblast-like cells and endothelial cells can spread, grow and multiply on PLA-printed discs and collagen tested PLA cages loaded with stromal-derived factor-1 (SDF-1); SDF-1-collagen supports cell growth of endothelial cells and induces the formation of neo-vessels. Another factor for thermoplastic matrix fillers is homogeneity, as it improves consistency. A low cycle fatigue behaviour of 3D-printed PLA-based composite material of porous scaffolds for bone defect replacement was studied (Senatov et al., 2016a). Factors such as the accumulated energy, the modulus of elasticity, the height of the samples, the elastic deformation and structural characteristics of porous scaffolds obtained by 3D printing of the FDM system were examined. PLA-based prints showed several defects such as a collapse of pores, height reductions, delamination, bending and shear of the printed layers. Additionally, the growth and propagation of these cracks during cyclic loading were observed. However, the introduction of dispersed hydroxyapatite particles reduced the rates of the accumulations of defects and the result showed that PLA with 15% hydroxyapatite porous scaffolds was able to function under cyclic loading at a stress of 21 MPa for a long time without change. Thus, substantial crack resistance is demonstrated and therefore there is the potential for use as implants for bone replacement. PLA-based scaffolds with porosity of 30 vol% ceramic particles showed an inhibition of the growth of cracks in polymer materials which is a typical behaviour of the mechanical properties of polymer composites (Senatov et al., 2016b). In particular, hydroxyapatite was found to play an important role in the strengthening and toughening of the polymer matrix as also suggested by other polymer composites. However, PLA scaffolds have certain limitations. These include low cell adhesion due to the hydrophobic property of PLA and inflammatory reactions in vivo caused by the lactic acid degradation product of high-molar-mass PLA and its accumulation due to inefficient removal from the surroundings of the scaffold's area and low impact toughness associated with its application (Bergsma et al., 1995; Donate et al., 2020; Liu et al., 2020a). In tissue engineering, it is therefore preferable to use a low-molar-mass PLA, which should provide sufficient structural support for cell adhesion as well as subsequent tissue development, while its rate of degradation should ideally match the rate of tissue regeneration. The low hydrophilicity is also an issue and that leads to lack of cell binding and also poor interaction between the printed PLA scaffold and the surrounding tissues (Wang et al., 2020). Another well-known PLA drawback is its lack of reactive side-chain groups, which prevents or complicates bulk modifications to improve its properties (Rasal et al., 2010). Often chemical or physical modification of PLA is needed to attain the desired properties for its intended biomedical applications and there are several ways to achieve this (Cuiffo et al., 2017).

FDM printing has also been used to manufacture PLA-based microneedles for drug delivery (Luzuriaga et al., 2018). This is a new class of transdermal microneedles for drug delivery purposes and promises a painless and hygienic alternative method to syringes (Aldawood et al., 2021). By piercing microscale pores through the stratum corneum, the drug is delivered to the bloodstream by controlling diffusion.

3.5.2 ADDITIVE MANUFACTURING OF PLA THROUGH SLS

SLS has been used for the fabrication of patient-specific scaffold implants with a complex geometry of interconnected pore structures for bone tissue regeneration purposes. A lot of research into scaffold fabrication is now focused on (bio)degradable polymers, including PLA, in an effort to replace non-biodegradable polymers. Biodegradable implants offer the possibility of being resorbed by the human body and will not require surgical explantation or permanently remain in the patient's body, unlike other non-biodegradable implants used for the treatment of bone defects (Wei et al., 2020). Several studies have reported the SLS method based application for the manufacture of PLA composite implants and tissue scaffolds, as this technique allows the incorporation of several materials into the 3D object (Antonov et al., 2005; Zhou et al., 2008; Bukharova et al., 2010; Duan et al., 2010; Lindner et al., 2011; Gayer et al., 2019). For instance, with the help of SLS, a mixture of PLA and poly(D-lactide) nanocomposites has been used for the manufacture of scaffolding from powder bed (Wu et al., 2018). The study evaluated several parameters such as energy density, preheating temperature, as well as scanning speed and distance by comparing the microstructures. The crystallisation morphology and microstructure analysis of samples generated with different process parameters were performed by polarised optical microscopy (POM) and scanning electron microscopy (SEM), respectively. The study highlighted several 3D printing structural defects associated with the microstructure of the specimens fabricated with each set of parameters, such as holes or gaps, arising from poor layer adhesion and sintering. The presence of this microporosity is generally known to affect the mechanical properties of printed parts and is one of the issues impeding any wider application of SLS in PLA. Poly(L-lactide) (PLLA) and hydroxyapatite composite have also been investigated in SLS technology (Tan et al., 2005). As can be seen from the presented SEM results of this study, the SLS method is capable of fabricating highly porous scaffolds for tissue engineering applications. Scaffolds made of poly(D,L-lactide) (PDLLA) and β-tricalcium phosphate (β-TCP) powder composite material (50/50 mass ratio) were successfully fabricated using SLS techniques (Lindner et al., 2011). The result of gel permeation chromatography (GPC) analysis indicated a decrease in the molar mass, which is a common phenomenon in polymer manufacturing processes including injection moulding. This result showed that the SLS technique melted the PDLLA polymer part of this composite material, indicating that the β-TCP is not affected by the laser. In addition, the X-ray diffraction analysis showed that the β-TCP is the only crystalline phase present in the specimens and no gaps were seen on a 1 μm microscopic scale between the polymer and the β-TCP component. This suggests that β-TCP has been homogeneously embedded in the PDLLA matrix although process related pores were seen at 50 μm.

By tailoring the material properties, the strength of scaffolds made with the SLS technology can be improved. Two batches of a PDLLA/β-TCP (50/50) composite powder with different polymer properties were prepared and sintered with SLS using identical processing parameters (Gayer et al., 2018). The object of the study was to investigate how much impact the material properties, such as molar mass, polymer particle size, filler particle size and internal viscosity, have on how a PLA and calcium carbonate (calcite) composite processable powder material can be produced by mechanical milling in the first place and how well this composite can be subsequently processed in the SLS technique. The results showed that the specimens manufactured from the batch with a smaller polymer particle size and low zero shear melt viscosity (Batch 1) had lower porosity and greater biaxial flexural strength (62 MPa) compared to specimens from the other batch (Figure 3.6) with a higher particle size and biaxial flexural strength of 23 MPa (Batch 2). This shows that, by optimising these parameters, specimens with higher mechanical strength can be fabricated with SLS.

Processing issues are one of the most important factors mentioned in the studies of PLA in the SLS technique. Therefore, among other things, a new preparation method for PLA/calcium carbonate composite powder for SLS application was investigated (Gayer et al., 2019). Composite powders with different compositions of about 75% by mass of polylactides (PLLA as well as PDLLA) and 25% by mass of calcium carbonate were prepared. Different PLA grades were also selected to represent the intrinsic viscosity range of 1.0–3.6 dl/g and the result revealed that the composite material with 1.0 dl/g, corresponding to PLA grade with low zero-shear viscosity (400 Pa·s) and with the particle size of 53 μm showed the best SLS machinability. Specimens with high strength of up to 75 MPa, due to low microporosity (about 2%), were sintered. This is due to the small polymer particle diameter as well as the small zero shear melt viscosity.

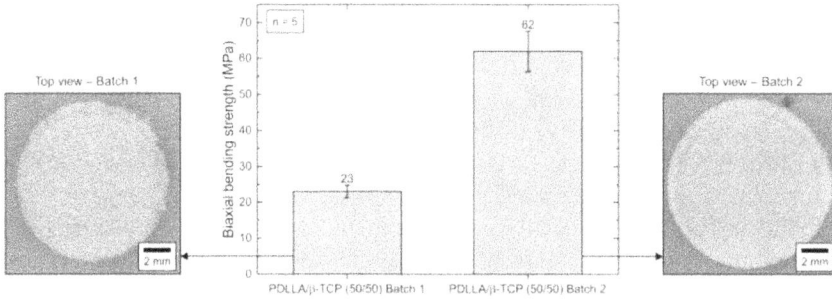

FIGURE 3.6   Biaxial flexural strength of the disc-shaped SLS test specimens with 10 mm diameter and 1 mm thickness. These are made of PDLLA/ β -TCP (50/50) Batch 1 and Batch 2.

Source: reprinted with permission from Gayer et al., Influence of the material properties of a poly(*D,L*-lactide)/β-tricalcium phosphate composite on the processability by selective **laser sintering**, *J. Mech. Behav. Biomed. Mater.* no 87 (2018):267–278. For more details, see the CC BY 4.0

A modified version of the SLS method has been developed, so-called surface SLS (SSLS) to broaden its scope to the application of thermoplastic polymers, such as PLA and PLGA (Antonov et al., 2006; Antonov et al., 2015; Santoro et al., 2016). The SSLS was also capable of providing a main of fabricated scaffolds with incorporated bioactive proteins, without them being destroyed during sintering. In principle, the SSLS method involves initiating the sintering process by adding a small amount (< 0.1% by mass) of carbon microparticles to the surfaces of the polymer particles. Another study demonstrated the viability of the SSLS method in the manufacture of PLA-based scaffolding (Kanczler et al., 2009). The in vivo studies of the suitability and biocompatibility of these scaffolds were examined. Scaffolds seeded with foetal femur-derived cells implanted in a murine critical size femur segmental defective model may aid bone regeneration.

3.5.3   ADDITIVE MANUFACTURING OF PLA THROUGH SL

The creation of 3D objects with the SL technique is a process that involves photo-polymerisation where a photosensitive polymer resin undergoes spatially controlled hardening upon interaction with a UV laser. Since SL techniques have the highest accuracy and resolution and offer the clearest details as well as the smoothest surface finish of all 3D printing techniques, it is no surprise that they have been the preferred method for orthopaedic applications (Williams et al., 2005; Melchels et al., 2010; Chartrain et al., 2018). PLA, due to its inherent processability, biocompatibility and (bio)degradability mechanical properties, has aroused great interest in the medical field in particular for its use in tissue engineering including implants tailored to a specific patient (Da Silva et al., 2018; Donate et al., 2020). Although not as much as other biodegradable resins, SL processing based on PLA has been explored mainly in applications related to tissue technology. In general, commercially available photopolymer resins for SL processing are multifunctional epoxy or (meth)acrylate monomers and lack the necessary biocompatibility and biodegradability, while PLA on the other hand possesses these properties but lacks photocurable derivatives where the C=C bonds in vinyl groups crosslink with different side groups and therefore cannot be processed with SL (Manapat et al., 2017; Bagheri and Jin, 2019). Consequently, PLA is modified by covalently end-functionalising with photo-crosslinkable unsaturated groups such as acrylate, methacrylate and fumarate to be readily polymerisable by the UV-laser-based SL process. The methacrylate-based monomers are often used, compared to other acrylate-based resins, due to the fact that they are less cytotoxic and have a high heat resistance (Manapat et al., 2017). This modification approach has been implemented as a strategy for developing PLA-based porous scaffolds with pre-designed high-resolution architectures (Melchels et al., 2009). The modification has been achieved by end-functionalising star-shaped oligo(D,L-lactide) with varying molar mass that was synthesised through ring-opening polymerisation (ROP) by reacting the terminal hydroxyl groups with methacryloyl chloride. The resulting prepolymers are subsequently photo-crosslinked in the presence of ethyl lactate as a non-reactive diluent. The effect of varying both of these factors (hydroxyapatite amount and pore properties) in oligo(D,L-lactide)/hydroxyapatite scaffolds on cell growth and differentiation was also evaluated (Tanodekaew et al., 2013). Oligolactide synthesised through ROP of D,L-lactide was functionalised with an excessive amount of methacrylate to form photocurable moieties. These prepolymers were then mixed with different percentages of hydroxyapatite to obtain oligo(D,L-lactide)/hydroxyapatite composites which are processed in the SL technique to fabricate scaffolds with different pore sizes and found that the scaffolds provide suitable conditions for cell growth support and differentiation, making them potentially suitable for bone tissue technology. In another example, oligolactide end-functionalised with fumaric acid monoethyl ester to provide photocurable functional groups have been reported (Grijpma et al., 2005). This process normally requires the reaction of excess fumaric acid, diethyl fumarate or fumaroyl chloride with the terminal hydroxy groups of the oligomers.

3.5.4   ADDITIVE MANUFACTURING OF PHAs THROUGH FDM

PHAs are thermoplastic polyesters that are naturally produced by the fermentation of numerous bacteria. PHA has excellent biocompatibility and biodegradability properties and is available from renewable resources, although it is used to a much lesser extent as a 3D print material compared to PLA. This is largely due to the physicochemical properties of PHA as different PHA types can contribute to varying mechanical and chemical properties based on their monomeric composition, but the high cost of producing PHA has also restricted its industrial applications (Wu et al., 2017). However, different PHAs, PHB and copolyester of 3-hydroxybutyrate and 3-hydroxyhexanoate, PHBHHx, as well as their composite with various biodegradable polymers, have shown excellent properties (Azad et al., 2020; Kovalcik et al., 2020; Mehrpouya et al., 2021). In particular, PHBV and PHB are by far the most studied of the PHAs as they are commercially available; however, PHBHHx has been attracting a lot of attention in recent years as it is now commercially produced by Kaneka Corporation (Tokyo, Japan) under the trade name PHBH™. A comprehensive list of other PHA manufacturers can be found elsewhere (Naser et al., 2021).

It is still the case that FDM technology does not attract major research efforts to produce 3D printing from PHAs. PHA is also a thermoplastic aliphatic polyester, similar to PLA; however, it is well documented that PHA is lagging behind PLA in terms of application and research in 3D printing, mainly due to higher production costs that have hampered early basic research. However, PHA composites have been widely used in the manufacture of various 3D objects due to commercial availability. Some of the print parameters have been explored and the ideal print temperature in FDM is reported to be around 110–170°C (Cano-Vicent et al., 2021). The relationship among abiotic degradation, processing conditions and print orientation such as parallel direction (horizontal) and perpendicular direction (vertical) of a commercial PLA/PHA filament in FDM machines has been investigated (Ausejo et al., 2018). Some of the challenges encountered in fabricating PHA filaments as a precursor material for 3D printers have been reported (Batchelor, 2016). PHA composites reinforced with siliceous sponge spicules have been reported as a biodegradable material of 3D printing filaments (Wu, 2018). Likewise, PHBV/palm fibre composite was used to make filaments with a diameter of 1.75 ± 0.05 mm by extrusion at 130–140°C and 50 rpm (Wu et al., 2017). The produced filament showed no phase separation, exhibiting a very good adhesion at the polymer-filler interface and better mechanical properties. It was also suggested that the mechanical properties of PHB are close to those of other biodegradable thermoplastic polyesters such as PLA (Naser et al., 2021).

One area that is generally attracting enormous research efforts in the field of 3D printing now is the manufacture of scaffolding for tissue engineering. Applications of PHA in scaffold-based tissue engineering have been explored due to PHA having biocompatibility and being a biodegradability thermo-processable polymer, making it a promising candidate for applications in tissue engineering (Lim et al., 2017). Using the FDM technique the thermal and rheological properties of commercial PHB, PHBV and PHBH™ scaffolds have been evaluated and confirmed as far as their suitability for tissue engineering (Kovalcik et al., 2020). It was also reported that porous 3D printed mesoporous bioactive glass and PHBHHx composite are potential scaffolds for bone regeneration (Zhao et al., 2014).

3.5.5 Additive Manufacturing of PHAs through SLS

SLS technology can be applied to all powdered polymers that melt but do not decompose under a laser beam as it subjects the materials to high temperatures. Both pure PHA and PHA composite have also been studied in the SLS techniques, albeit not to the same degree as FDM techniques as has been highlighted in several reviews on this topic (Chiulan et al., 2017; Kovalcik, 2021; Lee et al., 2021; Mehrpouya et al., 2021). Typically, suitable resins for SLS 3D printing are polyamides (nylon-12), epoxy resins, polystyrenes, polyaryletherketones, acrylic resins and thermoplastic elastomers. SLS enables the construction of scaffolding with complex internal and external geometries due to the layer-by-layer addition process. Hence, most of the reported studies of PHA in the SLS technique concern 3D scaffolds printing in the field of tissue engineering (Duan et al., 2010; Pereira et al., 2012a; Lupone et al., 2022). In SLS, the curing process of the resin depends on several factors, including resin content, temperature and the light source as well as its intensity. The correct setting of parameters, such as preheating temperature, construction orientation and appropriate laser settings such as the power, laser diameter, scanning speed and scanning distance are all important factors. How variation in energy density, by changing the speed and average power of the laser beam, affects the mechanical properties of a polymeric material have been investigated (Beal et al., 2009). This study revealed that laser power had a greater influence on density and mechanical properties compared to scanning speed. A more recent study of PHAs has concluded that the porosity and mechanical properties of the scaffold could be modified by changing the SLS process parameters such as the laser energy density, which causes variations in the relative densities, couplings between the layers and the amount of residual powder remaining inside the pores (Diermann et al., 2018). Interconnected PHBV scaffolds with porosities of up to 80% were fabricated without using pre-designed internal pore architectures, and the SLS method was used to evaluate how the scan spacing and thickness of the powder layer can affect parameters such as mechanical properties, dimensional deviations and morphology of PHA scaffolds for bone tissue engineering. The variation in the thickness of the powder layer had a higher impact on the values of the mechanical properties of the PHA scaffold compared to the variation in the scan spacing. In study of an SLS-based technique, pure PHB in powder form was used to fabricate porous structures with controlled pore size and the thermal stability during the process heating was evaluated (Pereira et al., 2012b). First, the PLA powder was used to print porous structures and the overflow powder from that first printing set was collected and reloaded into the machine to print a second set of porous structure. The same process was repeated one more time to produce a third set of prints. At the end of this third run, the powder was subjected to 32.5 h of SLS process cycles. Finally, the result showed that neither the PHB thermal properties nor the chemical composition had changed even after the powder had been subjected to these long SLS print processing times. Despite the low thermal stability of PHB in the melt, no signs of degradation were found, when the same powder was reused for additional 3D printing runs. This contrasts with conventional SLS powders, such as nylon. PHB scaffolding was also manufactured by AM via SLS (Saska et al., 2018). The morphology, porosity, thermal and mechanical properties of the resulting PHB scaffold were characterised. The result showed that the SLS method is capable of producing PHB scaffolds that are analogue to the design model, with good mechanical properties and a hierarchical structure with interconnected pores and inherent porosity. In addition, the results showed that the sintering process did not affect the thermal properties of PHB. PHA composites results have also been reported in the literature. By optimising several different parameters such as the layer thickness as well as the laser power and the scanning distance, calcium phosphate (Ca-P)/PHBV nanocomposite scaffolds of high quality were fabricated (Duan et al., 2011). Several other Ca-P/PHBV nanocomposite scaffolds manufactured via SLS for bone tissue were fabricated, including one that combines Ca-P/PHBV and carbonated hydroxyapatite (CHAp)/PLLA (Duan and Wang, 2010; Duan et al., 2010).

3.5.6 ADDITIVE MANUFACTURING OF PHAs THROUGH SL

There are very few studies on SL printing of PHAs. The main obstacle to the application of SL in PHA is its structure. PHA does not have photopolymerisable functional groups which are one of the desired characteristics for the success of the curing reaction, as the SL techniques require UV light-induced polymerisation. Prepolymers, oligomers with viscosity lower than 5 Pas and the contents of photosensitive active terminating functional groups are typical photopolymer materials used for SL-based 3D printing (Foli et al., 2020). The most commonly used commercial resins include acrylate, methacrylate, vinyl ether and epoxy prepolymers to ensure cross-linking. However, these materials are not biodegradable, and although modifications to the PHA structure can be achieved to enable effective SA curing by introducing UV-curable end groups, this modification can nevertheless have implications that could affect photochemistry as well as physical chemistry and could either increase or decrease the biodegradability of the polymer. Several studies have explored the application of PHAs for the manufacture of tissue-engineered scaffolds (Zhao et al., 2003; Lim et al., 2017; Dwivedi et al., 2020). This interest in PHA can be explained by its versatile mechanical properties which render it suitable for both soft and hard tissue printing, although only a few studies have dealt with the application of SL in tissue engineering. Using SL technology in tissue engineering, a human heart valve was fabricated from polyhydroxyoctanoate (PHO) and poly(4-hydroxybutyrate) (P4HB) blend (Sodian et al., 2002). A complex 3D CAD model derived from X-ray computed tomography was first created and then printed with SL. The thermoplastic and elastic nature of PHO and P4HB offered the flexibility needed to make an anatomically shaped heart valve that resembles the model. A solvent-free, two-stage photo source resin formulation method for preparing bis-methacryl terminated PHB diol oligomers has been suggested (Foli et al., 2020). As the transesterification of poly(3-hydroxybutyrate) with 1,4-butanediol is completed in the first step, it is then followed by the reaction with 2-isocyanatoethyl methacrylate in the second step. The obtained PHB-diol was diluted with propylene carbonate at 90°C to get the desired viscosity. Finally, after the addition of IRGACURE® 819 as a photoinitiator, this optimisation enables the PHB-diol to be photocurable under UV light.

3.5.7 ADDITIVE MANUFACTURING OF PCL THROUGH FDM

PCL is a commonly used polymer in the material extrusion-based 3D printers and there are many works concerning both the fabrication and the characterisation of PCL filaments and their 3D printed specimens (Hutmacher et al., 2001; Zein et al., 2002; Korpela et al., 2013; Aho et al., 2019). However, most of the studies concerning PCL present results for PCL with various other biodegradable composites polymers (Zhao et al., 2003; Shor et al., 2007; Nyberg et al., 2017; Haq et al., 2019a; Cakmak et al., 2020; Liu et al., 2020b). One commonly used method to enhance the properties of FDM printed parts is changing the print process parameters. The effect of process parameters such as nozzle temperature, volumetric flow rate, bed temperature, print speed, infill density, layer height and shell perimeter and flow rate on 3D-printed PCL parts was examined to find the optimal conditions (Mehraein, 2018). The study evaluated seven nozzle temperatures (from 115 to 175°C) and six different flow rates (100% [2.2 mm³/s] to 145% [3.19 mm³/s]). The nozzle temperature of 165°C and the flow rate of 135°C in a defect-free PCL FDM printed test sample with a layer height of 0.1 mm, shell perimeter of 2 and infill density of 90% gives specimens enhanced mechanical properties. PCL is widely used in the fabrication of 3D scaffolds due to its biocompatibility, rheological properties, loadbearing capacity, blend compatibility and slow degradation rate in vivo (2–3 years which allows time for bone remodelling). It generates fewer acidic degradation products (compared to other polyesters) and is also FDA approved (Ramanath et al., 2008; Mkhabela and Ray, 2014; Abdelfatah et al., 2021; Yang et al., 2021a). The application of medical grade 3D PCL as a stand for bone tissue technology was introduced a little more than a decade ago but has now been implanted in over 20,000 patients (Teoh et al., 2019). Pioneering research on PCL-based filaments for the FDM method was used for the fabrication of PCL-based scaffolding (Hutmacher et al., 2001). In this case, FDM was used for the design and fabrication of 3D scaffolding with a fully interconnected pore network and a porosity of 61% from plain PCL polymer pellets with an average number-average molar mass (M_n) of 80,000. Two PCL pallets with the same porosity (61%) and in two matrix orientations, a lay-down pattern of 0/60/120° and a lay-down pattern of 0/72/144/36/108°, a three-angle architecture and five-angle architecture respectively, were fabricated.

The mechanical properties of these PCL scaffolds and their in vitro biocompatibility were evaluated. The results show that both PCL scaffolds have a typical honeycomb behaviour by being deformed during the compression test and have good mechanical properties as well as excellent biocompatibility. This excellent mechanical property of the scaffolding is expected to be maintained for a period of up to 6 months, while it gradually deteriorates thereafter until it completely decomposes within 2 years. PCL-based scaffolds with different channel sizes (160–700 µm) and porosity (48–77%) were also used. These scaffolds were characterised by several different techniques, and it was concluded that their mechanical properties largely depend on their porosity, irrespective of their pattern and channel size. These porous PCL scaffolds had fully interconnected channel networks as well as mechanical deformation stress-strain behaviours that exhibited mechanical properties very similar to those observed for implantation site tissues; in vitro studies demonstrated excellent biocompatibility with human fibroblast and periosteal cells.

Pure PCL lacks cell adhesion due to its hydrophobic nature, and therefore modification by mixing with other polymer materials, including inorganic particles, provides a way to generally overcome these shortcomings and increase hydrophilicity and cell affinity (Shor et al., 2007; Fu et al., 2012; Turnbull et al., 2018). In particular, the addition of hydroxyapatite is stated not only to increase hydrophilicity but also to improve the mechanical property; the properties of PCL/hydroxyapatite composite can be altered by optimising the processing parameters.

PCL with hydroxyapatite filaments with varying content of hydroxyapatite from 5 to 25 wt% was produced and used to manufacture leg stands with hole sizes of 1200 µm. The mechanical properties of these filaments, as well as the 3D-printed racks, were evaluated. This showed a similar pattern with both tensile strength and fracture elongation rates decreasing, due to the poor mechanical properties of HA. A filament with 30 wt% hydroxyapatite was left out of the study as it was too brittle to be used as a filament.

FDM has also been used as a manufacturing method in the fabrication of a PCL-based 3D-printed drug delivery carrier. The FDM method is very versatile and allows the printing of drugs that can offer tailored delivery with variable dosage, release profiles and different shapes, sizes and geometric models such as hollow structures, spheres, pyramids, cubes, torus or even customised medicine that is not available in current compaction form (Goyanes et al., 2016; Holländer et al., 2016; Kempin et al., 2017; Tan et al., 2018; Đuranović et al., 2021; Windolf et al., 2022). PCL is a drug delivery vehicle well known not only for having desirable biodegradability properties and biocompatibility but also because of its high aggregation tendency and drug permeability. It has been used to prepare the delivery of various drugs (Dash and Konkimalla, 2012; Malikmammadov et al., 2018). An indomethacin loaded PCL filament with systematically varied drug content was prepared (Holländer et al., 2016). The filament was then with the help of the FDM method used for the printing of T-shaped indomethacin containing intrauterine devices. The study established the viability of FDM methods for the fabrication of controlled-release implantable devices and that both the filament as well as the 3D-printed prototypes were affected by the amount of drug loaded. Nevertheless, a diffusion-based drug release mechanism was observed with minor effects from the biodegradation of PCL. A combination of FDM 3D printing with salt leaching techniques was used for the fabrication of Cefazolin-containing scaffolds from a medical grade PCL (Visscher et al., 2018). The scaffolds had been intentionally manufactured to have a double micro- and macroporosity, where the microporosity properties enabled loading and the subsequent sustained release of the drug. It was also compared to the drug delivery pattern for several different polymers, such as PCL, PLLA, ethylcellulose and Eudragit® RS polymer. Fluorescent dye quinine was used as a model drug, which was loaded into the polymer during the filament preparation by hot melt extrusion (HME) to facilitate the visualisation of the drug distribution in the filament. The filament is then used to fabricate hollow-shaped implants using the FDM method. The results showed that PCL has the fastest release rate and, in addition, the release rate is largely dependent on both the type of polymer and the amount of drug.

It is also a common occurrence to use PCL with various composites for drug delivery purposes. This is mainly to improve some of the issues regarding the drug release: for instance, diffusion is not the dominant drug release mechanism for PCL 3D-printed drug delivery devices, as the drugs are tightly embedded once added to the polymer matrix (Holländer et al., 2016). Some other concerns are the high processing temperature and the addition of organic solvents to prevent clogging of the nozzles, both of which can interfere with drug efficacy. The addition of suitable biodegradable polymeric materials eliminates some of the problems that limit the PCL application in drug delivery. The effect of plasticisers on the thermal and mechanical properties of PCL is well documented as their addition can provide a way to improve the overall processability and enhance 3D printability (Alhijjaj et al., 2016; Haq et al., 2017; Viidik et al., 2021). PCL was combined with water-soluble chitosan to make 3D-printed drug delivery implants (Yang et al., 2022). The PCL was impregnated with ibuprofen as the model drug, while the chitosan was dissolved to form controlled and effective release channels. The addition of chitosan and the drug acts as a plasticiser and nucleation agent thereby reducing the crystallinity and tensile strength of the material by weakening the intermolecular forces of PCL. Thus, the need to use organic solvents is eliminated and the implant can be printed at a lower temperature. By optimising the drug and chitosan content of the 3D-printed implant, a delayed release of 99% was obtained which lasted for 5 days. The implant could be used as catheters or scaffolding to support the wound-healing process.

3.5.8 ADDITIVE MANUFACTURING OF PCL THROUGH SLS

Various studies have reported the fabrication of PCL scaffolds using the SLS technique. For instance, the SLS technique is used to fine-tune the porosity and thus study factors (such as scaffold geometry and process conditions) that affect the mechanical properties to optimise the best conditions allowing maximum fold architectures that mimic the mechanical properties of human trabecular bone via SLS Porous PCL scaffold architectures that mimic the mechanical properties of human trabecular bone have been computationally designed and then fabricated via SLS (Williams et al., 2005). SLS is also used for the fabrication of cartilage tissue scaffolds from PCL hybridised with collagen hydrogel (Chen et al., 2019). It is also quite common to use PCL composite for tissue engineering applications, such as PCL composites filled with different volume fractions of β-TCP to provide the appropriate mechanical strength required for a scaffold, (Chung et al., 2010; Eshraghi and Das, 2012). The SLS technology has also found application in the manufacture of medicines, in particular PCL based drug delivery systems. The technique offers the manufacture of medicines with controlled release dosing, which can be a promising technology that enables the development of tailored medication. This is an area that is gaining increased attention (Awad et al., 2020; Kulinowski et al., 2021). The SLS has enabled the fabrication of PCL impregnated with progesterone, ibuprofen, chemotherapy drugs, copovidone and paracetamol (Salmoria et al., 2012; Salmoria et al., 2016; Salmoria et al., 2017a; Gueche et al., 2021; Yang et al., 2021b). The properties of resorbable porous PCL/ibuprofen implants made by SLS that could be used for bone regeneration and the control of the inflammatory process shortly after implantation have been studied. The samples were sintered in SLS equipment using a CO_2 laser with a beam diameter of 250 mm. A total drug release of 75% was observed after 26.40 h, which showed a non-linear release profile with respect to time and with a possible combination of diffusion and erosion mechanisms (Salmoria et al., 2016).

3.5.9 ADDITIVE MANUFACTURING OF PCL THROUGH SL

Like PHA, PCL is not photocurable in SL, so further modifications are needed to produce photocurable prepolymers. Many methacrylted compositions with photocurable groups have found the potential for SL 3D printing (Matsuda et al., 2000; Williams et al., 2005; Elomaa et al., 2011). Pioneering work suggested the synthesis of photocurable PCL based liquids for use in SL (Matsuda et al., 2000). Photocurable and biodegradable liquid copolymers of ε-caprolactone and trimethylene carbonate were prepared by ring-opening copolymerisation and then subjected to coumarin derivatisation at their hydroxyl terminal. Photocrosslinkable PCL was also synthesised by ring-opening polymerisation of ε-caprolactone and methacrylation at the hydroxyl terminal of the resulting PCL prepolymer oligomers with varying molecular structure (Elomaa et al., 2011). SL 3D technology was then used to make porous scaffolding from the photocurable PCL macromer resin containing IRGACURE® 369 photoinitiator, vitamin E inhibitor to prevent premature crosslinking and also Orasol® Orange G dye to better control the cure depth. The scaffolding was exactly similar to the CAD design model and had a high porosity (70.5 ± 0.8%), with a pore size of 465 μm and the interconnection of the pores was complete throughout the 3D structures. In addition, no shrinkage occurred after the end of the curing process because no changes were observed on the scaffold dimensions. The effect of prepolymer molar mass on the properties of cross-linked PCL with different molar masses was evaluated (Green et al., 2018). However, this study has gone a step further by also examining the effect of the functionality of the polymer. The researcher compared a high-molar-mass PCL with low functional groups with that of low molar mass and a high degree of functionality. The result indicated that the former (PCL with low functional groups) exhibited a higher level of strain at break and lower modulus, while the opposite result was seen for the later (low-molar-mass polymer with a high degree of functionality). Another study of photocurable PCL methacrylate resin for use in SL examined the relationship between the polymer degradation rate and the degree of methacrylation (Field et al., 2021). The degree of methacrylation that occurs during the photocuring process affects the crosslinking density of the cured material, which in turn affects both the mechanical properties and the degradation rate. By varying the degree of methacrylation from 17–77%, a crosslinked PCL methacrylate resin of Young's modulus from 0.12–3.51 MPa could be obtained. This also affects the degradation, as complete degradation was observed in 17 days for the resin with the lowest degree of methacrylation, while non-significant degradation was seen in 21 days for the one with the highest degree of methacrylation. It is also worth noting here that β-carotene was used as a biocompatible photoinitiator, unlike the IRGACURE® 369 reported in previous cases (Elomaa et al., 2011).

3.6 CONCLUSIONS

As this is a diverse subject area and one that has accumulated a significant body of research in the last few years, our aim in this section was never to offer a comprehensive overview of the pertinent topics, but rather to provide some contextual material with a number of examples. When it comes to AM, as expected, each of the methods has its own advantages and disadvantages. It is therefore quite common to buy several different 3D printers in the same manufacturing environment. Nevertheless, FDM is still the most widely used technique for rapid prototyping while SL is used for prototyping and high-resolution manufacturing. Commercial 3D materials are increasing in both demand and supply, but there is still limited availability of materials that are compatible with some of the AM processes. Some of the biodegradable materials require an effort to improve their processability with 3D printers, but this may in turn change their properties and thus affect their biodegradability. An example of this is the stereolithography 3D printing method, whereby photocurable liquid resin is a requirement and PLA lacks photocurable moieties and therefore cannot be processed with SL. However, some bio-based acrylate photopolymer resins are also emerging.

Although currently biodegradable polymers account for only a small proportion of the total global consumption of thermoplastics, their potential to contribute to a more sustainable future is nevertheless significant. Notwithstanding, the question is often raised as to whether biodegradable polymers will ever fully solve the problem of plastic waste and completely replace non-biodegradable plastics. Presumably, this will probably never be the case, but it should not dampen the current research momentum. The use of biodegradable polymers in medical applications, such as scaffolds, coronary stents and drug delivery systems, as well as everyday items in our lives, such as food packaging, coffee capsules, straws and shopping bags is a testament to how far this has come in recent years. This gives at least some hope to facilitate the ongoing transition towards a circular economy, as it reduces the amount of non-biodegradable plastic going to landfills. Most of these (bio)degradable items are made from the (bio)degradable polymers described, in this section, such as PLA, PHA and PCL using one of the processing techniques discussed here, such as extrusion, injection moulding or 3D printing. The study of their physicochemical, mechanical and structural properties in order to optimise the manufacturing conditions is an active research topic in the polymer processing field. These important research challenges will be crucial for both the identification of new biodegradable polymers and the modification of existing polymers to increase their biodegradability in the future.

REFERENCES

Abdelfatah, J., Paz, R., Alemán-Domínguez, M.E., Monzón, M., Donate, R., and Winter, G. 2021. Experimental analysis of the enzymatic degradation of polycaprolactone: microcrystalline cellulose composites and numerical method for the prediction of the degraded geometry. *Materials* 14(9): 2460.

Abe, M.M., Martins, J.R., Sanvezzo, P.B., Macedo, J.V., Branciforti, M.C., Halley, P., Botaro, V.R., and Brienzo, M. 2021. Advantages and disadvantages of bioplastics production from starch and lignocellulosic components. *Polymers* 13(15): 2484.

Aho, J., Bøtker, J.P., Genina, N., Edinger, M., Arnfast, L., and Rantanen, J. 2019. Roadmap to 3D-printed oral pharmaceutical dosage forms: feedstock filament properties and characterization for fused deposition modeling. *J. Pharm. Sci. Res.* 108(1): 26–35.

Akhoundi, B., Nabipour, M., Hajami, F., and Shakoori, D. 2020. An experimental study of nozzle temperature and heat treatment (annealing) effects on mechanical properties of high-temperature polylactic acid in fused deposition modeling. *Polym. Eng. Sci.* 60(5): 979–987.

Akindoyo, J.O., Beg, M.D.H., Ghazali, S., Heim, H.P., and Feldmann, M. 2017. Effects of surface modification on dispersion, mechanical, thermal and dynamic mechanical properties of injection molded PLA-hydroxyapatite composites. *Compos. Part A Appl. Sci.* 103: 96–105.

Aldawood, F.K., Andar, A., and Desai, S. 2021. A comprehensive review of microneedles: types, materials, processes, characterizations and applications. *Polymers* 13(16): 2815.

Alhijjaj, M., Belton, P., and Qi, S. 2016. An investigation into the use of polymer blends to improve the printability of and regulate drug release from pharmaceutical solid dispersions prepared via fused deposition modeling (FDM) 3D printing. *Eur. J. Pharm. Biopharm.* 108: 111–125.

Al-Itry, R., Lamnawar, K., and Maazouz, A. 2012. Improvement of thermal stability, rheological and mechanical properties of PLA, PBAT and their blends by reactive extrusion with functionalized epoxy. *Polym. Degrad. Stab.* 97(10): 1898–1914.

Amrita, Manoj, A., and Panda, R.C. 2022. Biodegradable filament for 3D printing process: A review. *Eng. Sci.* 18: 11–19.

Anderson, K.S., Schreck, K.M., and Hillmyer, M.A. 2008. Toughening polylactide. *Polym. Rev.* 48(1): 85–108.

Andrady, A.L., and Neal, M.A. 2009. Applications and societal benefits of plastics. *Philos. Trans. R. Soc. Lond., B, Biol. Sci.* 364(1526): 1977–1984.

Antonov, E.N., Bagratashvili, V.N., Howdle, S.M., Konovalov, A.N., Popov, V.K., and Panchenko, V.Y. 2006. Fabrication of polymer scaffolds for tissue engineering using surface selective laser sintering. *Laser Phys.* 16(5): 774–787.

Antonov, E.N., Bagratashvili, V.N., Whitaker, M.J., Barry, J.J., Shakesheff, K.M., Konovalov, A.N., Popov, V.K., and Howdle, S.M. 2005. Three-dimensional bioactive and biodegradable scaffolds fabricated by surface-selective laser sintering. *Adv. Mater.* 17(3): 327–330.

Antonov, E.N., Krotova, L.I., Minaev, N.V., Minaeva, S.A.E., Mironov, A.V., Popov, V.K., and Bagratashvili, V.N. 2015. Surface-selective laser sintering of thermolabile polymer particles using water as heating sensitizer. *Quantum Electron.* 45(11): 1023.

Arrieta, M.P., Fortunati, E., Dominici, F., López, J., and Kenny, J.M. 2015. Bionanocomposite films based on plasticized PLA-PHB/cellulose nanocrystal blends. *Carbohydr. Polym.* 121: 265–275.

Arrieta, M.P., Samper, M.D., Aldas, M., and López, J. 2017. On the use of PLA-PHB blends for sustainable food packaging applications. *Materials* 10(9): 1008.

Ashter, S.A. 2016. *Introduction to Bioplastics Engineering.* Oxford: William Andrew Publishing.

ASTM 52900-21. 2021. *Additive Manufacturing – General Principles – Fundamentals and Vocabulary.* West Conshohocken: ASTM International.

Ausejo, J.G., Rydz, J., Musiol, M., Sikorska, W., Sobota, M., Włodarczyk, J., Adamus, G., Janeczek, H., Kwiecień, I., Hercog, A., and Johnston, B. 2018. A comparative study of three-dimensional printing directions: The degradation and toxicological profile of a PLA/PHA blend. *Polym. Degrad. Stab.* 152: 191–207.

Awad, A., Fina, F., Goyanes, A., Gaisford, S., and Basit, A.W. 2020.3D Printing: Principles and pharmaceutical applications of selective laser sintering. *Int. J. Pharm.* 586: 119594.

Azad, M.A., Olawuni, D., Kimbell, G., Badruddoza, A.Z.M., Hossain, M., and Sultana, T. 2020. Polymers for extrusion-based 3D printing of pharmaceuticals: A holistic materials – Process perspective. *Pharmaceutics* 12(2): 124.

Bagheri, A., and Jin, J. 2019. Photopolymerization in 3D printing. *ACS Appl. Polym. Mater.* 1(4): 593–611.

Bahramian, B., Fathi, A., and Dehghani, F. 2016. A renewable and compostable polymer for reducing consumption of non-degradable plastics. *Polym. Degrad. Stab.* 133: 174–181.

Barnett, E., and Gosselin, C. 2015. Weak support material techniques for alternative additive manufacturing materials. *Addit. Manuf.* 8: 95–104.

Batchelor, W.M. 2016. *PHA Biopolymer Filament for 3D Printing.* PhD Thesis. Williamsburg: College of William and Mary.

Beal, V.E., Paggi, R.A., Salmoria, G.V., and Lago, A. 2009. Statistical evaluation of laser energy density effect on mechanical properties of polyamide parts manufactured by selective laser sintering. *J. Appl. Polym. Sci.* 113(5): 2910–2919.

Behm, J.E., Waite, B.R., Hsieh, S.T., and Helmus, M.R. 2018. Benefits and limitations of three-dimensional printing technology for ecological research. *BMC Ecol.* 18(1): 1–13.

Beltrán, F.R., Infante, C., de la Orden, M.U., and Urreaga, J.M. 2019. Mechanical recycling of poly(lactic acid): Evaluation of a chain extender and a peroxide as additives for upgrading the recycled plastic. *J. Cleaner Prod.* 219: 46–56.

Bergsma, J.E., de Bruijn, W.C., Rozema, F.R., Bos, R.R.M., and Boering, G. 1995. Late degradation tissue response to poly(*L*-lactide) bone plates and screws. *Biomaterials* 16(1): 25–31.

Berman, B. 2012. 3-D printing: The new industrial revolution. *Bus. Horiz.* 55(2): 155–162.

Berry, E., Brown, J.M., Connell, M.C., Craven, M., Efford, N.D., Radjenovic, A., and Smith, M.A. 1997. Preliminary experience with medical applications of rapid prototyping by selective laser sintering. *Med. Eng. Phys.* 19(1): 90–96.

Biron, M. 2016. EcoDesign. In *Material Selection for Thermoplastic Parts*, ed. Biron, M., pp. 603–653. Oxford: William Andrew Publishing.

Bledzki, A.K., and Jaszkiewicz, A. 2010. Mechanical performance of biocomposites based on PLA and PHBV reinforced with natural fibres – A comparative study to PP. *Compos. Sci. Technol.* 70(12): 1687–1696.

Bledzki, A.K., Jaszkiewicz, A., and Scherzer, D. 2009. Mechanical properties of PLA composites with man-made cellulose and abaca fibres. *Compos. Part A Appl. Sci. Manuf.* 40(4): 404–412.

Borkar, T., Goenka, V., and Jaiswal, A.K. 2021. Application of poly-ε-caprolactone in extrusion-based bioprinting. *Bioprinting* 21: e00111.

Bossu, J., Le Moigne, N., Dieudonné-George, P., Dumazert, L., Guillard, V., and Angellier-Coussy, H. 2021. Impact of the processing temperature on the crystallization behavior and mechanical properties of poly[*R*-3-hydroxybutyrate-*co*-(*R*-3-hydroxyvalerate)]. *Polymer* 229: 123987.

Bucci, D.Z., Tavares, L.B.B., and Sell, I. 2005. PHB packaging for the storage of food products. *Polym. Test.* 24(5): 564–571.

Bucci, D.Z., Tavares, L.B.B., and Sell, I. 2007. Biodegradation and physical evaluation of PHB packaging. *Polym. Test.* 26(7): 908–915.

Bugnicourt, E., Cinelli, P., Lazzeri, A., and Alvarez, V.A. 2014. Polyhydroxyalkanoate (PHA): Review of synthesis, characteristics, processing and potential applications in packaging. *eXPRESS Polym. Lett.* 8(11): 791–808.

Bukharova, T.B., Antonov, E.N., Popov, V.K., Fatkhudinov, T.K., Popova, A.V., Volkov, A.V., Bochkova, S.A., Bagratashvili, V.N., and Gol'Dshtein, D.V. 2010. Biocompatibility of tissue engineering constructions from porous polylactide carriers obtained by the method of selective laser sintering and bone marrow-derived multipotent stromal cells. *Bull. Exp. Biol. Med.* 149(1): 148–153.

Burniol-Figols, A., Skiadas, I.V., Daugaard, A.E., and Gavala, H.N. 2020. Polyhydroxyalkanoate (PHA) purification through dilute aqueous ammonia digestion at elevated temperatures. *J. Chem. Technol. Biotechnol.* 95(5): 1519–1532.

Cakmak, A.M., Unal, S., Sahin, A., Oktar, F.N., Sengor, M., Ekren, N., Gunduz, O., and Kalaskar, D.M. 2020. 3D printed polycaprolactone/gelatin/bacterial cellulose/hydroxyapatite composite scaffold for bone tissue engineering. *Polymers* 12(9): 1962.

Cano-Vicent, A., Tambuwala, M.M., Hassan, S.S., Barh, D., Aljabali, A.A., Birkett, M., Arjunan, A., and Serrano-Aroca, Á. 2021. Fused deposition modelling: current status, methodology, applications and future prospects. *Addit. Manuf.* 47: 102378.

Carrasco, F., Pagès, P., Gámez-Pérez, J., Santana, O.O., and Maspoch, M.L. 2010. Processing of poly(lactic acid): characterization of chemical structure, thermal stability and mechanical properties. *Polym. Degrad. Stab.* 95(2): 116–125.

Castro-Aguirre, E., Iniguez-Franco, Samsudin, F.H., Fang, X., and Auras, R. 2016. Poly(lactic acid) – Mass production, processing, industrial applications, and end of life. *Adv. Drug Delivery Rev.* 107: 333–366.

Chartrain, N.A., Williams, C.B., and Whittington, A.R. 2018. A review on fabricating tissue scaffolds using vat photopolymerization. *Acta Biomater.* 74: 90–111.

Chen, S., Zhu, L., Wen, W., Lu, L., Zhou, C., and Luo, B. 2019. Fabrication and evaluation of 3D printed poly(*L*-lactide) scaffold functionalized with quercetin-polydopamine for bone tissue engineering. *ACS Biomater. Sci. Eng.* 5(5): 2506–2518.

Chen, Y., Tang, T., and Ayranci, C. 2022. Moisture-induced anti-plasticization of polylactic acid: Experiments and modeling. *J. Appl. Polym. Sci.* 139(24): 52369.

Chiulan, I., Frone, A.N., Brandabur, C., and Panaitescu, D.M. 2017. Recent advances in 3D printing of aliphatic polyesters. *Bioengineering* 5(1): 2.

Chung, H., Jee, H., and Das, S. 2010. Selective laser sintering of PCL/TCP composites for tissue engineering scaffolds. *J. Mech. Sci. Technol.* 24: 241–244.

Corneillie, S., and Smet, M. 2015. PLA architectures: The role of branching. *Polym. Chem.* 6(6): 850–867.

Correa, J.P., Molina, V., Sanchez, M., Kainz, C., Eisenberg, P., and Massani, M.B. 2017. Improving ham shelf life with a polyhydroxybutyrate/polycaprolactone biodegradable film activated with nisin. *Food Packag. Shelf Life* 11: 31–39.

Costanzo, A., Spotorno, R., Candal, M.V., Fernández, M.M., Müller, A.J., Graham, R.S., Cavallo, D., and McIlroy, C. 2020. Residual alignment and its effect on weld strength in material-extrusion 3D-printingof polylactic acid. *Addit. Manuf.* 36: 101415.

Cui, Z., Nelson, B., Peng, Y., Li, K., Pilla, S., Li, W.J., Turng, L.S., and Shen, C. 2012. Fabrication and characterization of injection molded poly(ε-caprolactone) and poly(ε-caprolactone)/hydroxyapatite scaffolds for tissue engineering. *Mater. Sci. Eng., C.* 32(6): 1674–1681.

Cuiffo, M.A., Snyder, J., Elliott, A.M., Romero, N., Kannan, S., and Halada, G.P. 2017. Impact of the fused deposition (FDM) printing process on polylactic acid (PLA) chemistry and structure. *Appl. Sci.* 7(6): 579.

Da Silva, D., Kaduri, M., Poley, M., Adir, O., Krinsky, N., Shainsky-Roitman, J., and Schroeder, A. 2018. Biocompatibility, biodegradation and excretion of polylactic acid (PLA) in medical implants and theranostic systems. *Chem. Eng. J.* 340: 9–14.

Dang, X.P. 2014. General frameworks for optimization of plastic injection molding process Parameters. *Simul. Modell. Pract. Theory.* 41: 15–27.

Dash, T.K., and Konkimalla, V.B. 2012. Poly-ε-caprolactone based formulations for drug delivery and tissue engineering: a review. *JRC* 158(1): 15–33.

de Koning, G.J.M., Lemstra, P.J., Hill, D.J.T., Carswell, T.G., and O'Donnell, J.H. 1992. Ageing phenomena in bacterial poly[(R)-3-hydroxybutyrate]: 1. A study on the mobility in poly[(R)-3-hydroxybutyrate] powders by monitoring the radical decay with temperature after γ-radiolysis at 77 K. *Polymer* 33(15): 3295–3297.

de Schutter, G., Lesage, K., Mechtcherine, V., Nerella, V.N., Habert, G., and Agusti-Juan, I. 2018. Vision of 3D printing with concrete – Technical, economic and environmental potentials. *Cem. Concr. Res.* 112: 25–36.

Deckard, C.R. 1989. Method and apparatus for producing parts by selective sintering. US Patent 5316580.

Dey, A., Roan Eagle, I.N., and Yodo, N. 2021. A review on filament materials for fused filament fabrication. *J. Manuf. Mater. Process.* 5(3): 69.

Diermann, S.H., Lu, M., Zhao, Y., Vandi, L.J., Dargusch, M., and Huang, H. 2018. Synthesis, microstructure, and mechanical behaviour of a unique porous PHBV scaffold manufactured using selective laser sintering. *J. Mech. Behav. Biomed. Mater.* 84: 151–160.

Din, M.I., Ghaffar, T., Najeeb, J., Hussain, Z., Khalid, R., and Zahid, H. 2020. Potential perspectives of biodegradable plastics for food packaging application-review of properties and recent developments. *Food Addit. Contam. Pt. A.* 37(4): 665–680.

Ding, W., Jahani, D., Chang, E., Alemdar, A., Park, C.B., and Sain, M. 2016. Development of PLA/cellulosic fiber composite foams using injection molding: Crystallization and foaming behaviors. *Compos. Part A Appl. Sci.* 83, 130–139.

Dizon, J.R.C., Espera Jr, A.H., Chen, Q., and Advincula, R.C. 2018. Mechanical characterization of 3D-printed polymers. *Addit. Manuf.* 20: 44–67.

Dodziuk, H. 2016. Applications of 3D printing in healthcare. *Polish. J. Cardio-Thorac. Surg.* 13(3): 283.

Domb, A.J., Kumar, N., and Ezra, A. 2011. *Biodegradable Polymers in Clinical use and Clinical Development.* Hoboken: John Wiley & Sons Ltd.

Donate, R., Monzón, M., and Alemán-Domínguez, M.E. 2020. Additive manufacturing of PLA-based scaffolds intended for bone regeneration and strategies to improve their biological properties. *E-Polymers* 20(1): 571–599.

Duan, B., Cheung, W.L., and Wang, M. 2011. Optimized fabrication of CA-P/PHBV nanocomposite scaffolds via selective laser sintering for bone tissue engineering. *Biofabrication* 3(1): 015001.

Duan, B., and Wang, M. 2010. Customized CA-P/PHBV nanocomposite scaffolds for bone tissue engineering: Design, fabrication, surface modification and sustained release of growth factor. *J. R. Soc. Interface.* 7(5): S615–S629.

Duan, B., Wang, M., Zhou, W.Y., Cheung, W.L., Li, Z.Y., and Lu, W.W. 2010. Three-dimensional nanocomposite scaffolds fabricated via selective laser sintering for bone tissue engineering. *Acta Biomater.* 6(12): 4495–4505.

Đuranović, M., Obeid, S., Madžarević, M., Cvijić, S., and Ibrić, S. 2021. Paracetamol extended release FDM 3D printlets: Evaluation of formulation variables on printability and drug release. *Int. J. Pharm.* 592: 120053.

Dwivedi, R., Pandey, R., Kumar, S., and Mehrotra, D. 2020. Polyhydroxyalkanoates (PHA): Role in bone scaffolds. *JOBCR* 10(1): 389–392.

Elomaa, L., Teixeira, S., Hakala, R., Korhonen, H., Grijpma, D.W., and Seppälä, J.V. 2011. Preparation of poly(ε-caprolactone)-based tissue engineering scaffolds by stereolithography. *Acta Biomater.* 7(11): 3850–3856.

Eshraghi, S., and Das, S. 2012. Micromechanical finite-element modeling and experimental characterization of the compressive mechanical properties of polycaprolactone-hydroxyapatite composite scaffolds prepared by selective laser sintering for bone tissue engineering. *Acta Biomater.* 8(8): 3138–3143.

Farah, S., Anderson, D.G., and Langer, R. 2016. Physical and mechanical properties of PLA, and their functions in widespread applications – A comprehensive review. *Adv. Drug Deliv. Rev.* 107: 367–392.

Fazzini, G., Paolini, P., Paolucci, R., Chiulli, D., Barile, G., Leoni, A., Muttillo, M., Pantoli, L., and Ferri, G. 2019. Print on air: FDM 3D printing without supports. In *II Workshop on Metrology for Industry 4.0 and IoT*, pp. 350–354. IEEE: Naples.

Ferretti, P., Leon-Cardenas, C., Santi, G.M., Sali, M., Ciotti, E., Frizziero, L., Donnici, G., and Liverani, A. 2021. Relationship between FDM 3D printing parameters study: Parameter optimization for lower defects. *Polymers* 13(13): 2190.

Field, J., Haycock, J.W., Boissonade, F.M., and Claeyssens, F. 2021. A tuneable, photocurable, poly(caprolactone)-based resin for tissue engineering – Synthesis, characterisation and use in stereolithography. *Molecules* 26(5): 1199.

Fiori, S. 2014. Industrial uses of PLA. In *Poly(lactic acid) Science and Technology: Processing, Properties, Additives and Applications*, eds. Jiménez, A., Peltzer, M., and Ruseckaite R., pp. 315–333. Cambridge: RSC Publishing.

Foli, G., Degli Esposti, M., Morselli, D., and Fabbri, P. 2020. Two-step solvent-free synthesis of poly(hydroxybutyrate)-based photocurable resin with potential application in stereolithography. *Macromol. Rapid Commun.* 41(11): 1900660.

Fu, X., Sammons, R.L., Bertóti, I., Jenkins, M.J., and Dong, H. 2012. Active screen plasma surface modification of polycaprolactone to improve cell attachment. *J. Biomed. Mater. Res. B Appl. Biomater.* 100(2): 314–320.

Gálvez, J., Correa Aguirre, J.P., Hidalgo Salazar, M.A., Mondragón, B.V., Wagner, E., and Caicedo, C. 2020. Effect of extrusion screw speed and plasticizer proportions on the rheological, thermal, mechanical, morphological and superficial properties of PLA. *Polymers* 12(9): 2111.

Gamon, G., Evon, P., and Rigal, L. 2013. Twin-screw extrusion impact on natural fibre morphology and material properties in poly(lactic acid) based biocomposites. *Ind. Crops Prod.* 46: 173–185.

Garcia-Garcia, D., Ferri, J.M., Boronat, T., López-Martínez, J., and Balart, R. 2016. Processing and characterization of binary poly(hydroxybutyrate) (PHB) and poly(caprolactone) (PCL) blends with improved impact properties. *Polym. Bull.* 73(12): 3333–3350.

Gayer, C., Abert, J., Bullemer, M., Grom, S., Jauer, L., Meiners, W., Reinauer, F., Vučak, M., Wissenbach, K., Poprawe, R., and Schleifenbaum, J.H. 2018. Influence of the material properties of a poly(*D,L*-lactide)/β-tricalcium phosphate composite on the processability by selective laser sintering. *J. Mech. Behav. Biomed. Mater.* 87: 267–278.

Gayer, C., Ritter, J., Bullemer, M., Grom, S., Jauer, L., Meiners, W., Pfister, A., Reinauer, F., Vučak, M., Wissenbach, K., and Fischer, H. 2019. Development of a solvent-free polylactide/calcium carbonate composite for selective laser sintering of bone tissue engineering scaffolds. *Mater. Sci. Eng.*, C. 101: 660–673.

Gkartzou, E., Koumoulos, E.P., and Charitidis, C.A. 2017. Production and 3D printing processing of bio-based thermoplastic filament. *Manuf. Rev.* 4: 1.

Goyanes, A., Det-Amornrat, U., Wang, J., Basit, A.W., and Gaisford, S. 2016.3D scanning and 3D printing as innovative technologies for fabricating personalized topical drug delivery systems. *JRC.* 234: 41–48.

Green, B.J., Worthington, K.S., Thompson, J.R., Bunn, S.J., Rethwisch, M., Kaalberg, E.E., Jiao, C., Wiley, L.A., Mullins, R.F., Stone, E.M., and Sohn, E.H. 2018. Effect of molecular weight and functionality on acrylated poly(caprolactone) for stereolithography and biomedical applications. *Biomacromolecules* 19(9): 3682–3692.

Grijpma, D.W., Hou, Q., and Feijen, J. 2005. Preparation of biodegradable networks by photo-crosslinking lactide, ε-caprolactone and trimethylene carbonate-based oligomers functionalized with fumaric acid monoethyl ester. *Biomaterials* 26(16): 2795–2802.

Gueche, Y.A., Sanchez-Ballester, N.M., Bataille, B., Aubert, A., Leclercq, L., Rossi, J.C., and Soulairol, I. 2021. Selective laser sintering of solid oral dosage forms with copovidone and paracetamol using a CO_2 laser. *Pharmaceutics* 13(2): 160.

Gupta, M.C., and Deshmukh, V.G. 1982. Thermal oxidative degradation of poly-lactic acid. *Colloid. Polym. Sci.* 260(5): 514–517.

Hahladakis, J.N., Velis, C.A., Weber, R., Iacovidou, E., and Purnell, P. 2018. An overview of chemical additives present in plastics: migration, release, fate, and environmental impact during their use, disposal and recycling. *J. Hazard. Mater.* 344: 179–199.

Haleem, A., and Javaid, M. 2019. 3D scanning applications in medical field: A literature-based review. *CEGH* 7(2): 199–210.

Haq, R.H.A., Rahman, M.N.A., Ariffin, A.M.T., Hassan, M.F., Yunos, M.Z., and Adzila, S. 2017. Characterization and mechanical analysis of PCL/PLA Composites for FDM Feedstock Filament. *IOP conference series: Mater. Sci. Eng.* 226: 012038.

Haq, R.H.A., Rahman, M.N.A.R.A., Arifin, A.M.T., Hassan, M.F., Taib, I., and Wahit, M.U. 2019a. Thermal properties of polycaprolactone (PCL) reinforced montmorillonite (MMT) and hydroxyapatite (ha) as an alternate of FDM composite filament. *J. Adv. Res. Fluid Mech. Therm. Sci.* 62(1): 112–121.

Haq, R.H.A., Taib, I., Rahman, M.N.A., Haw, H.F., Abdullah, H., Ahmad, S., Ariffin, A.M.T., and Hassan, M.F. 2019b. Mechanical properties of PCL/PLA composite sample produce from 3D printer and injection molding. *Int. J. Integr. Eng.* 11(5): 102–108.

Harris, A.M., and Lee, E.C. 2008. Improving mechanical performance of injection molded PLA by controlling crystallinity. *J. Appl. Polym. Sci.* 107(4): 2246–2255.

Helanto, K., Talja, R., and Rojas, O.J. 2021. Talc reinforcement of polylactide and biodegradable polyester blends via injection-molding and pilot-scale film extrusion. *J. Appl. Polym. Sci.* 138(41): 51225.

Hench, L., and Jones, J. 2005. *Biomaterials, Artificial Organs, and Tissue Engineering.* Sawston: Woodhead Publishing.

Herrera, N., Olsén, P., and Berglund, L.A. 2020. Strongly improved mechanical properties of thermoplastic biocomposites by PCL grafting inside holocellulose wood fibers. *ACS Sustain. Chem. Eng.* 8(32): 11977–11985.

Holländer, J., Genina, N., Jukarainen, H., Khajeheian, M., Rosling, A., Mäkilä, E., and Sandler, N. 2016. Three-dimensional printed PCL-based implantable prototypes of medical devices for controlled drug delivery. *J. Pharm. Sci.* 105(9): 2665–2676.

Hsueh, M.H., Lai, C.J., Wang, S.H., Zeng, Y.S., Hsieh, C.H., Pan, C.Y., and Huang, W.C. 2021. Effect of printing parameters on the thermal and mechanical properties of 3D-printed PLA and PETG, using fused deposition modeling. *Polymers* 13(11): 1758.

Hull, C.W. 1986. Apparatus for production of three-dimensional objects by stereolithography. US Patent 4575330A.

Hutmacher, D.W., Schantz, T., Zein, I., Ng, K.W., Teoh, S.H., and Tan, K.C. 2001. Mechanical properties and cell cultural response of polycaprolactone scaffolds designed and fabricated via fused deposition modeling. *J. Biomed. Mater. Res.* 55(2): 203–216.

Idram, I., Bintara, R.D., Lai, J.Y., Essomba, T., and Lee, P.Y. 2019. Development of mesh-defect removal algorithm to enhance the fitting of 3D-printed parts for comminuted bone fractures. *J. Med. Biol. Eng.* 39(6): 855–873.

Jacobsen, S., and Fritz, H.G. 1999. Plasticizing polylactide – The effect of different plasticizers on the mechanical properties. *Polym. Eng. Sci.* 39(7): 1303–1310.

Jamshidian, M., Tehrany, E.A., Imran, M., Jacquot, M., and Desobry, S. 2010. Poly-lactic acid: Production, applications, nanocomposites, and release studies. *Compr. Rev. Food Sci.* 9(5): 552–571.

Jia, H., Gu, S.Y., and Chang, K. 2018. 3D printed self-expandable vascular stents from biodegradable shape memory polymer. *Adv. Polym. Technol.* 37(8): 3222–3228.

Jiang, J., Xu, X., and Stringer, J. 2018a. Support structures for additive manufacturing: A review. *J. Manuf. Mater. Process.* 2(4): 64.

Jiang, N., Yu, T., and Li, Y. 2018b. Effect of hydrothermal aging on injection molded short jute fiber reinforced poly(lactic acid) (PLA) composites. *J. Polym. Environ.* 26(8): 3176–3186.

Jo, W., Kwon, O.-C., and Moon, M.-W. 2018. Investigation of influence of heat treatment on mechanical strength of FDM printed 3D objects. *Rapid Prototyping J.* 24(3): 637–644.

Kabir, E., Kaur, R., Lee, J., Kim, K.-H., and Kwon, E.E. 2020. Prospects of biopolymer technology as an alternative option for non-degradable plastics and sustainable management of plastic wastes. *J. Clean. Prod.* 258: 120536.

Kanczler, J.M., Mirmalek-Sani, S.H., Hanley, N.A., Ivanov, A.L., Barry, J.J., Upton, C., Shakesheff, K.M., Howdle, S.M., and Antonov, E.N. 2009. Biocompatibility and osteogenic potential of human fetal femur-derived cells on surface selective laser sintered scaffolds. *Acta Biomater.* 5(6): 2063–2071.

Kempin, W., Franz, C., Koster, L.C., Schneider, F., Bogdahn, M., Weitschies, W., and Seidlitz, A. 2017. Assessment of different polymers and drug loads for fused deposition modeling of drug loaded implants. *Eur. J. Pharm. Biopharm.* 115: 84–93.

Kodama, H. 1981. Automatic method for fabricating a three-dimensional plastic model with photo-hardening polymer. *Rev. Sci. Instrum.* 52(11): 1770–1773.

Koller, M. 2020. *The Handbook of Polyhydroxyalkanoates.* Boca Raton: CRC Press.

Kopinke, F.-D., Remmler, M., Mackenzie, K., Möder, M., and Wachsen, O. 1996. Thermal decomposition of biodegradable polyesters – II. Poly(lactic acid). *Polym. Degrad. Stab.* 53(3): 329–342.

Korpela, J., Kokkari, A., Korhonen, H., Malin, M., Närhi, T., and Seppälä, J. 2013. Biodegradable and bioactive porous scaffold structures prepared using fused deposition modeling. *J. Biomed. Mater. Res. B Appl. Biomater.* 101(4): 610–619.

Kovalcik, A. 2021. Recent advances in 3D printing of polyhydroxyalkanoates: A review. *Eurobiotech J.* 5(1): 48–55.

Kovalcik, A., Sangroniz, L., Kalina, M., Skopalova, K., Humpolíček, P., Omastova, M., Mundigler, N., and Müller, A.J. 2020. Properties of scaffolds prepared by fused deposition modeling of poly(hydroxyalkanoates). *Int. J. Biol. Macromol.* 161: 364–376.

Kulinowski, P., Malczewski, P., Pesta, E., Łaszcz, M., Mendyk, A., Polak, S., and Dorożyński, P. 2021. Selective laser sintering (SLS) technique for pharmaceutical applications – Development of high dose controlled release printlets. *Addit. Manuf.* 38: 101761.

Lay, M., Thajudin, N.L.N., Hamid, Z.A.A., Rusli, A., Abdullah, M.K., and Shuib, R.K. 2019. Comparison of physical and mechanical properties of PLA, ABS and nylon 6 fabricated using fused deposition modeling and injection molding. *Compos. B. Eng.* 176: 107341.

Lee, C.-Y., and Liu, C.-Y. 2019. The influence of forced-air cooling on a 3D printed PLA part manufactured by fused filament fabrication. *Addit. Manuf.* 25: 196–203.

Lee, C.H., Padzil, F.N.B.M., Lee, S.H., Ainun, Z.M.A., and Abdullah, L.C. 2021. Potential for natural fiber reinforcement in PLA polymer filaments for fused deposition modeling (FDM) additive manufacturing: A review. *Polymers* 13(9): 1407.

Lemu, H.G., and Kurtovic, S. 2011. 3D printing for rapid manufacturing: Study of dimensional and geometrical accuracy. In *IFIP International Conference on Advances in Production Management Systems*, pp. 470–479. Berlin: Springer.

Li, Z., Wang, Z., Gan, X., Fu, D., Fei, G., and Xia, H. 2017. Selective laser sintering 3D printing: a way to construct 3D electrically conductive segregated network in polymer matrix. *Macromol. Mater. Eng.* 302(11): 1700211.

Li, Z., Yang, J., and Loh, X.J. 2016. Polyhydroxyalkanoates: Opening doors for a sustainable future. *NPG Asia Mater.* 8(4): e265–e265.

Lim, J., You, M., Li, J., and Li, Z. 2017. Emerging bone tissue engineering via polyhydroxyalkanoate (PHA)-based scaffolds. *Mater. Sci. Eng. C.* 79: 917–929.

Lim, L.-T., Auras, R., and Rubino, M. 2008. Processing technologies for poly(lactic acid). *Prog. Polym. Sci.* 33(8): 820–852.

Lindner, M., Hoeges, S., Meiners, W., Wissenbach, K., Smeets, R., Telle, R., Poprawe, R., and Fischer, H. 2011. Manufacturing of individual biodegradable bone substitute implants using selective laser melting technique. *J. Biomed. Mater. Res. A.* 97(4): 466–471.

Liu, B., Shi, Q., Hu, L., Huang, Z., Zhu, X., and Zhang, Z. 2020a. Engineering digital polymer based on thiol-maleimide Michael coupling toward effective writing and reading. *Polym. Chem.* 11(10): 1702–1707.

Liu, H., He, H., and Huang, B. 2020b. Favorable thermoresponsive shape memory effects of 3D printed poly(lactic acid)/poly(ε-caprolactone) blends fabricated by fused deposition modelling. *Macromol. Mater. Eng.* 305(11): 2000295.

Liu, S., Qin, S., He, M., Zhou, D., Qin, Q., and Wang, H. 2020c. Current applications of poly(lactic acid) composites in tissue engineering and drug delivery. *Compos. B. Eng.* 199: 108238.

Liu, X., Wang, T., Chow, L.C., Yang, M., and Mitchell, J.W. 2014. Effects of inorganic fillers on the thermal and mechanical properties of poly(lactic acid). *Int. J. Polym. Sci.* 2014: 827028.

Ljungberg, N., and Wesslen, B. 2002. The effects of plasticizers on the dynamic mechanical and thermal properties of poly(lactic acid). *J. Appl. Polym. Sci.* 86(5): 1227–1234.

Long, J., Gholizadeh, H., Lu, J., Bunt, C., and Seyfoddin, A. 2017. Application of fused deposition modelling (FDM) method of 3D printing in drug delivery. *Curr. Pharm. Des.* 23(3): 433–439.

Lovera, D., Márquez, L., Balsamo, V., Taddei, A., Castelli, C., and Müller, A.J. 2007. Crystallization, morphology, and enzymatic degradation of polyhydroxybutyrate/polycaprolactone (PHB/PCL) blends. *Macromol. Chem. Phys.* 208(9): 924–937.

Luna, C.B.B., Siqueira, D.D., Ferreira, E.D.S.B., Araújo, E.M., and Wellen, R.M.R., 2021. Effect of injection parameters on the thermal, mechanical and thermomechanical properties of polycaprolactone (PCL). *J. Elastomers Plast.* 53(8): 1045–1062.

Lupone, F., Padovano, E., Casamento, F., and Badini, C. 2022. Process phenomena and material properties in selective laser sintering of polymers: A review. *Materials* 15(1): 183.

Luzuriaga, M.A., Berry, D.R., Reagan, J.C., Smaldone, R.A., and Gassensmith, J.J. 2018. Biodegradable 3D printed polymer microneedles for transdermal drug delivery. *Lab Chip* 18(8): 1223–1230.

Lyu, J.S., Lee, J.-S., and Han, J. 2019. Development of a biodegradable polycaprolactone film incorporated with an antimicrobial agent via an extrusion process. *Sci. Rep.* 9(1): 1–11.

Ma, P., Cai, X., Wang, W., Duan, F., Shi, D., and Lemstra, P.J., Lemstra. 2014a. Crystallization behavior of partially crosslinked poly(β-hydroxyalkonates)/poly(butylene succinate) blends. *J. Appl. Polym. Sci.* 131(21).

Ma, P., Hristova-Bogaerds, D.G., Zhang, Y., and Lemstra, P.J. 2014b. Enhancement in crystallization kinetics of the bacterially synthesized poly(β-hydroxybutyrate) by poly(butylene succinate). *Polym. Bull.* 71(4): 907–923.

Malikmammadov, E., Tanir, T.E., Kiziltay, A., Hasirci, V., and Hasirci, N. 2018. PCL and PCL-based materials in biomedical applications. *J. Biomater. Sci. Polym. Ed.* 29(7–9): 863–893.

Manapat, J.Z., Chen, Q., Ye, P., and Advincula, R.C. 2017.3D printing of polymer nanocomposites via stereolithography. *Macromol. Mater. Eng.* 302(9): 1600553.

Martin, D.P., and Williams, S.F. 2003. Medical applications of poly-4-hydroxybutyrate: a strong flexible absorbable biomaterial. *Biochem. Eng. J.* 16(2): 97–105.

Mathew, A.P., Oksman, K., and Sain, M. 2006. The effect of morphology and chemical characteristics of cellulose reinforcements on the crystallinity of polylactic acid. *J. Appl. Polym. Sci.* 101(1): 300–310.

Matsuda, T., Mizutani, M., and Arnold, S.C. 2000. Molecular design of photocurable liquid biodegradable copolymers. 1. Synthesis and photocuring characteristics. *Macromolecules* 33(3): 795–800.

Matzinos, P., Tserki, V., Kontoyiannis, A., and Panayiotou, C., 2002. Processing and characterization of starch/polycaprolactone products. *Polym. Degrad. Stab.* 77(1): 17–24.

Mazurchevici, A.D., Nedelcu, D., and Popa, R. 2021. Additive manufacturing of composite materials by FDM technology: A review. *Indian J. Eng. Mater. Sci.* (IJEMS) 27(2): 179–192.

Mazzanti, V., Malagutti, L., and Mollica. F 2019. FDM 3D printing of polymers containing natural fillers: A review of their mechanical properties. *Polymers* 11(7): 1094.

Mehraein, H. 2018. *Impact of Process Parameters on Mechanical Properties of 3D Printed Polycaprolactone (PCL) parts*. PhD Thesis. Wichita: Wichita State University.

Mehrpouya, M., Vahabi, H., Barletta, M., Laheurte, P., and Langlois, V. 2021. Additive manufacturing of polyhydroxyalkanoates (PHAs) biopolymers: Materials, printing techniques, and applications. *Mater. Sci. Eng. C.* 127: 112216.

Melchels, F.P., Feijen, J., and Grijpma, D.W. 2009. A poly(D,L-lactide) resin for the preparation of tissue engineering scaffolds by stereolithography. *Biomaterials* 30(23–24): 3801–3809.2010.

Melchels, F.P., Feijen, J., and Grijpma, D.W. 2010. A review on stereolithography and its applications in biomedical engineering. *Biomaterials* 31(24): 6121–6130.

Mkhabela, V.J., and Ray, S.S. 2014. Poly(ε-caprolactone) nanocomposite scaffolds for tissue engineering: A brief overview. *J. Nanosci. Nanotechnol.* 14(1): 535–545.

Montano-Herrera, L., Pratt, S., Arcos-Hernandez, M.V., Halley, P.J., Lant, P.A., Werker, A., and Laycock, B. 2014. In-line monitoring of thermal degradation of PHA during melt-processing by near infrared spectroscopy. *New Biotechnol.* 31(4): 357–363.

Moradi, M., Moghadam, M.K., and Malekshahi Beiranvand, Z. 2021, CO_2 Laser engraving of injection moulded polycarbonate: Experimental investigation, *Lasers Eng.* 48(4–6): 293–303.

Muthuraj, R., Valerio, O., and Mekonnen, T.H. 2021. Recent developments in short- and medium-chain-length polyhydroxyalkanoates: Production, properties, and applications. *Int. J. Biol. Macromol.* 187: 422–440.

Mysiukiewicz, O., Barczewski, M., Skórczewska, K., and Matykiewicz, D. 2020. Correlation between processing parameters and degradation of different polylactide grades during twin-screw extrusion. *Polymers* 12(6): 1333.

Naftulin, J.S., Kimchi, E.Y., and Cash, S.S. 2015. Streamlined, inexpensive 3D printing of the brain and skull. *PLoS One* 10(8): e0136198.

Naser, A.Z., Deiab, I., and Darras, B.M. 2021. Poly(lactic acid) (PLA) and polyhydroxyalkanoates (PHAs), green alternatives to petroleum-based plastics: A review. *RSC Adv.* 11(28): 17151–17196.

Ngo, T.D., Kashani, A., Imbalzano, G., Nguyen, K.T.Q., and Hui, D. 2018. Additive manufacturing (3D printing): A review of materials, methods, applications and challenges. *Compos. B. Eng.* 143: 172–196.

Nguyen, T.K., and Lee, B.-K. 2018. Post-processing of FDM parts to improve surface and thermal properties. *Rapid Prototyp. J.* 24(7): 1091–1100.

Niaounakis, M. 2015. *Biopolymers: Processing and Products*. Norwich: William Andrew.

Nyberg, E., Rindone, A., Dorafshar, A., and Grayson, W.L. 2017. Comparison of 3D-printed poly-ε-caprolactone scaffolds functionalized with tricalcium phosphate, hydroxyapatite, Bio-Oss, or decellularized bone matrix. *Tissue Eng. Pt. A.* 23(11–12): 503–514.

Okan, M., Aydin, H.M., and Barsbay, M. 2019. Current approaches to waste polymer utilization and minimization: A review. *J. Chem. Technol. Biotechnol.* 94(1): 8–21.

Oropallo, W., and Piegl, L.A. 2016. Ten challenges in 3D printing. *Eng. Comput.* 32(1): 135–148.

Panaitescu, D.M., Lupescu, I., Frone, A.N., Chiulan, I., Nicolae, C.A., Tofan, V., Stefaniu, A., Somoghi, R., and Trusca, R. 2017. Medium chain-length polyhydroxyalkanoate copolymer modified by bacterial cellulose for medical devices. *Biomacromolecules* 18(10): 3222–3232.

Park, J.-M., Jeon, J., Koak, J.-Y., Kim, S.-K., and Heo, S.-J. 2021. Dimensional accuracy and surface characteristics of 3D-printed dental casts. *J. Prosthet. Dent.* 126(3): 427–437.

Parulekar, Y., and Mohanty, A.K. 2007. Extruded biodegradable cast films from polyhydroxyalkanoate and thermoplastic starch blends: fabrication and characterization. *Macromol. Mater. Eng.* 292(12): 1218–1228.

Pereira, T.F., Oliveira, M.F., Maia, I.A., Silva, J.V., Costa, M.F., and Thiré, R.M. 2012b. 3D printing of poly(3-hydroxybutyrate) porous structures using selective laser sintering. *Macromol. Symp.* 319: 64–73.

Pereira, T.F., Silva, M.A.C., Oliveira, M.F., Maia, I.A., Silva, J.V.L., Costa, M.F., and Thiré, R.M.S.M. 2012a. Effect of process parameters on the properties of selective laser sintered poly(3-hydroxybutyrate) scaffolds for bone tissue engineering: This paper analyzes how laser scan spacing and powder layer thickness affect the morphology and mechanical properties of sls-made scaffolds by using a volume energy density function. *Virtual Phys. Prototyp.* 7(4): 275–285.

Petersen, E.E., and Pearce, J. 2017. Emergence of home manufacturing in the developed world: return on investment for open-source 3-D printers. *Technologies* 5(1): 7.

Petersen, K., Nielsen, P.V., Bertelsen, G., Lawther, M., Olsen, M.B., Nilsson, N.H., and Mortensen, G. 1999. Potential of biobased materials for food packaging. *Trends Food Sci. Technol* 10(2): 52–68.

Pilla, S., Kramschuster, A., Yang, L., Lee, J., Gong, S., and Turng, L.-S. 2009. Microcellular injection-molding of polylactide with chain-extender. *Mater. Sci. Eng. C.* 29(4): 1258–1265.

Pollack, S., Venkatesh, C., Neff, M., Healy, A.V., Hu, G., Fuenmayor, E.A., Lyons, J.G., Major, I., and Devine, D.M. 2019. Polymer-based additive manufacturing: Historical developments, process types and material considerations. In *Polymer-Based Additive Manufacturing*, pp. 1–22. Berlin and Heidelberg: Springer.

Przybysz, M., Marć, M., Klein, M., Saeb, M.R., and Formela, K. 2018. Structural, mechanical and thermal behavior assessments of PCL/PHB blends reactively compatibilized with organic peroxides. *Polym. Test.* 67: 513–521.

Puppi, D., and Chiellini, F. 2020. Biodegradable polymers for biomedical additive manufacturing. *Appl. Mater. Today* 20: 100700.

Quiles-Carrillo, L., Duart, S., Montanes, N., Torres-Giner, S., and Balart, R., 2018. Enhancement of the mechanical and thermal properties of injection-molded polylactide parts by the addition of acrylated epoxidized soybean oil. *Mater. Des.* 140: 54–63.

Rahim, T.N.A.T., Abdullah, A.M., and Akil, H.M. 2019. Recent developments in fused deposition modeling-based 3D printing of polymers and their composites. *Polym. Rev.* 59(4): 589–624.

Ramanath, H.S., Chua, C.K., Leong, K.F., and Shah, K.D. 2008. Melt flow behaviour of poly-ε caprolactone in fused deposition modelling. *J. Mater. Sci. Mater. Med.* 19(7): 2541–2550.

Rasal, R.M., Janorkar, A.V., and Hirt, D.E. 2010. Poly(lactic acid) modifications. *Prog. Polym. Sci.* 35(3): 338–356.

Rayegani, F., and Onwubolu, G.C. 2014. Fused deposition modelling (FDM) process parameter prediction and optimization using group method for data handling (GMDH) and differential evolution (DE). *Int. J. Adv. Manuf. Technol.* 73(1): 509–519.

Rayna, T., and Striukova, L. 2016. From rapid prototyping to home fabrication: How 3D printing is changing business model innovation. *Technol. Forecast. Soc. Change* 102: 214–224.

Reddy, C.S.K., Ghai, R., and Kalia, V. 2003. Polyhydroxyalkanoates: An overview. *Bioresour. Technol.* 87(2): 137–146.

Rengier, F., Mehndiratta, A., Von Tengg-Kobligk, H., Zechmann, C.M., Unterhinninghofen, R., Kauczor, H-U., and Giesel, F.L. 2010. 3D printing based on imaging data: Review of medical applications. *Int. J. Comput. Assist. Radiol. Surg.* 5(4): 335–341.

Rimini, S., Vincent, J., and Wake, N. 2022. Computer-aided design principles for anatomic modeling. In *3D Printing for the Radiologist*, ed. Wake, N., pp. 45–59. Amsterdam: Elsevier.

Ritz, U., Gerke, R., Götz, H., Stein, S., and Rommens, P.M. 2017. A new bone substitute developed from 3D-prints of polylactide (PLA) loaded with collagen I: An *in vitro* study. *Int. J. Mol. Sci.* 18(12): 2569.

Rodríguez-Panes, A., Claver, J., and Camacho, A.M. 2018. The influence of manufacturing parameters on the mechanical behaviour of PLA and ABS pieces manufactured by FDM: A comparative analysis. *Materials* 11(8): 1333.

Rosenboom, J.-G., Langer, R., and Traverso, G. 2022. Bioplastics for a circular economy. *Nat. Rev. Mater.* 7: 117–137.

Ruellan, A., Ducruet, V., and Domenek, S. 2014. Plasticization of poly(lactide). In *Poly(lactic acid) Science and Technology. Processing, Properties, Additives and Applications*, eds. Jimenez, A., Peltzer, M., and Ruseckaite, R., pp. 124–170. Cambridge: RSC Publishing.

Rydz, J., Włodarczyk, J., Gonzalez Ausejo, J., Musioł, M., Sikorska, W., Sobota, M., Hercog, A., Duale, K., and Janeczek, H. 2020. Three-dimensional printed PLA and PLA/PHA dumbbell-shaped specimens: Material defects and their impact on degradation behavior. *Materials* 13(8): 2005.

Saini, P., Arora, M., and Kumar, M.R. 2016. Poly(lactic acid) blends in biomedical applications. *Adv. Drug Deliv. Rev.* 107: 47–59.

Salmoria, G.V., Cardenuto, M.R., Roesler, C.R.M., Zepon, K.M.m and Kanis, L.A. 2016. PCL/ibuprofen implants fabricated by selective laser sintering for orbital repair. *Procedia CIRP* 49: 188–192.

Salmoria, G.V., Klauss, P., Zepon, K., Kanis, L.A., Roesler, C.R.M., and Vieira, L.F. 2012. Development of functionally-graded reservoir of PCL/PG by selective laser sintering for drug delivery devices: this paper presents a selective laser sintering-fabricated drug delivery system that contains graded progesterone content. *Virtual. Phys. Prototyp.* 7(2): 107–115.

Salmoria, G.V., Sibilia, F., Henschel, V.G., Fare, S., and Tanzi, M.C. 2017a. Structure and properties of polycaprolactone/ibuprofen rods prepared by melt extrusion for implantable drug delivery. *Polym. Bull.* 74(12): 4973–4987.

Salmoria, G.V., Vieira, F.E., Ghizoni, G.B., Marques, M.S., and Kanis, L.A. 2017b. 3D printing of PCL/fluorouracil tablets by selective laser sintering: Properties of implantable drug delivery for cartilage cancer treatment. *ROM* 2(3): 1–7.

Santoro, M., Shah, S.R., Walker, J.L., and Mikos, A.G. 2016. Poly(lactic acid) nanofibrous scaffolds for tissue engineering. *Adv. Drug Deliv. Rev.* 107: 206–212.

Saptarshi, S.M., and Zhou, C. 2019. Basics of 3D printing: engineering aspects. In *3D printing in orthopaedic surgery*, eds. Dipaola, M., and Wodajo, F.M., pp. 17–30. Amsterdam: Elsevier.

Saska, S., Pires, L.C., Cominotte, M.A., Mendes, L.S., de Oliveira, M.F., Maia, I.A., da Silva, J.V.L., Ribeiro, S.J.L., and Cirelli, J.A. 2018. Three-dimensional printing and *in vitro* evaluation of poly(3-hydroxybutyrate) scaffolds functionalized with osteogenic growth peptide for tissue engineering. *Mater. Sci. Eng. C.* 89: 265–273.

Schyns, Z.O., and Shaver, M.P. 2021. Mechanical recycling of packaging plastics: A review. *Macromol. Rapid Commun.* 42(3): 2000415.

Senatov, F.S., Niaza, K.V., Stepashkin, A.A., and Kaloshkin, S.D. 2016a. Low-cycle fatigue behavior of 3D-printed PLA-based porous scaffolds. *Compos. B. Eng.* 97: 193–200.

Senatov, F.S., Niaza, K.V., Zadorozhnyy, M.Y., Maksimkin, A.V., Kaloshkin, S.D., Estrin, and Y.Z. 2016b. Mechanical properties and shape memory effect of 3D-printed PLA-based porous scaffolds. *J. Mech. Behav. Biomed. Mater.* 57: 139–148.

Seng, C.T., Eh Noum, S.Y., Sivanesan, S.K., and Yu, L.J. 2020. Reduction of hygroscopicity of PLA filament for 3D printing by introducing nano silica as filler. *AIP Conf. Proceed.* 2233:020024.

Shor, L., Güçeri, S., Wen, X., Gandhi, M., and Sun, W. 2007. Fabrication of three-dimensional polycaprolactone/hydroxyapatite tissue scaffolds and osteoblast-scaffold interactions *in vitro*. *Biomaterials* 28(35): 5291–5297.

Signori, F., Coltelli, M.-B., and Bronco, S. 2009. Thermal degradation of poly(lactic acid) (PLA) and poly(butylene adipate-*co*-terephthalate) (PBAT) and their blends upon melt processing. *Polym. Degrad. Stab.* 94(1): 74–82.

Sikora, J., Majewski, Ł., and Puszka, A. 2020. Modern biodegradable plastics – Processing and properties: Part I. *Materials* 13(8): 1986.

Sikora, J.W., Majewski, Ł., and Puszka, A. 2021. Modern biodegradable plastics – Processing and properties Part II. *Materials* 14(10): 2523.

Sodian, R., M. Loebe, A., Hein, D.P., Martin, S.P., Hoerstrup, E.V., Potapov, Hausmann, H., Lueth, T., and Hetzer, R. 2002. Application of stereolithography for scaffold fabrication for tissue engineered heart valves. *ASAIO J.* 48(1): 12–16.

Stratasys Inc. 1991. Fast, precise, safe prototypes with FDM. *Sol. Freef. Fabricat. Symp. Aust.* 50: 53–60.

Syrlybayev, D., Zharylkassyn, B., Seisekulova, A., Akhmetov, M., Perveen, A., and Talamona, D. 2021. Optimisation of strength properties of FDM printed parts – A critical review. *Polymers* 13(10): 1587.

Tábi, T., Ageyeva, T., and Kovács, J.G. 2021. Improving the ductility and heat deflection temperature of injection molded poly(lactic acid) products: A comprehensive review. *Polym. Test.* 101: 107282.

Tan, D.K., Maniruzzaman, M., and Nokhodchi, A. 2018. Advanced pharmaceutical applications of hot-melt extrusion coupled with fused deposition modelling (FDM) 3D printing for personalised drug delivery. *Pharmaceutics* 10(4): 203.

Tan, K.H., Chua, C.K., Leong, K.F., Cheah, C.M., Gui, W.S., Tan, W.S., and Wiria, F.E. 2005. Selective laser sintering of biocompatible polymers for applications in tissue engineering. *Biomed. Mater. Eng.* 15(1–2): 113–124.

Tanodekaew, S., Channasanon, S., Kaewkong, P., and Uppanan, P. 2013. PLA-HA scaffolds: preparation and bioactivity. *Proc. Eng.* 59: 144–149.

Taubner, V., and Shishoo, R. 2001. Influence of processing parameters on the degradation of poly(*L*-lactide) during extrusion. *J. Appl. Polym. Sci.* 79(12): 2128–2135.

Ten, E., Jiang, L., Zhang, J., and Wolcott, M.P. 2015. Mechanical performance of polyhydroxyalkanoate (PHA)-based biocomposites. *Biocomposites* 39–52.

Teoh, S.-H., Goh, B.-T., and Lim, J. 2019. Three-dimensional printed polycaprolactone scaffolds for bone regeneration success and future perspective. *Tissue Eng. Part A.* 25(13–14), 931–935.

Thellen, C., Coyne, M., Froio, D., Auerbach, M., Wirsen, C., and Ratto, J.A. 2008. A processing, characterization and marine biodegradation study of melt-extruded polyhydroxyalkanoate (PHA) films. *J. Polym. Environ.* 1: 1–11.

Thomas, D.S., and Gilbert, S.W. 2014. Costs and cost effectiveness of additive manufacturing. *NIST Special Publication* 1176: 12.

Tlegenov, Y., Hong, G.S., and Lu, W.F. 2018. Nozzle condition monitoring in 3D printing. *Robot. Comput. Integr. Manuf.* 54: 45–55.

Tripathi, N., Misra, M., and Mohanty, A.K. 2021. Durable polylactic acid (PLA)-based sustainable engineered blends and biocomposites: Recent developments, challenges, and opportunities. *ACS Engineering Au* 1(1): 7–38.

Tronvoll, S.A., Welo, T., and Elverum, C.W. 2018. The effects of voids on structural properties of fused deposition modelled parts: A probabilistic approach. *Int. J. Adv. Manuf.* 97(9): 3607–3618.

Turnbull, G., Clarke, J., Picard, F., Riches, P., Jia, L., Han, F., Li, B., and Shu, W. 2018. 3D bioactive composite scaffolds for bone tissue engineering. *Bioact. Mater.* 3(3): 278–314.

Valerga, A.P., Batista, M., Fernandez-Vidal, S.R., and Gamez, A.J. 2019. Impact of chemical post-processing in fused deposition modelling (FDM) on polylactic acid (PLA) surface quality and structure. *Polymers* 11(3): 566

Valerga, A.P., Batista, M., Salguero, J., and Girot, F. 2018. Influence of PLA filament conditions on characteristics of FDM parts. *Materials* 11(8): 1322.

Vanheusden, C., Samyn, P., Goderis, B., Hamid, M., Reddy, N., Ethirajan, A., Peeters, R., and Buntinx, M. 2021. Extrusion and injection molding of poly(3-hydroxybutyrate-*co*-3-hydroxyhexanoate) (PHBHHx): Influence of processing conditions on mechanical properties and microstructure. *Polymers* 13(22): 4012.

Viidik, L., Vesala, J., Laitinen, R., Korhonen, O., Ketolainen, J., Aruväli, J., Kirsimäe, K., Kogermann, K., Heinämäki, J., Laidmäe, I., and Ervasti, T. 2021. Preparation and characterization of hot-melt extruded polycaprolactone-based filaments intended for 3D-printing of tablets. *Eur. J. Pharm. Sci.* 158: 105619.

Visscher, L.E., Dang, H.P., Knackstedt, M.A., Hutmacher, D.W., and Tran, P.A. 2018. 3D printed polycaprolactone scaffolds with dual macro-microporosity for applications in local delivery of antibiotics. *Mater. Sci. Eng. C.* 87: 78–89.

Vlachopoulos, J., and Strutt, D. 2003. Polymer processing. *Mater. Sci. Technol.* 19(9): 1161–1169.

Vroman, I., and Tighzert, L. 2009. Biodegradable polymers. *Materials* 2(2): 307–344.

Wahit, M.U., Akos, N.I., and Laftah, W.A. 2012. Influence of natural fibers on the mechanical properties and biodegradation of poly(lactic acid) and poly(ε-caprolactone) composites: A review. *Polym. Compos.* 33(7): 1045–1053.

Wang, Y., Lei, M., Wei, Q., Wang, Y., Zhang, J., Guo, Y., and Saroia, J. 2020. 3D printing biocompatible l-Arg/GNPs/PLA nanocomposites with enhanced mechanical property and thermal stability. *J. Mater. Sci.* 55(12): 5064–5078.

Wei, S., Ma, J.-X., Xu, L., Gu, X.-S., and Ma, X.-L. 2020. Biodegradable materials for bone defect repair. *Mil. Med. Res.* 7(1): 1–25.

Williams, J.M., Adewunmi, A., Schek, R.M., Flanagan, C.L., Krebsbach, P.H., Feinberg, S.E., Hollister, S.J., and Das, S. 2005. Bone tissue engineering using polycaprolactone scaffolds fabricated via selective laser sintering. *Biomaterials* 26(23): 4817–4827.

Windolf, H., Chamberlain, R., and Quodbach, J. 2022. Dose-independent drug release from 3D printed oral medicines for patient-specific dosing to improve therapy safety. *Int. J. Pharm.* 616: 121555.

Wu, C.-S. 2018. Characterization, functionality, and application of siliceous sponge spicules additive-based manufacturing biopolymer composites. *Addit. Manuf.* 22: 13–20.

Wu, C.-S., Liao, H.-T., and Cai, Y.-X. 2017. Characterisation, biodegradability and application of palm fibre-reinforced polyhydroxyalkanoate composites. *Polym. Degrad. Stab.* 140: 55–63.

Wu, J., Yang, R., Zheng, J., and Wang, X. 2018. Super heat deflection resistance stereocomplex crystallisation of PLA system achieved by selective laser sintering. *Micro Nano Lett.* 13(11): 1604–1608.

Yang, L., Li, S., Li, Y., Yang, M., and Yuan, Q. 2019. Experimental investigations for optimizing the extrusion parameters on FDM PLA printed parts. *J. Mater. Eng. Perform.* 28(1): 169–182.

Yang, X., Wang, Y., Zhou, Y., Chen, J., and Wan, Q. 2021a. The application of polycaprolactone in three-dimensional printing scaffolds for bone tissue engineering. *Polymers.* 13(16): 2754.

Yang, Y., Wu, H., Fu, Q., Xie, X., Song, Y., Xu, M., and Li, J. 2022. 3D-printed polycaprolactone-chitosan based drug delivery implants for personalized administration. *Mater. Des.* 214: 110394.

Yang, Y., Xu, Y., Wei, S., and Shan, W. 2021b. Oral preparations with tunable dissolution behavior based on selective laser sintering technique. *Int. J. Pharm.* 593: 120127.

Yang, Z.T., Yang, J.X., Fan, J.H., Feng, Y.H., and Huang, Z.X. 2021c. Preparation of super-toughened poly(*L*-lactide) composites under elongational flow: A strategy for balancing stiffness and ductility. *Compos. Sci. Technol.* 208: 108758.

Yao, T., Ye, J., Deng, Z., Zhang, K., Ma, Y., and Ouyang, H. 2020. Tensile failure strength and separation angle of FDM 3D printing PLA material: Experimental and theoretical analyses. *Compos. B. Eng.* 188: 107894.

Zein, I., Hutmacher, D.W., Tan, K.C., and Teoh, S.H. 2002. Fused deposition modeling of novel scaffold architectures for tissue engineering applications. *Biomaterials* 23(4): 1169–1185.

Zekavat, A.R., Jansson, A., Larsson, J., and Pejryd, L. 2019. Investigating the effect of fabrication temperature on mechanical properties of fused deposition modeling parts using X-ray computed tomography. *Int. J. Adv. Manuf. Technol.* 100(1): 287–296.

Zhang, C., Lu, L., Li, W., Li, L., and Zhou, C. 2016. Effects of crystallization temperature and spherulite size on cracking behavior of semi-crystalline polymers. *Polym. Bull.* 73(11): 2961–2972.

Zhang, M., and Thomas, N.L. 2011. Blending polylactic acid with polyhydroxybutyrate: the effect on thermal, mechanical, and biodegradation properties. *Adv. Polym. Technol.* 30(2): 67–79.

Zhang, X.C. 2016. *Science and Principles of Biodegradable and Bioresorbable Medical Polymers: Materials and Properties.* Sawston: Woodhead publishing.

Zhao, K., Deng, Y., Chen, J.C., and Chen, G.-Q. 2003. Polyhydroxyalkanoate (PHA) scaffolds with good mechanical properties and biocompatibility. *Biomaterials* 24(6): 1041–1045.

Zhao, S., Zhu, M., Zhang, J., Zhang, Y., Liu, Z., Zhu, Y., and Zhang, C. 2014. Three dimensionally printed mesoporous bioactive glass and poly(3-hydroxybutyrate-*co*-3-hydroxyhexanoate) composite scaffolds for bone regeneration. *J. Mater. Chem. B.* 2(36): 6106–6118.

Zhou, W.Y., Lee, S.H., Wang, M., Cheung, W.L., and Ip, W.Y. 2008. Selective laser sintering of porous tissue engineering scaffolds from poly(*L*-lactide)/carbonated hydroxyapatite nanocomposite microspheres. *J. Mater. Sci. Mater. Med.* 19(7): 2535–2540.

Zhou, Y., Fan, M., and Chen, L. 2016. Interface and bonding mechanisms of plant fibre composites: An overview. *Compos. B. Eng.* 101: 31–45.

Zhu, C., Nomura, C.T., Perrotta, J.A., Stipanovic, A.J., and Nakas, J.P. 2012. The effect of nucleating agents on physical properties of poly-3-hydroxybutyrate (PHB) and poly-3-hydroxybutyrate-*co*-3-hydroxyvalerate (PHB-*co*-HV) produced by burkholderia cepacia ATCC 17759. *Polym. Test.* 31(5): 579–585.

4 Comprehensive Characterisation of Complex Polymer Systems

Khadar Duale

The term "complex polymers" unites architecturally complex compounds that have different polymer structures: hyperbranched polymers, bottlebrushes, block copolymers, amphiphilic networks, dendrimers, as well as stars-and-rings polymers (Huang et al., 2015; Alvaradejo et al., 2019; Michalski et al., 2019; Cook et al., 2020; García-Gallego et al., 2020). In recent years, these complex polymers have found an increasingly wide range of applications for new emerging technology and biomedical applications designed to create novel materials with unique physical, chemical and biological properties (Ulery et al., 2011; Jin et al., 2012; Loh, 2014; Feig et al., 2018; Van Bochove et al., 2019; Cook et al., 2020; García-Gallego et al., 2020).

Usually, these polymer compounds contain segmented copolymers of a variety of molecules and have molar mass, molecular architecture as well as chemical structures and composition type of distribution (Kilz et al., 2000). Likewise, they may also possess different chain length functionality, a variety of end-groups and a sequence of monomer units in the polymer chain; they, therefore, exhibit a wide range in size from nanometres up to micrometres (Cook et al., 2020). Furthermore, their method of synthesis can also vary, depending on the choice of monomer, polymer composition, method and polymerisation conditions (Nicolaÿ et al., 2007; Setijadi et al., 2009; Huang et al., 2015; Michalski et al., 2019; García-Gallego et al., 2020).

As these multicomponent macromolecules combine the distinct properties of their constituent polymer units and host multiple different functional moieties on a single discrete polymer, their chemical structure is determined by their molar mass and functionality but also by the sequence of monomer units in the polymer chain (Pasch et al., 2003). Obviously, this poses a much more insurmountable challenge to the standard analytical techniques and an appropriate analytical method to investigate their apparent molecular heterogeneity is therefore required (Pasch et al., 2003; Uliyanchenko et al., 2012; Pasch, 2013; Crotty et al., 2016; Knol et al., 2021).

Comprehensive characterisation of these polymer materials becomes even more convoluted as the size of related structures increases, and their complexity also increases. Both molar mass and molar-mass distribution are two of the factors affecting the properties of polymers (Nunes et al., 1982). For such complex polymers, it is important to analyse and exactly determine the thermal properties carefully

DOI: 10.1201/9780429352799-5

(Menczel et al., 2009). This is, for example, of great significance in determining the crystallinity and the degree of cross-linking and thus defining the degree of long-range ordering in the polymer, which strongly influences its properties (Sakamoto et al., 2013). For example, star-shaped polymers have a lower melt viscosity than linear polymers of the same molar mass while, in turn, the thermal and mechanical properties of star-shaped polymers are affected more by arm length than molar mass (Kim et al., 2004; Doganci, 2021). It is therefore important to establish the physical properties of the complex polymer since the information obtained from these analyses can then be linked directly to the molar mass and molar-mass distribution factors of the polymer. This eventually affects the processability of the polymer and influences, among other things, the interfacial interaction between the polymer matrix and fillers and other additives during the different processing steps and ultimately the performance of the product (Nadgorny et al., 2017; Zhu et al., 2021). An example of this is melt and solution spinning. In the melt spinning method, a polymer with low molar mass and broad molar-mass distribution is usually preferred, while a polymer with very high molar mass and narrow molar-mass distribution is preferred in the manufacture of fibres by solution spinning (Talebi et al., 2010).

The topography and surface morphology of these complex polymers are studied using a combination of multidisciplinary techniques such as transmission electron microscopy, optical microscopy, scanning probe techniques and X-ray diffraction (Macko et al., 2002). For safe applications of complex polymers in areas relating to human health as well as the environment, knowledge of their structure, properties, performance and function is essential. Atomic force microscopy (AFM) and scanning electron microscope (SEM) are techniques that allow visualisation of the morphology of the sample surface, particle size and fragments with superior spatial resolution and that also present quantitative measurements of their mechanical and electrical properties at smaller scales (Rydz et al., 2019; Wu et al., 2020).

The ability to accurately characterise polymer distributions is one of the most challenging aspects of standard polymer analysis using liquid chromatography (LC), but this is a task that becomes much more intricate when dealing with complex polymer systems (Trathnigg, 2000; Knol et al., 2021). LC is an essential tool commonly used for analysing the molecular structure, heterogeneity and distribution. Although LC is a powerful tool, there are still some issues that limit its potential. For instance, the resolution in LC is not high enough to separate complex polymers into individual oligomers over a wide molar mass range and there are no standards available for complex polymers. Furthermore, LC methods often use refractive index detection systems, which are not accurate enough for the analysis of complex polymers since the increase in the refractive index is often not known (Knol et al., 2021). Several improvements have taken place: LC has gone from high performance to ultra-high performance in the last few years, with increased pressure and reduced particle sizes enabling faster separation times or higher resolution in the same analysis time, when compared to conventional columns (Dong, 2019). Other improvements include those in the areas of instrumentation, sample preparation and the introduction of multidimensional liquid chromatography (Schoenmakers et al., 2014; Stoll et al., 2017). The multidimensional liquid chromatography (MDLC) methods have been explored in the characterisation of complex polymers (Berek and Šišková, 2010). In the first step, the macromolecules are separated according to their chemical composition, without taking their molar mass into account, whereas in the second step, the fractions obtained are sent online in their entirety to the size exclusion column and separated here according to their molar mass. Liquid chromatography under critical conditions (LCCC) is another separation method used to separate block or graft copolymers by the length of individual blocks (Abdul-Karim et al., 2017; Wesdemiotis, 2017). This allows multiple adsorption interactions between the polymer and stationary phase where, for example, the separation of block copolymers from their homopolymer precursors is based solely on their different functionalities (Macko et al., 2002).

Gel permeation chromatography (GPC) which is a type of size exclusion chromatography (SEC) is a widely used liquid chromatographic separation method for the routine characterisation of polymers and is the workhorse for many polymer laboratory analyses, providing efficient and rapid separation of the polymer chains in terms of their molar mass averages and distributions according to their chemical composition (Moore, 1964; Berek, 2000; Trathnigg, 2000). In contrast to other separation techniques that depend on chemical interactions, SEC relies on the size of the polymer (hydrodynamic volume) in solution to separate macromolecules and therefore often does not allow assessment of the chemical composition and architecture of structurally more complex polymers (Gaborieau et al., 2011). This is because the size of macromolecules in solution is related to molecular properties such as chemical structure, physical architecture and molar mass (Kilz et al., 2015). In addition, the SEC technique rarely produces simultaneous data on more than a single molecular property, and for complex polymers even molar mass averages and distributions obtained using SEC can only be viewed as semi-quantitative values (Berek, 2000). Another common problem with both LC and gas chromatography (GC) is that they cannot be analysed with some complex polymers that require the formation of solutions due to a lack of suitable solvents, as well as poor solubility, due to their size, or this can even arise from the fact that they are strongly cross-linked during synthesis (Gies et al., 2008; Ferreira et al., 2011). The lack of total dissolution complicates and possibly hinders direct analysis that can be performed by conventional chromatographic methods. This difficulty in dissolving the sample during dissolution and obtaining a clear suspension provokes the use of temperature (heat, microwave irradiation and ultrasonic treatment) or chemical additives (acid, base and salt), but these methods may result in the degradation of the polymer chain prior to SEC analysis, which is not desirable (Gaborieau et al., 2011). It is also worth mentioning that this cross-linking causes additional dimensional stability to the polymers, enhancement of mechanical properties and delayed hydrolysis and hence alters their biodegradability (Liu et al., 2011). The coupling of SEC to nuclear magnetic resonance (NMR) spectroscopy is another method that can also provide a wealth of both qualitative and quantitative data, but it is very costly (Hiller et al., 2014).

Mass spectrometry (MS) techniques have been extensively used to apply chemical composition and structure analysis of polymers (Gruendling et al., 2010). In particular, MS techniques equipped with soft ionisation techniques that create fewer fragments are advantageous. Here, both electrospray ionisation mass spectrometry (ESI-MS) and matrix-assisted laser desorption/ionisation mass spectrometry (MALDI-MS) have been increasingly gaining mention in polymer structure analysis research (Li, 2009; Gruendling et al., 2010; Kowalczuk et al., 2016; Wesdemiotis, 2017). Each technique has its own merits, but MALDI has been utilised extensively and is a critical tool applied for quantitative structural analysis, including material characterisation and degradation studies of complex polymers as it generates intact, singly charged gas-phase ions of high molar mass analytes (Li, 2009). Compared with MALDI, ESI-MS generated several series of multiply charged ions. While this can simplify the mass measurement of high mass macromolecules by reducing their mass/charge ratio (m/z) to a range of values available for mass spectrometry analysis, however, this can also lead to additional mass spectral complexity when mixtures of different species are ionised (Nohmi and Fenn, 1992). The time-of-flight (TOF) detection system has played an important role in the dominance of the MALDI technique because the molecules that have been put into the gas phase by MALDI can now be separated precisely according to their mass, thus providing non-averaging detailed structural information about the individual molecules. In terms of detection of both small and large molecules, MALDI also offers a superior wide mass range compared to ESI-MS, with lower and upper mass limits of approximately 400 and 350,000 Da respectively (Vlek et al., 2012). Nevertheless, samples with very small molar masses below 400 Da can often be analysed by laser desorption/ionisation (LDI) techniques. Moreover, sample preparation techniques are central to MALDI analysis in order to optimise the quality of mass spectra. Solvent-free sample preparation is also viable, which can be very useful for the analysis of insoluble analytes (Karas et al., 1988; Dolan et al., 2004). The mass spectrometry techniques offer the possibility of performing tandem mass spectrometry (MS/MS) studies, whereby mass selected parent ions are collided with a colloidal gas to further fragment and decompose into structurally indicative fragments that can be assigned to individual oligomers.

In complex polymers, the increasing structural or architectural complexity makes it difficult for MS and MS/MS methods to provide accurate information (Crotty et al., 2016). MALDI has some known intensity problems with large molecules. Due to ion suppression, selective ionisation, sweet spots and hot spots, it can be difficult to obtain quantitative information from signal intensities and this also impacts reproducibility (Szaéjli et al., 2008). Furthermore, MS techniques are limited to measuring only ion intensity and mass-to-charge (m/z) ratios. One way to address this is to use the mass spectrometry apparatus hyphenated with chromatographic separation techniques (Pasch, 2013; Crotty et al., 2016). Hyphenation is a technique that combines or connects two different analytical instruments (Hirschfeld, 1980). The combination of a powerful separation technique and the accuracy of the spectroscopic characterisation method has proven to be among the most powerful methods for molecular characterisation of complex polymer systems (Murgasova et al., 2002; Gruendling et al., 2010; Pasch, 2013). In addition, gas-phase ion mobility separations as well as various mass analysers such as ion traps, quadrupole and time-of-flight are also explored (Kanu et al., 2008; Gruendling et al., 2010; Crotty et al., 2016). Ion mobility (IM) mass spectrometry is an analytical technique that separates ions in the gas phase based on size, shape and charge with mass analysis and has been used to study macromolecular isomers and isobars as well as polymers with different branching architectures (Kanu et al., 2008; Lanucara et al., 2014; Alexander et al.,

2018). Sequence analysis of macrocyclic polyester copolymers applying ESI-IM and MALDI-TOF-MS was also reported (Alexander et al., 2018). The MALDI analysis showed sodiated oligomers without end groups as this is due to the macrocyclic architectures of the detected ions. ESI-MS mass spectra for the polyesters were more convoluted due to the superposition of singly and multiply charged ions. The combination of the MS/MS technique with separation of IM allowed the deconvolution of the spectra. Since the singly charged ions drift more slowly through the IM region than multiple charge ion species, they are therefore separated in a unique mobility region to make spectra without multicharged ions.

In another example, a hyphenated technique using a combination of mass spectrometry techniques such as MALDI or ESI-MS with liquid chromatography separation techniques (LC or SEC) analysis of complex polymers to perform fragmentation patterns and end-groups analysis was reported (Barqawi et al., 2013). In this study, the unique capability of multidimensional chromatographic methods in combination with MS techniques to analyse structural arrangements of both symmetrical triblock copolymers (AnBAn) (triethylene oxide-*b*-polyisobutylene-*b*-triethylene oxide) and non-symmetrical trisegment copolymers (AnBAm) is demonstrated. These complex polymers are biocompatible and are known to form segmental polymeric micelles or polymersomal membranes and therefore have potential applications for drug delivery.

The final polymers were separated from intermediate polymers first, using high-performance liquid chromatography (HPLC), which is followed by the second-dimension measurements with 2D-LC/SEC techniques. The subsequent analysis of the different segmental polymers together with the molar mass was then performed by direct coupling of 2D-LC/SEC to a MALDI-TOF via a spray module. This final step generated a series of time-correlated MALDI-MS spectra, representing symmetrical and non-symmetrical triblock copolymers.

Nowadays, copolymers consisting of more than three different monomers or oligomers are not an exception. Therefore, the increasing number of synthetic polymers with tailor-made functionality and molecular heterogeneity require an in-depth understanding of their complex molecular structures, thermal properties and degradation products. However, in order for these new materials to be thoroughly assessed before their use in various applications, their structure and composition must be methodically examined and their characteristic properties and comparison – depending on the type/mechanism of degradation – as well as environmental assessment completed. Such extensive characterisations of these complex polymer systems pose a challenge to conventional analytical techniques currently available. The characterisation of these systems is usually done in many different ways and the choice of instrument is highly dependent on the sample and the specific properties of the polymer under consideration. Hyphenation between different separation techniques such as LC and ion mobility with mass spectrometry techniques has contributed immensely to this area. This has extended the possibilities to analyse complex polymers and enables e.g. separation of isobaric co-eluting compounds, branched and linear species. Unfortunately, even the most powerful analytical techniques have trouble characterising complex mixtures of such closely related molecules in great detail. The notion of characterising all aspects of a complex polymer structure in sufficient detail to predict its performance is appealing and perhaps not unrealistic, but it will never be realised without paying a high price, and it is therefore debatable whether or not it is economically feasible. This may be the case for medical polymers used in the medical and pharmaceutical sector, where they have to fulfil certain strict safety and quality requirements and the cost is greater than the benefit.

REFERENCES

Abdul-Karim, R., Musharraf, S.G., and Malik, M.I. 2017. Synthesis and characterization of novel biodegradable di-and tri-block copolymers based on ethylene carbonate polymer as hydrophobic segment. *J. Polym. Sci. Part A: Polym. Chem.* 55(11): 1887–1893.

Alexander, N.E., Swanson, J.P., Joy, A., and Wesdemiotis, C. 2018. Sequence analysis of cyclic polyester copolymers using ion mobility tandem mass spectrometry. *Int. J. Mass Spectrom.* 429: 151–157.

Alvaradejo, G.G., Nguyen, H.V.T., Harvey, P., Gallagher, N.M., Le, D., Ottaviani, M.F., Jasanoff, A., Delaittre, G., and Johnson, J.A. 2019. Polyoxazoline-based bottlebrush and brush-arm star polymers via romp: Syntheses and applications as organic radical contrast agents. *ACS Macro Lett.* 8(4): 473–478.

Barqawi, H., Schulz, M., Olubummo, A., Saurland, V., and Binder, W.H. 2013. 2D-LC/SEC-(MALDI-TOF)-MS characterization of symmetric and nonsymmetric biocompatible PEO m–PIB–PEO n block copolymers. *Macromolecules* 46(19): 7638–7649.

Berek, D. 2000. Coupled liquid chromatographic techniques for the separation of complex polymers. *Prog. Polym. Sci.* 25(7): 873–908.

Berek, D., and Šišková, A. 2010. Comprehensive molecular characterization of complex polymer systems by sequenced two-dimensional liquid chromatography. Principle of operation. *Macromolecules* 43(23): 9627–9634.

Cook, A.B., and Perrier, S. 2020. Branched and dendritic polymer architectures: Functional nanomaterials for therapeutic delivery. *Adv. Funct. Mater.* 30(2): 1901001.

Crotty, S., Gerişlioğlu, S., Endres, K.J., Wesdemiotis, C., and Schubert., U.S. 2016. Polymer architectures via mass spectrometry and hyphenated techniques: A Review. *Anal. Chim. Acta.* 932: 1–21.

Doganci, M.D. 2021. Effects of star-shaped PCL having different numbers of arms on the mechanical, morphological, and thermal properties of PLA/PCL blends. *J. Polym. Res.* 28(1): 1–13.

Dolan, A.R., and Wood, T.D. 2004 Analysis of polyaniline oligomers by laser desorption ionization and solventless MALDI. *J. Am. Soc. Mass. Spectrom.* 15(6): 893–899.

Dong, M.W. 2019. *HPLC and UHPLC for Practicing Scientists.* Hoboken: John Wiley & Sons Ltd.

Feig, V.R., Tran, H., and Bao, Z. 2018. Biodegradable polymeric materials in degradable electronic Devices. *ACS Cent. Sci.* 4(3): 337–48.

Ferreira, J., Syrett, J., Whittaker, M., Haddleton, D., Davis, T.P., and Boyer, C. 2011. Optimizing the generation of narrow polydispersity 'arm-first'star polymers made using raft polymerization. *Polym. Chem.* 2(8): 1671–177.

Gaborieau, M., and Castignolles, P. 2011. Size-exclusion chromatography (SEC) of branched polymers and polysaccharides. *Anal. Bioanal. Chem.* 399(4): 1413–1423.

García-Gallego, S., Stenström, P., Mesa-Antunez, P., Zhang, Y., and Malkoch, M. 2020. Synthesis of heterofunctional polyester dendrimers with internal and external functionalities as versatile multipurpose platforms. *Biomacromolecules* 21(10): 4273–4279.

Gies, A.P., Kliman, M., McLean, J.A., and Hercules, D.M. 2008 Characterization of branching in aramid polymers studied by MALDI-ion mobility/mass spectrometry. *Macromolecules* 41 (22): 8299–8301.

Gruendling, T., Weidner, S., Falkenhagen, J., and Barner-Kowollik, C. 2010. Mass spectrometry in polymer chemistry: A state-of-the-art up-date. *Polym. Chem.* 1(5): 599–617.

Hiller, W., Sinha, P., Hehn, M., and Pasch, H. 2014. Online LC-NMR–from an expensive toy to a powerful tool in polymer analysis. *Prog. Polym. Sci.* 39(5): 979–1016.

Hirschfeld, T. 1980. The hyphenated methods. *Anal. Chem.* 52(2): 297A–312A.

Huang, Y., Wang, D., Zhu, X., Yan, D., and Chen, R. 2015. Synthesis and therapeutic applications of biocompatible or biodegradable hyperbranched polymers. *Polym. Chem.* 6(15): 2794–2812.

Jin, H., Huang, W., Zhu, X., Zhou, Y., and Yan, D. 2012. Biocompatible or biodegradable hyperbranched polymers: from self-assembly to cytomimetic applications. *Chem. Soc. Rev.* 41(18): 5986–5997.

Kanu, A.B., Dwivedi, P., Tam, M., Matz, L., and Hill Jr, H.H. 2008. Ion mobility-mass spectrometry. *J. Mass Spectrom.* 43(1): 1–22.

Karas, M., and Hillenkamp, F. 1988. Laser desorption ionization of proteins with molecular masses exceeding 10,000 daltons. *Anal. Chem.* 60(20): 2299–2301.

Kilz, P., and Pasch, H. 2000. Coupled liquid chromatographic techniques in molecular characterization. *EAC.* 9: 7495.

Kilz, P., and Radke, W. 2015. Application of two-dimensional chromatography to the characterization of macromolecules and biomacromolecules. *Anal. Bioanal. Chem.* 407(1): 193–215.

Kim, E.S., Kim, B.C., and Kim, S.H. 2004. Structural effect of linear and star-shaped poly(*L*-lactic acid) on physical properties. *J. Polym. Sci. B: Polym. Phys.* 42(6): 939–946.

Knol, W.C., Pirok, B.W., and Peters, R.A. 2021. Detection challenges in quantitative polymer analysis by liquid chromatography. *J. Sep. Sci.* 44(1): 63–87.

Kowalczuk, M., and Adamus, G. 2016. Mass spectrometry for the elucidation of the subtle molecular structure of biodegradable polymers and their degradation products. *Mass Spectrom. Rev.* 35(1): 188–198.

Lanucara, F., Holman, S.W., Gray, C.J., and Eyers, C.E. 2014. The power of ion mobility-mass spectrometry for structural characterization and the study of conformational dynamics. *Nat. Chem.* 6(4): 281–294.

Li, L. 2009. *MALDI Mass Spectrometry for Synthetic Polymer Analysis*. Hoboken: John Wiley & Sons Ltd.

Liu, H., and Zhang, J. 2011. Research progress in toughening modification of poly(lactic acid). *J. Polym. Sci., Part B: Polym. Phys.* 49(15): 1051–1083.

Loh, X.J. 2014. Supramolecular host-guest polymeric materials for biomedical applications. *Materials Horizons* 1(2): 185–95.

Macko, T., Hunkeler, D., and Berek, D. 2002. Liquid chromatography of synthetic polymers under critical conditions. The case of single eluents and the role of θ conditions. *Macromolecules* 35(5): 1797–1804.

Menczel, J.D., and Prime, R.B. 2009. *Thermal Analysis of Polymers: Fundamentals and Applications*. Hoboken: John Wiley & Sons Ltd.

Michalski, A., Brzezinski, M., Lapienis, G., and Biela, T. 2019. Star-shaped and branched polylactides: Synthesis, characterization, and properties. *Prog. Polym. Sci.* 89: 159–212.

Moore, J.C. 1964. Gel permeation chromatography. I. A new method for molecular weight distribution of high polymers. *J. Polym. Sci. Part A: General Papers* 2: 835–843.

Murgasova, R., and Hercules, D.M. 2002. Polymer characterization by combining liquid chromatography with MALDI and ESI mass spectrometry. *Anal. Bioanal. Chem.* 373(6): 481–489.

Nadgorny, M., Gentekos, D.T., Xiao, Z., Singleton, S.P., Fors, B.P., and Connal, L.A. 2017. Manipulation of molecular weight distribution shape as a new strategy to control processing parameters. *Macromol. Rapid Commun.* 38(19): 1700352.

Nicolaÿ, R., Marx, L., Hémery, P., and Matyjaszewski, K. 2007. Synthesis of multisegmented degradable polymers by atom transfer radical cross-coupling. *Macromolecules* 40(26): 9217–9223.

Nohmi, T., and Fenn, J.B. 1992. Electrospray mass spectrometry of poly(ethylene glycols) with molecular weights up to five million. *J. Am. Chem. Soc.* 114(9): 3241–3246.

Nunes, R.W., Martin, J.R., and Johnson, J.F. 1982. Influence of molecular weight and molecular weight distribution on mechanical properties of polymers. *Polym. Eng. Sci.* 22(4): 205–228.

Pasch, H. 2013. Hyphenated separation techniques for complex polymers. *Polym. Chem.* 4(9): 2628–2650.

Pasch, H., and Schrepp, W. 2003. *Analysis of Complex Polymers, in MALDI-TOF Mass Spectrometry of Synthetic Polymers*. Berlin: Springer Berlin Heidelberg.

Rydz, J., Šišková, A., and Andicsová Eckstein, A. 2019. Scanning electron microscopy and atomic force microscopy: topographic and dynamical surface studies of blends, composites, and hybrid functional materials for sustainable future. *Adv. Mater. Sci. Eng.* 2019: 6871785.

Sakamoto, Y., and Tsuji, H. 2013. Stereocomplex crystallization behavior and physical properties of linear 1-arm, 2-arm, and branched 4-arm poly(*L*-lactide)/poly(*D*-lactide) blends: effects of chain directional change and branching. *Macromol. Chem. Phys.* 214(7): 776–86.

Schoenmakers, P., and Aarnoutse, P. 2014. Multi-dimensional separations of polymers. *Anal. Chem.* 86(13): 6172–6179.

Setijadi, E., Tao, L., Liu, J., Jia, Z., Boyer, C., and Davis, T.P. 2009. Biodegradable star polymers functionalized with β-cyclodextrin inclusion complexes. *Biomacromolecules* 10(9): 2699–2707.

Stoll, D.R., and Carr, P.W. 2017. Two-dimensional liquid chromatography: A state of the art tutorial. *Anal. Chem.* 89(1): 519–31.

Szaéjli, E., Feheér, T., and Medzihradszky, K.F. 2008 Investigating the quantitative nature of MALDI-TOF MS. *Mol. Cell. Proteomics* 7(12): 2410–2418.

Talebi, S., Duchateau, R., Rastogi, S., Kaschta, J., Peters, G.W., and Lemstra, P.J. 2010. Molar mass and molecular weight distribution determination of UHMWPE synthesized using a living homogeneous catalyst. *Macromolecules* 43(6): 2780–2788.

Trathnigg, B. 2000. Size-exclusion chromatography of polymers. In *Encyclopedia of Analytical Chemistry*, ed. Meyers, R.A., pp. 8008–8034. Chichester: John Wiley & Sons Ltd.

Ulery, B.D., Nair, L.S., and Laurencin, C.T. 2011. Biomedical applications of biodegradable polymers. *J. Polym. Sci., Part B: Polym. Phys.* 49(12): 832–864.

Uliyanchenko, E., Van der Wal, S., and Schoenmakers, P.J. 2012. Challenges in polymer analysis by liquid chromatography. *Polym. Chem.* 3(9): 2313–2335.

Van Bochove, B., and Grijpma, D.W. 2019. Photo-crosslinked synthetic biodegradable polymer networks for biomedical applications. *J. Biomater. Sci. Polym. Ed.* 30(2): 77–106.

Vlek, A.L., Bonten, M.J., and Boel, C.E. 2012. Direct matrix-assisted laser desorption ionization time-of-flight mass spectrometry improves appropriateness of antibiotic treatment of bacteremia. *PloS One* 7(3): 32589.

Wesdemiotis, C. 2017. Multidimensional mass spectrometry of synthetic polymers and advanced materials. *Angew. Chem. Int. Ed.* 56(6): 1452–1464.

Wu, F., Misra, M., and Mohanty, A.K. 2020. Tailoring the toughness of sustainable polymer blends from biodegradable plastics via morphology transition observed by atomic force microscopy. *Polym. Degrad. Stab.* 173: 109066.

Zhu, J., Abeykoon, C., and Karim, N. 2021. Investigation into the effects of fillers in polymer processing. *Int. J. Lightweight Mater. Manuf.* 4(3): 370–382.

5 Current Trends in the Area of Blends and Composites

Marta Musioł

CONTENTS

5.1 BLENDS

Polymeric materials are still being development in search of materials with improved properties such as mechanical, physicochemical etc. as well as in terms of biological properties. New application paths require new materials that are on the one hand easy to process and inexpensive and on the other hand safe for human health and the environment. One polymer regardless of its origin is unable to meet different requirements. Optimising the properties of the plastic materials can be achieved by physical mixing of dissimilar polymers (Goonoo et al., 2015). Obtaining polymer blends by a physical process is much cheaper than, e.g., copolymerisation (chemical approaches). During the preparation of the blend, some chemical interactions can but this is not necessarily always the case. Melt blending is the preferable method for production of new materials by the industry. From the technological point of view the factors which determine performance are morphology, miscibility and compatibility (Muthuraj et al., 2018).

DOI: 10.1201/9780429352799-6

5.1.1 Morphology

Polymer morphology is a physical phenomenon related to organisation and form on a size scale lower than the whole sample but bigger than the atomic arrangement. It can be classified depending on the degree of order as crystalline, semi-crystalline and amorphous. But also, it contains information concerning structural domains association and distribution in the macrostructure without excluding the shape and size of additives and fillers (Sawyer et al., 1987; Khalifeh, 2020). Preparation of the blend from dissimilar polymers can create various morphology. Miscibility and compatibility of blend components have a significant impact on this phenomenon. The blend of crystalline and amorphous polymers in the melt can have a single phase suggesting that they are miscible. However, in the solid-state amorphous regions are mixed with the crystalline phase (Inoue, 2003). Incompatibility between the polymer pairs in the blend leads to dispersion and more pronounced changes in morphology, for these blends are observed in comparison to neat polymers.

Many features have an influence on the morphology of a blend of incompatible polymers. These include the elastic properties and viscosities of both components, but also the processing method itself, which can cause coalescence of slight drops into larger ones and increase of instability of elongated streak-like structures (Van Der Vegt and Elmendorp, 1986).

Crystalline and amorphous polymers during the formation of miscible blend in the melt exhibit a single phase, and when it becomes solid, the amorphous region is observed with a pure crystalline phase. Exclusion of non-crystalline phase from crystalline regions during crystallisation depends on the growth rate of the crystalline ingredient and the diffusion rate of the non-crystalline ingredient, which affects the morphology of the polymer blends. In the case when the upper critical solution temperature is lower than the melting temperature for a crystalline/amorphous polymer blends the phase separation disrupts crystallisation. Blend composition, rate of solvent evaporation and polymer/polymer interaction parameter (χ_{12}) are responsible for structure formation during solvent evaporation. Faster casting is needed for the bi-continuous structure formation when the χ_{12} is larger (Inoue, 2003). By selecting the appropriate conditions, a homogeneous blend can be obtained. However, the heating of those blends above the glass transition temperature (T_g) caused the phase separation by spinodal decomposition (Kammer and Kummerloewe, 1990). Spinodal decomposition occurs when one phase spontaneously separates to two phases without the presence of a nucleation step (Favvas and Mitropoulos, 2008). Phase separation with simultaneous chemical reactions is quite often used to obtain a material with certain properties (Inoue, 2003).

Blending of the polymers pairs which are usually incompatible causes the dispersion. The dispersion scale can be reduced by the blending until an equilibrium is reached. However, it depends on process conditions and on the properties of the polymers themselves (Van Der Vegt and Elmendorp, 1986). Dispersed particles of immiscible polymers during extrusion are alternately exposed to shear stress and allowed to relax. Viscosity of both polymers and their interfacial tension has an influence on the ability to return to the particle's original shape. If the size of particles is larger, the deformation of the spherical shape during shearing is bigger. Changes in processing parameters may lead to variations in the morphology of the blends (Inoue, 2003). The processing itself has a significant impact on the morphology of the blend surface.

5.1.2 Miscibility

Physicochemical properties of blends are determined by the miscibility of polymers. This phenomenon can help to define the polymer pair behaviour during blending by the usage of an appropriate model. Polymer pairs need to fulfil the following condition (Equation 1) to obtain the complete miscibility.

$$\Delta G_m = \Delta H_m - T\Delta S_m < 0 \qquad \text{(Eq. 1)}$$

Where: ΔG_m – Gibb's free energy of mixing, ΔH_m – enthalpy of mixing, T – temperature and ΔS_m – entropy of mixing.

For miscible polymer blends the free energy of mixing has a negative value; the $T\Delta S_m$ must be positive because during mixing there is an increase of ΔS_m. It follows that ΔH_m value has an influence on the sign of Gibb's free energy of mixing. According to Eq. 1, to obtain a complete miscible blend the ΔH_m value must be less than entropic contribution.

The Flory-Huggins lattice theory is considered as the starting point for theoretical interpretations of polymer blends. It takes into account the specific properties of polymers, enabling the use of this model to describe the behaviour of polymer pairs upon blending (Andersen, 2004; Imre et al., 2013; Kataoka et al., 2014; Goonoo et al., 2015). The degree of miscibility has a significant influence on the properties of blends. Glass transition temperature can provide the information to qualify the blend as completely miscible, partially miscible or completely immiscible. If one T_g value is observed then the blend is considered miscible, but if two distinguishable T_g are observed it is immiscible. The shifting of T_g values of one blend component and its approach to the T_g value of the second one gives information about partial miscibility (Imre and Pukánszky, 2013; Korycki et al., 2021).

5.1.3 Compatibility

Compatibility is a term with a rather practical meaning. It denotes the ability of the individual constituent substances in a blend of immiscible polymers to exhibit interfacial adhesion. Hydrogen bonding belongs to the favourable specific interaction enabled to achieve compatibility of the blend (Sabzi and Boushehri, 2006). To obtain the blend of immiscible polymers with improved properties compatibilisation is used. There are two methods to enhance blending performance: non-reactive and reactive. Non-reactive strategies use the copolymers prepared to achieve the miscibility with both immiscible blended polymers. When one block from block copolymer is mixing with one prepared blend component the second one mixes with another. Blended polymers connect through the bridges formed by interaction with a compatibiliser which improves their miscibility. Since biodegradable polymers usually have functional groups, it is possible to use reactive compatibility strategies. Reactive functional groups of blended polymers react with functional groups of the compatibiliser during melt blending. Obtained block or graft copolymers play the role of proper compatibilisers (Muthuraj et al., 2018; Ferri et al., 2020).

5.1.4 Preparation of Biodegradable Blends

Biodegradable polymers are becoming more and more common. Their biodegradability has a significant impact on the place of their application. They are mainly used in short-lived products, especially as packaging. Their importance in medicine is also not overestimated, where their biocompatibility together with biodegradability gives great possibilities of applications (Haider et al., 2019; Silva et al., 2020). Most applications require specific properties that can only be achieved with the blending of two polymers. The selection of appropriate processing conditions as well as the type of compatibiliser during blending of biodegradable polymers improves the properties of the blend obtained. Preparation of the poly(3-hydroxybutyrate-*co*-hydroxyvalerate) (PHBV) with poly(butylene succinate) (PBS) by the co-dissolving and casting resulted in an incompatible blend, regardless of the percentage of the polymers contained (Qiu et al., 2003).

The miscibility of blend it is not a prerequisite for better properties of the final material. Biodegradable blends obtained by the mixing of poly(propylene carbonate) (PPC) with PBS turned out to be immiscible, confirmed by dynamic mechanical analysis (DMA) and scanning electron microscopy (SEM) analysis. However, when the amount of PBS increases the strength at break and yield strength of the blends improved. Also, introduction of PBS caused a significant increase in blend thermal stability in comparison to neat PPC. The improved properties of the PPC/PBS blends without obtaining miscibility can provide a wide range of application as biodegradable materials (Pang et al., 2008).

Improving miscibility of PBS/poly(ethylene succinate) (PES) blends happened as a result of their annealing. PBS/PES blends were prepared by the solution-casting method from the chloroform. After evaporation of the solvent the blend samples were annealed under nitrogen at 150°C. Obtained results were indicated the transesterification between PES and PBS caused by the annealing process. Differential scanning calorimetry (DSC) analysis confirmed the miscibility of annealed blend during T_g investigation (Kataoka et al., 2014).

Different methods of blend preparation were applied for obtaining poly(1,4-butylene adipate-*co*-1,4-butylene terephthalate) (PBAT)/poly(hydroxy ether of bisphenol A) (PH) materials. Samples prepared by casting from chloroform or with addition of annealing showed increase in miscibility between polymers. For both methods immiscible or partially miscible blends were obtained; the usage of the annealing process did not make it possible to achieve this state. When the melt blend method was used, DSC analysis indicated that a single T_g was observed for the PBAT/PH blend in all compositional ranges, confirming the achievement of miscibility (Sangroniz et al., 2018).

Reactive melt blending can be conducted by one or more processing steps. Polylactide (PLA) was blended with poly(trimethylene terephthalate) (PTT) with reactive compatibiliser poly(ethylene-*co*-glycidyl methacrylate) (PEGMA) to improve the material properties. A one-step blending procedure used to obtain PLA/PTT/PEGMA blend caused dispersion of PEGMA in PLA matrix, limiting the occurrence of the compatibilisation reaction. Tensile properties of blend prepared by this procedure were unimproved. However, the use of two-step blending resulted in enhancing elongation at the break. During the first step the PEGMA was mixed with PTT and then in the second step the PLA was added. Changes in the producing procedure caused no interfacial detachment between phases of PLA and PTT. It follows that there was a marked enhancement of the interfacial adhesion between the PLA and PTT phases. The size of PTT dispersed phase has also been reduced. The increase in thermal properties of the PLA/PTT/PEGMA blend obtained during two-step procedure was observed. Therefore, previously mentioned materials show the possibilities of a large range of applications (Kultravut et al., 2019).

Addition of a reactive compatibiliser in the form of multifunctional epoxy oligomers to the PBAT/PLA blend improved its impact toughness and tensile qualities. Joncryl® ADR-4370S is a chain extender used in this blend as the compatibiliser influences the mechanical properties as expected. In the thermal stability no significant changes were observed regardless of compatibiliser quantity (Nunes et al., 2019, Wang et al., 2019).

Joncryl® ADR-4370F was incorporated in PBAT/thermoplastic poly(propylene carbonate) polyurethane (PPCU) blend. Obtained material shows good miscibility inducted by the presence of a chain extender causing intermolecular reactions between PBAT and PPCU. The improved miscibility has a positive effect on the remaining properties of the blend. A blend in the form of a film obtained by the blowing films technique after melt blending showed good mechanical properties with high tensile strength. Materials were tested for applications as fresh vegetable and fruit packages; the PBAT/PPCU/Joncryl® ADR-4370F blend exhibited good gas barrier performance and low water vapour permeability (Zhang et al., 2021a).

Using a different type of chain extender (Joncryl® ADR-4468) in the PBAT/PPC blend also improved mechanical properties. The melt blending method was used for blend preparation. Micrographs of the PBAT/PPC/Joncryl® ADR-4468 blend show that the fracture surface of the PBAT/PPC flattering point to strong interfacial adhesion between polymers. This evidences an increase of PBAT/PPC compatibility additionally confirmed by DSC where the T_g of PBAT rises. In addition to improving compatibility, the addition of chain extender into the PBAT/PPC blend resulted in an increase in thermal stability (Zhao et al., 2021).

In situ compatibilisation of PBAT/PLA blends was also conducted by the addition of a free radical initiator – dicumyl peroxide (DCP). Addition of DCP initiates the formation of the network structures and branching between PBAT and PLA. After compatibilisation the T_g of PBAT phase increases insignificantly while the T_g of PLA remains unchanged. Improvement in the mechanical properties was observed. Presence of the PBAT and DCP had no effect on the hydrolytic degradation of PLA in the blend (Ma et al., 2014).

Modification of PBAT properties by blending with different polymers or additives can greatly expand its application possibilities. PBAT/polyglycolide (PGA) blends with changing content of PGA were prepared by melt blending at 230°C for 8 min. Then the compression-moulding was used to obtain the sheets from prepared blends. The influence of the PGA amount on the blend properties like microstructure, melt viscosity, crystallisation properties, tensile property or barrier performance was investigated. Presence of PGA in the blend caused significant changes in the mechanical properties; tensile yield strength and modulus increased with an increase of PGA content. Those two parameters are very important to determine the deformation resistance of polymeric materials. But it must be mentioned that tensile ductility in respect of elongation at break for PBAT/PGA blends dropped significantly in comparison to neat PBAT. The application of Joncryl® ADR-4370S as a chain extender aims to improve the blend properties by getting better compatibility. Ternary blends with Joncryl® ADR-4370S were also prepared by melt blending, keeping the same conditions as for binary blends. The SEM results obtained for PBAT/PGA/ Joncryl® ADR-4370S blend indicated great decrease of the PGA size domain dispersed in the PBAT matrix relative to the PBAT/PGA blend. The observed phenomenon is due to the in situ reaction among PBAT, PGA and Joncryl® ADR-4370S taking place during melt-blending. This proves that the chain extender works as intended. Addition of chain extender to the PBAT/PGA blend improved also the tensile ductility. Despite the negative impact of an increase in chain extender content on tensile yield strength and modulus it is considered that the obtained materials show higher processing stability and better stiffness-ductility balance (Shen et al., 2021).

4,4'-Methylenebis(phenyl isocyanate) (MDI) can be used as a compatibiliser for PGA/PBAT blend. Materials for further investigation were obtained by the extrusion. The compatibilisation reaction was prepared in situ. The influence of PBAT and MDI content in the blend on materials properties was investigated. Obtained results indicated on the chain extension reaction, PBAT and PGA hydroxyl end groups react with diisocyanate forming the urethane bonds. In situ compatibilisation in this case leads to obtaining the mixture of PGA-crosslink-PBAT, branched PBAT and PGA. Addition of MDI caused the increase of complex viscosity of blends, and SEM results confirmed the improvement of blend compatibility. Also, the presence of the MDI enhanced the mechanical properties of the blend. Increase of PBAT content in the blend had also significant influence on the material properties, especially evident when the PBAT content exceeded 60%. During processing, until this amount of PBAT was reached the dispersed phase morphology changed from a spherical structure to in situ microfibre. Enhancement of compatibilisation of polymers in the blends caused the decrease of crystallisation temperature, crystallinity and crystallisation size which has a beneficial effect on improving toughness. The developed method of obtaining blends led to obtaining materials not only of enhanced toughness but also of increased thermal stability (Wang et al., 2021).

Melt blending of PLA with epoxidised poly(styrene-*b*-butadiene-*b*-styrene) (ESBS) resulted, at the interface, in the creation of graft copolymer. Possibly the reaction took place between the terminal carboxyl groups of PLA and epoxy groups of ESBS. In situ peroxy-formic acid method was used to obtain ESBS with a dissimilar epoxidation degree. The obtained blends show an improvement in mechanical properties. However, depending on the degree of epoxidation, the differences become pronounced, especially after exceeding 25.5%. For the samples PLA/ESBS which contain 10% of ESBS with epoxidation degree at 35.8%, the impact strength was 6.7 times higher than for neat PLA. This same sample exhibited the increase of elongation from 3.5% observed for neat PLA to 218.8% for this blend. Phase homogeneity and good compatibility were demonstrated by the received PLA/ESBS blends, which is most likely responsible for a significant improvement in the impact toughness and elongation of the materials obtained (Wang et al., 2016). Reactive compatibilisation was applied for obtaining the PLA/natural rubber blend. Because PLA and natural rubber show the poor compatibility caused by the difference in polarity the dynamic vulcanisation had been used to receive the PLA/natural rubber blend. This term is used to describe selective vulcanisation during melt, mixing the rubber phase with thermoplastic polymer. Two-phase material is obtained when in plastic matrix the crosslinked elastomer is dispersed. Higher amounts of natural rubber cause the increase of PLA-natural rubber graft detected by Fourier transform infrared spectroscopy (FTIR), which suggests that at the PLA/natural rubber interfaces reactive compatibilisation occurred. The use of a 35% natural rubber additive in the PLA/natural rubber blend resulted in the improvement of the mechanical properties, including an increase in the Izod impact strength from 2.75 kJ/m^2 for neat PLA to 58.3 kJ/m^2 for this blend composition (Yua et al., 2014).

The solvent technique was used to obtain dialdehyde cellulose/PLA blend. Dimethylformamide (DMF) was used as a solvent and poly(ethylene glycol) (PEG) as compatibiliser to allow interaction between dialdehyde cellulose and PLA. Dialdehyde cellulose was synthesised by the oxidation of microcrystalline cellulose to dialdehyde cellulose with NaIO4. The use of a compatibiliser resulted in a decrease in the decomposition temperature and increased mass residual during thermogravimetric analysis (TGA; Şirin et al., 2021).

Thermoplastic starch (TPS) was obtained in the presence of a plasticiser subjected to shear-pressure temperature. Application of urea as a plasticiser caused the TPS retrogradation suppression. TPS is biodegradable and rather cheap material, however the mechanical properties of it are poor. Blending of TPS with other polymers can help to overcome these disadvantages. Poly(ε-caprolactone) (PCL), due to its properties, can improve the properties of the blend obtained by mixing it with TPS. The high price of PCL also has a significant impact on the search for opportunities to lower the price by making blends with cheaper materials such as TPS. Urea plasticised thermoplastic starch (UTPS)/PCL blends were obtained by the extrusion in the same conditions as neat UTPS to avoid changes in the thermal history of the materials obtained. The different composition of the obtained mixtures enables optimisation in terms of mechanical and thermal properties. DSC analysis shows that UTPS is a nucleating agent for the PCL. Thermal stability (from TGA) increased for UTPS with increasing PCL content. Increasing content in the blend above 50% deteriorates the UTPS dispersion in the PCL matrix. Elastic modulus and maximum tensile strength of blends were improved versus neat UTPS with increasing of the PCL content. Due to its properties, this type of material can be used in agriculture for the controlled release of nutrients into the soil (Corrêa et al., 2017).

Thermoplastic starch, as a component of PCL/TPS blend, was prepared by the two-step process. The starch, glycerol and distilled water were mixed and kept in 65 °C till the viscosity was raised. From this mixture the film was formed, which was then cut and homogenised by melt-mixing. After pressing, the compression-moulded TPS-material was obtained. The blends with PCL were prepared by melt-mixing in these same conditions, which was used for TPS preparation. The influence of blend compositions on their properties was evaluated. DSC results indicated the increase of crystallinity (X_c) for the sample with 30% of TPS which is attributed to enhanced nucleation at the interface. However, as the TPS content increases further, the crystallinity begins to decline. In the blend with 70% of TPS X_c decreased below the value for neat PCL, which follows from co-continuous structure of that blend with partially fine PCL domains. Hardness of all the blends was determined by a microindentation test. Based on comparison of predictive models to the experimentally determined micromechanical properties it was found that obtained results indicated the strong interfacial adhesion between components and very good compatibility (Nevoralová et al., 2020).

Blending of PLA with PCL caused the improvement of PLA properties without loss of the biodegradability feature. PLA/PCL blends with various composition ratios were prepared by melt mixing and injection moulding. Two methods of injection moulding were used to obtain standard tensile bars (conventional and microcellular) for mechanical test and thermal analysis. Results gained by DSC analysis shows that processing conditions do not have an influence on the blends thermal properties. Additionally, the PCL melting temperature did not change regardless of the composition of the blend. The same applies to the degree of crystallinity. Presence of the PCL in the PLA/PCL blend caused the increase of PLA decomposition temperature determined by TGA. However, the derivative thermogravimetric (DTGA) curves showed two separate peaks. The blends modulus and tensile strength decreased with increasing PCL content in the PLA/PCL blend, and the elongation greatly increased compared to neat PLA. Addition of PCL to the PLA caused the changes of the material from brittle to ductile failure. Changing the composition of the blend caused clear changes in the properties of the obtained material, which led to a wide field of applications of these materials, depending on the composition (Zhao and Zhao, 2016).

Different poly(vinyl alcohol) (PVA) types were used to obtain the blends with starch and glycerol by the extrusion when using various temperature profiles. The prepared blends were tested for their mechanical properties. Depending on the PVA type used and temperature profile, the great volatility of the mechanical properties was observed. In this context, the use of different PVA and processing conditions gives great possibilities of influencing the mechanical properties of the obtained materials (Zanela et al., 2016).

5.1.5 APPLICATION OF BIODEGRADABLE BLENDS

The possibilities offered by the blending of biodegradable polymers as well as the use of various types of compatibilisers significantly increase the application area of new materials. Research mainly focuses on improving mechanical properties while maintaining biodegradability.

Ecovio® from BASF Company which is a commercial PBAT/PLA blend has been tested for applications in the cosmetics industry. A prototype packaging for cosmetics was tested for life expectancy during ageing in cosmetic media at elevated temperature. Obtained results indicated that PBAT/PLA blend exhibits adequate stability during the experiment. This means that the material is suitable for use in the cosmetics industry as packaging (Sikorska et al., 2017).

For packaging application, the blend of PLA and cellulose acetate or chitosan were prepared and investigated. The blends were obtained in the form of film on the press after mixing in the twin-screw extruder. Blends were processed without any additional additives or plasticisers. For the PLA/chitosan blend with the amount of chitosan below 20%, no chitosan phase in the foil was observed, pointing to good dispersion of chitosan in the PLA matrix. In the PLA/cellulose acetate blends the non-anchoring of the polymer spheres was visible which results from the lack of use of compatibilisers. However, due to the processing temperature which was above the melting point of cellulose acetate, the particles attained diameters close to a micron. Morphology of the samples has a significant influence on mechanical properties so the PLA/chitosan samples with 10% of chitosan showed the best mechanical strength. Previously mentioned research indicated that the plasticisers or other additives are not needed to obtain the blends with potential packaging application (Claro et al., 2016).

Preparation of filaments for fused deposition modelling (FDM) requires many attempts to optimise the material also by using blends with different compositions. PGA/PBAT blends with addition of multi-functional epoxy chain extender styrene-glycidyl methacrylate (Joncryl® ADR-4370) were prepared with different polymer mass ratio. Blends have been tested and assessed for mechanical and thermal properties. The most optimal composition in terms of properties was selected and extruded into filaments. Adequate thermal stability and mechanical properties were shown by the blend of PGA/PBAT with 85/15 in mass ratio. A desktop single-screw filament extruder was used to fabricate the selected blend into filament. Mechanical test of the 3D-printed samples and for comparison the injection-moulded sample show that the materials obtained by both methods indicated similar impact, flexural and tensile properties. Diamond-triply periodic minimal surfaces structures possessed excellent energy absorption capacity and the graded thickness had an influence on the prevention of large stress fluctuations. PGA/PBAT blend was printed by FDM in the form of pore architecture with radially graded structures and two sheet thicknesses and they were highly equivalent with geometric models. Obtained results indicated that the PGA/PBAT with 85/15 in mass ratio enables their creation by the FDM complex diamond-triply periodic minimal surfaces structures. Modification of blend composition and applied additives open a great opportunity for preparation of FDM filaments (Zhang et al., 2021b).

The controlled release of plant protection products is now becoming an increasingly common approach to growing crops in agriculture. This approach not only works in savings resulting from the reduction of the frequency of agrochemicals use but also their amount released into the soil, which has a significant impact on environmental protection. The controlled release system used in agriculture has to meet several requirements, the most important being the rate of release. This property is closely related to the growing season of crops. Therefore, various modifications of biodegradable polymers and their subsequent blending are used. Poly(*L*-lactide-*co*-glycolide)-*b*-poly(ethylene glycol) (PLLGA-PEG) terpolymers obtained by ring-opening polymerisation were blended with dextrin-g-PCL or maltodextrin-*g*-PCL. Both grafted copolymers were synthesised by copolymerisation of dextrin or maltodextrin with ε-caprolactone (CL). PLLGA-PEG/maltodextrin-*g*-PCL or PLLGA-PEG/dextrin-*g*-PCL blends were received by the foil casting after mixing of solutions of all ingredients. Metazachlor and pendimethalin which are often used herbicides were used as a plant protection product in controlled release systems. The blends with plant protection products were subjected to the degradation process under various conditions in order to evaluate the operation of the gradual herbicide release system. Results obtained for PLLGA-PEG/dextrin-*g*-PCL are very promising. This system releases the 90% of loaded herbicide into the soil within a 3-month experiment period. Presence of modified dextrin in the blend has a significant influence on the beginning phase of degradation in soil due to its susceptibility to enzyme action; it accelerates significantly. This was not observed for samples containing maltodextrin which could be due to differences in water absorption of both compounds. The controlled-release system is designed to release the herbicide unchanged into the soil. PLLGA-PEG/dextrin-*g*-PCL meets the aforementioned requirements, however, the process itself runs much better for metazachlor then pendimethalin. Nevertheless, it results from the properties of the herbicide itself, which dissolves much worse in water than metazachlor (Lewicka et al., 2020).

Disposable masks which are often used not only for protection from the virus – like during the pandemic period of COVID-19 – but also in the city are often used to protect human health from dust suspended in the air. Those masks are mainly made from polypropylene which, with their considerable wear, as during a pandemic, causes a serious environmental problem. So far, the method of recycling this type of waste has not been indicated, hence the interest in introducing biodegradable masks to the market. A PLA/thermoplastic polyurethane (TPU) blend was used to obtain melt-blown nonwovens materials. To improve the tensile strength and miscibility the polymeric chain extender Joncryl® ADR-4400 was used. PLA/TPU blends in various compositions were prepared via a twin-screw extruder, and 1% of J Joncryl® ADR-4400 was added to each blend. Obtained materials were cooled and pelletised, to prepare for further processes. A meltblown machine was used for the spinning process of the PLA/TPU blends. Meltblown nonwoven materials were manufactured successfully from all blends. For materials with a TPU content of up to 15%, an improvement in mechanical properties was observed. Above this value some defects on webs appeared. PLA/TPU blends seem to be good candidates for application in the toughened meltblown nonwoven in the field of disposal filtration for production of masks. However, compatibility of polymers used still needs some improvement (Rahman et al., 2022).

Shape memory effect is an ability of the materials to return to their original shape after deformation by the action of an inducement (changes in temperature, pH etc.). The most popular materials which show this effect are polymers. Due to the use of shape memory polymers (SMP)s mainly in the areas of medicine, research on the development of new materials focuses on biodegradable polymers. SMP possess the permanent netpoints which allow it to return to its original shape after the action of an appropriate stimulus. In the case of biodegradable polymers material is generally heating above a transition temperature; then the SMP is deformed to the temporary shape and cooled below this temperature. However, when reheated above the T_g, the SPM reverts to a permanent shape. The development of a blend of biodegradable polymers having a shape memory effect requires the selection of materials that the blend has permanent netpoints. However, there are polymers that can be programmed below the T_g and yet they return to their original shape.

For example, the covalently crosslinked PCL networks have this property which is called *reversible plasticity shape memory*. This effect was also observed where noncovalent interactions occur, which are able to sufficiently stabilise the SMP permanent shape during cold drawing and thermal recovery. The self-healing polymers appearing more and more often attract a lot of attention due to their properties. The self-healing effect, like shape memory, can be caused by various factors; however, also in this case, heating is most often used. Due to its low melting point, PCL is often used as a healing agent. The use of the aforementioned effects may lead to the production of materials exhibiting shape memory-assisted self-healing functionality. This is due to the fact that in order for self-healing to occur, the edges of the damage must be brought closer to each other, but it requires the use of external force. Shape memory-assisted self-healing is a healing mechanism that uses the shape memory effect to bring damaged/fractured surfaces into close contact prior to self-healing, removing the need for external force. To investigate the shape memory-assisted self-healing ability the blend of PCL and TPU was used. PCL/TPU blends in different composition were prepared by melt compounding. Then the pressing was used to obtain the film samples. The prepared blends contained up to 60% PCL; they had the ability to reversible plasticity shape memory and shape memory-assisted self-healing. Cracked specimens were subjected to gentle heating (90°C/30 min) during which, due to reversible plasticity shape memory, the edges of the cracks came closer together and then healed by redistribution of the molten PCL. The composition of the blend largely determined the degree of healing. In the blends where amount of PCL was below 30% the almost-whole properties were restored. Blends with 50 and 60% of PCL, however, did not show the better properties, which is ascribed to their propensity to neck during the opening of the crack. Also, those blends at elevated temperatures show poor mechanical properties. Obtained material with proper composition can be used as a replacement for pure TPU with self-healing properties and applied in, e.g., automotive and sporting goods. However, some niche areas like biomedical or microelectronic also can be the places of application for these materials (Bhattacharya et al., 2020).

Biomedical application belongs to the most popular field for biodegradable materials. Shape memory as a material property is of particular importance in this area. The blends of PLA/PCL were obtained by the extrusion, which is the usual solution followed in the industry. Different compositions of the blends were obtained to investigate among others the shape memory behaviour. All of the blends were immiscible which has been demonstrated by the DSC analysis and morphology studies. Blending of those two polymers had no significant effect on the changes in values of the melting temperature (T_m) for both PCL and PLA homopolymers and T_g of PCL. Different phase separation which was observed by field emission scanning electron microscopes (FE-SEM) resulted from various blend compositions also pointing to the immiscibility of the blends. Despite the fact that the blends turned out to be incompatible, the material with 30% of PCL showed the shape memory effect. Depending on the quantity of PCL added the blends showed the wide range of mechanical properties. Obtaining multifunctional blends by the easily scalable process can find application not only in the biomedical field but also in packaging (Navarro-Baena et al., 2016).

Biodegradable PLA possesses a lot of additional advantages like nontoxicity and biocompatibility which makes it a material commonly used in the field of medical research. According to many studies, PLA also exhibits excellent shape memory properties. The two phases, amorphous and crystalline, are responsible for the shape memory effect of PLA. Each of the phases has a role in this phenomenon, the crystalline phase is responsible for storing energy when the product is in a temporary shape, while the amorphous phase is responsible for the change that can be released above the T_g. Increasing the applicability of PLA in this area can be caused by modifying the material by, for example, blending it with other polymers. Such modifications affect the major disadvantages of PLA as SMP by reducing internal brittleness and T_g. Addition of PEG improved the PLA mechanical properties, and its biodegradability and non-toxicity enable the use of this material in the medical field. Cosolvent of dichloromethane and ethyl alcohol were used to dissolve a PLA/PGA blend with different mass ratio. On the polytetrafluoroethylene moulds the blend solution was cast and leaves to fully evaporate solvents in an aim to obtaining the film. The different content of PEG in the blends made it possible to find the optimal composition that improves the properties of the materials and does not lead to phase separation. An increase in the PEG content above 15% caused phase separation and hence decrease in the elongation at the break of obtained blends. The shape memory effect for these blends was investigated and the result of composition changes was assessed. In order to accurately determine the influence of the blend composition on the properties of SMP, several key factors were tested: strain temperature, tensile rate as well as recovery temperature and stretch strain. The conducted research has shown that the fixing ratio decreased accordingly with the PEG increase and this can cause an increase in the recoverable part of the deformation. The shape recovery ratio decreases during the PEG introduction. However, the use of an appropriate amount of PEG additive causes the shape recovery factor values to remain at a high level. Significant deterioration of properties was observed for samples containing more than 15% of PEG. Addition of PEG cause the plasticising effect and leads to the shape recovery at lower temperature of PLA/PEG blends than for neat PLA. However, the higher temperature of action leads to a more complete shape recovery by the blends. Shape recovery by the tested samples also depended on the environment in which the tests were conducted. The researches were conducted by a DMA test, and in the water bath some difference between results can be explained by a type of sample heating. For DMA when the air is used for increasing temperature the heating effect is worse than in the water bath. Taking into account all the obtained results and bearing in mind the differences in test conditions, it can be concluded that PLA/PEG blends with the appropriate PEG content are excellent candidates for shape memory materials (Guo et al., 2018).

SMP is developing very dynamically in the field of medical applications, but it requires searching for new materials or modifying the available ones. Blending belongs to the simpler modification methods. SMPs were obtained by blending of poly(*D,L*-lactide-*co*-ε-caprolactone) (PDLLA-PCL) and stereocomplex polylactide (sc-PLA). PDLLA-PCL was synthesised with a different ratio of *D,L*-lactide and ε-caprolactone by bulk ring-opening polymerisation. To obtain the PDLLA-PCL/sc-PLA blend, the PDLLA-PCL, poly(*D*-lactide) (PDLA) and poly(*L*-lactide)

(PLLA) were dissolved in dichloromethane and mixed. Compression moulding with prior solution casting was used for film preparation. For the shape memory effect of the blends the content of CL unit in the PDLLA-PCL was significant. Content ratio of CL in PDLLA-PCL copolymers has an influence on the PDLLA-PCL/sc-PLA blend shape recovery temperature. Changes in the PDLLA-PCL compositions also had an impact on the other properties like storage moduli, T_g, shape memory transition temperature (T_{trans}) etc. DSC results indicated that the highly random distributions of monomers in the synthesised PDLLA-PCL molecular chains were obtained. This conclusion was drawn on the basis of the presence of only one single narrow T_g on DSC curves. T_{trans} of PDLLA-PCL/sc-PLA blend can be easily tuned by controlling the CL content ratio of switching segments of PDLLA-PCL, because the increase of CL content cause the decrease of blend T_{trans}. Hence, the PDLLA-PCL/sc-PLA blend with 14% content of CL in PDLLA-PCL is able to be adjusted to fit the human body temperature. Decreasing the CL ratio could result in increase of elastic modulus, shape-recovery ratio and switching temperature, which significantly extends the possibilities of PDLLA-PCL/sc-PLA blends applications (Hashimoto et al., 2021).

SMPs in the classic form can remember two shapes: a temporary shape and the one that requires programming. However, in the polymer research area the triple-shape memory polymers or even multiple-shape memory polymers are becoming more and more popular due to the ability to remember more than two shapes. Several T_g, as well as the appearance of two or more separate molecular switches resulting from the polymer chain structure, are responsible for the possibility of shape memory effect of several shapes' occurrence. In order to develop materials with triple-shape memory the blends of poly(L-lactide-co-glycolide) (PLLGA) with oligo(butylene succinate) (OBS), oligo(butylene succinate-co-butylene citrate) (OBSC) and oligo(butylene citrate) (OBC) were prepared. PLLGA was synthesised by copolymerisation in bulk and the blends, with different amounts of bioresorbable aliphatic oligoesters with side hydroxyl groups obtained by the extrusion. All of obtained blends were investigated to check the dual shape-memory property by the tensile testing machine with a thermostated chamber. The influence of oligomers additions on the blends thermal properties shows the T_g lowering. Also, some of the blends possess only one T_g; in the blends with higher oligomer content two values are observed. However, the presence of oligomers has an influence also on the mechanical properties and shape memory behaviour. PLLGA blends with 20% of butylene citrate units and OBSC shows the triple shape memory behaviour (see Figure 5.1).

Hydrogen bonds formed between oxygen atoms of the PLLGA and side OH groups of the oligomer are responsible for memory of the first shape. Entanglement of the polymer chains and phase separation phenomena make it possible to demonstrate by the materials the memory of the second shape. It follows that blends with a high content of oligomers with side hydroxyl groups – and additionally those which are compatible – present the memory of many shapes. These types of materials can easily find application in the medical field. The material containing less than 20 wt% of OBSC with its relatively good mechanical properties and high flexibility is especially promising for manufacturing bioresorbable implants (Smola-Dmochowska et al., 2020).

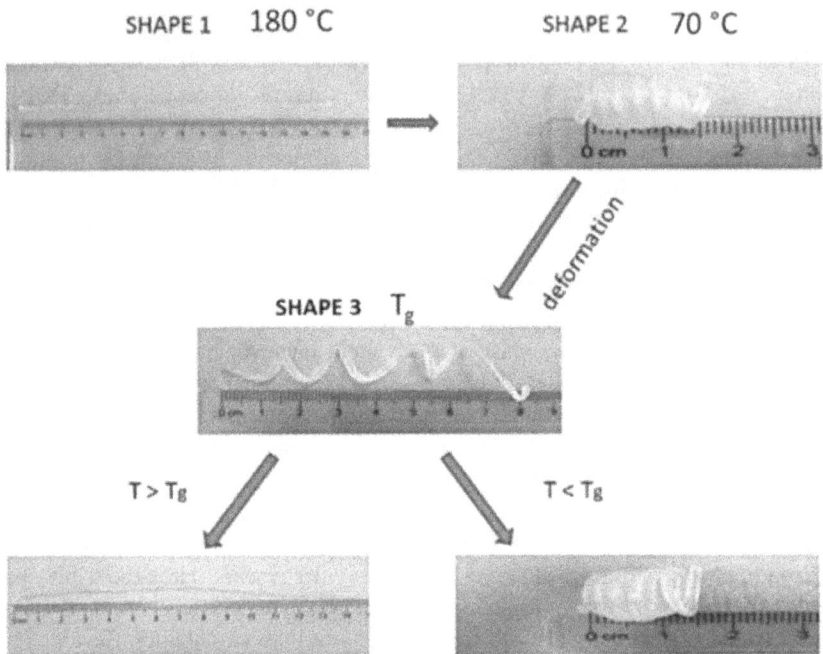

FIGURE 5.1 The triple-shape memory effect of blend PLLGA/OBSC 80/20.

Source: reprinted with permission from Smola-Dmochowska et al., Triple-shape memory behaviour of modified lactide/glycolide copolymers, *Polymers* no 12:2984 (2020). For more details, see the CC BY 4.0

5.1.6 DEGRADATION OF BIODEGRADABLE BLENDS

Blending of biodegradable polymers can cause some differences in the degradation rate, and the addition of some compatibilisers also influence on that profile. That why the investigation of degradation under different conditions is of great importance in the prediction study. The obtained results not only give an overview of the material behaviour after use and its impact on the natural environment but also on its processing and shelf life.

PBAT/PLA and PBAT/poly(3-hydroxybutyrate) (PHB) blends and the neat polymers were obtained using a twin-screw extruder, and then the films were gained from melt by the cast film die. Blends were prepared in the 50 wt%/50 wt% composition. All samples were subjected to the degradation test under laboratory composting conditions based on ASTM D5338, where the normalised carbon dioxide (CO_2) was evaluated (ASTM D5338-15(2021), 2021). Results obtained by the attenuated total reflectance FTIR (ATR-FTIR) analysis indicated the selectivity of degradation during composting for different blend phases. Both PLA and PHB polymers degrade faster in the blend in comparison to PBAT, in connection with films made from blends after incubation in compost from a porous network. The previous results together with the observed decrease in molar mass also led to a decrease in the mechanical properties of the samples during the degradation. Changing the composition of the blend has a clear effect on the degradation profile under composting conditions. By introducing changes in the composition at the design stage of the blend, one can influence the speed of its degeneration and, consequently, the time of use (Tabasi and Ajji, 2015).

Biodegradable polymer blends are available as a commercial plastic, for example the Ecovio® from BASF Company. Ecovio® is a blend of PBAT and with 25% of PLA (Musioł et al., 2022). This commercial blend as a material suitable for cosmetic packaging applications has been tested for compostability. The samples were incubated in the industrial composting conditions. The degradation test proceeded for 6 weeks and after this period of time the pits and cracks were visible on the samples surface. The influence of the sample thickness of the degradation process was also evaluated. For the thicker sample, layer-by-layer degradation was observed, and for the thinner sample, holes appeared all over the surface. Changes in the molar mass which were observed after degradation indicated the hydrolytic degradation under these conditions, which causes a molar mass decrease. The thickness of the sample also had the effect of reducing the molar mass which was more pronounced for the thicker sample (Sikorska et al., 2017).

Laboratory degradation studies in the soil at 70°C at the specified humidity level (70%) were conducted for PBAT/PLA blend. The aim of these investigation was to determine the changes in the microstructure within degradation time. Positron annihilation lifetime spectroscopic (PALS) was used to detect numerical concentration in the amorphous region and nano-scale free volume size to help the interaction understanding between properties of the blend and its microstructure. Blends were prepared by the solution casting with the different mass ratio. PALS analysis of the samples after specified degradation times show that the quantity of free volume holes decreases gradually after a slight increase at the beginning; the size after significant decrease starts to increase. These results provide information about the degradation process and its impact on properties and indicate the directions of further development of the blend modification (Chen et al., 2021). PBAT/PLA belongs to the blends used commercially; manufacturers introduce additives that improve processing but also those that reduce the price of the material. These blends are marketed in the form of foil from which disposable bags are made. Commercial additives have a significant influence on the degradation profile of PBAT/PLA blend. For example, talc, which belongs to popular commercial additives, exhibits the dichotomy of the wetting properties. This feature has a great influence on the degradation profile in the dependence of the environment's relative humidity. Commercial PBAT/PLA blends were incubated in the different environments both in the water at 70°C and in the two types of composting industrial conditions. In all media both blend components show progress in degradation over the course of the experiment but at different rate. Also, the thickness variations of the investigated foil with these same compositions have an effect on the degradation profile (Musioł et al., 2018).

PLA/PCL blends obtained by extrusion process were subjected to degradation tests in vitro under hydrolytic conditions and under composting conditions. Abiotic degradation was conducted in a phosphate buffer solution at 37°C and composting was performed according to the ISO 20200 standard. For the previously mentioned conditions of degradation tests for blends under composting, the faster degradation was observed. Additionally, the presence of PCL in the blend slows down blend degradation during composting (ISO 20200:2015; Navarro-Baena, et al., 2016).

Biodegradation under composting conditions of PCL/TPS was investigated for different blend compositions. The average temperature of the compost during processing is at a level of the melting temperature region of PCL. This causes those crystalline parts of PCL not to be a hindrance for enzymes, so the biodegradation goes rather quickly. The difference in the amount of TPS in the blend has had a significant impact on the initial stage of this process. However, in the later composting phase all of the materials, even neat PCL, achieved complete biodegradation almost simultaneously. On the other hand, in the case of biodegradation in soil, the amount of PCL was decisive for its rate. Here the molar mass of PCL and its crystallinity had a significant influence. Relatively high PCL molar mass caused the limitations in the rate of degradation in soil. The greater amount of PCL in the blend restricted enzymes' access to TPS and thus slowed down the process. The obtained results indicate that the selection of the appropriate blend composition can lead to obtaining the material for applications in specific conditions (Nevoralová et al., 2020).

To receive the better material properties for the preparation of TPS/PCL blend by extrusion, the thermoplastic starch plasticised with urea was used, and as compatibiliser PCL was grafted with maleic anhydride. PCL-g-(maleic anhydride) for blend preparation was obtained by reactive extrusion. Investigated materials were then incubated in the soil aimed to evaluate degradation rate in these conditions. TPS/PCL blends shown the good mechanical properties with stretchable characteristics for use as mulching film. The amount of TPS in the blends had an influence not only on mechanical properties but also on degradation rate. At the beginning the TPS degraded with release of urea. Changes in the composition of the blends increased the possibilities of their use, while not blocking the biodegradation of the foils obtained from them (Corrêa et al., 2022).

Prototypes cosmetic packages obtained by the three-dimensional (3D) printing from PLA/PHA filaments show the difference in the degradation profile in comparison to the packages made from only PLA. The degradation tests were conducted under laboratory and industrial composting conditions and additionally under natural weathering conditions. A weathering test is aimed at checking how the degradation of the tested material is progressing in external weather conditions, in order to evaluate its progress for the product abandoned, which did not end up in the appropriate disposal system. Additionally, the degradation of clean packaging and packaging containing a small amount of paraffin, simulating a cosmetic residue, was compared. Under laboratory composting conditions the presence of paraffin accelerated the degradation rate. The presence of PHA under these conditions also increases the degradation rate compared to pure PLA. This is because PLA degrades under simple hydrolysis, although there are known fungi that can cleave their main chain. However, the effect of the presence of paraffin outweighs the effect induced by PHA. Between the composting systems (BIODEGMA, compost pile), the degradation rate of prototype packages was different, mainly caused by the difference in the humidity. Increase in environment humidity caused the increase of PLA degradation rate. Under natural weathering conditions the presence of paraffin slowed down degradation due to its water-repellent activity. The degradation of the tested materials under atmospheric conditions is slow and is mainly influenced by temperature and humidity (Rydz et al., 2019).

One of the applications of the possibility using of biodegradable materials is the disposable packaging industry. Due to a significant amount of plastic waste, in which packaging constitutes the vast majority, it was necessary to develop materials that are safe for the environment. Biodegradable polymers are used in a wide range of environmental protection as a way to reduce waste in landfills. One such area is disposable packaging, especially for food. The use of blends of these polymers increases the possibilities of their application, from rigid packaging to flexible films, depending on the needs. Addition of active compounds with antimicrobial and antioxidant properties further increases the possibilities of blends used in packages for fresh foodstuffs susceptible to microbial spoilage. As a prototype material for food packaging with antimicrobial properties the PLA/PHBV blends with 15 wt% of plasticiser (PEG 1000) and 2% of various phenolic acids were prepared. Ferulic, p-coumaric or protocatechuic acid were added to the blend due to their antimicrobial properties. Blends with PLA/PHBV polymer ratio of 75:25 were processed by melt blending and compression moulding. Presence of the phenolic acids had an influence on the film thickness. Protocatechuic acid caused the reduction in the thickness and the ferulic acid in opposition caused the increase of this value, which may arise from established interchain forces between the antimicrobial agents and polymers. Due to the development of these materials as food packaging, which will be subjected to organic recycling after use, studies have been carried out on the degradation of these blends. Film samples were placed in the reactor with 1 kg of synthetic solid residue and the disintegration test was carried out based on the ISO 20200 standard. Biodegradation tests were conducted according the method adapted from the ISO 14855 standard (ISO 14855-1:2012). The amount of generated CO_2 from film samples was measured during aerobic biodegradation under controlled composting conditions. The presence of antimicrobial agents did not inhibit the degradation under the aforementioned conditions but only slightly extended its time. All tested materials completely disintegrated within 35 days. The induction period of biodegradation was extended by the presence of phenolic acids especially when protocatechuic acid was used. Without the phenolic acids the neat blend was biodegraded in full after 20 days of incubation in compost. The obtained results indicate that the developed materials are suitable for food packages, which will be disposed of after use by organic recycling (ISO 20200:2015; Hernández-García et al., 2022).

The application capabilities offered by the richness of compositions, additives or the method of processing blends can be additionally enriched with the introduction of fillers, obtaining composites that increase their possibilities even more. The introduction of various types of fillers into the matrices containing homopolymers and those containing blends, however, requires the selection of appropriate processing conditions and/or special additives increasing the adhesion of the matrix fibre.

5.2 COMPOSITES

Composite materials are already widely used; by definition they are made up of two or more components. Depending on the size of the particles, they can be divided into macro-, micro- and nanocomposites. What distinguishes composites from other materials is that their components do not dissolve in each other and therefore form separate phases. The reinforcing phase is usually fibres or particles that are suspended in a so-called matrix. Due to their structure, composites allow for mass reduction without losing the strength and stiffness of the products obtained after replacing the conventional metal alloys used so far. These properties mean that even an increase in price in relation to classic materials is not a problem, because in addition, composites often also have improved resistance to impact and corrosion as well as having improved thermal conductivity. Not only the presence of the filler itself but also its quantity, size, orientation and shape have a huge impact on the properties of composites, including stiffness. For composites in which the matrix is a polymer, three types are distinguished depending on the type of filling: nanocomposites and composites reinforced with short and continuous fibres. Due to differences between the matrix and the filler in the modulus and strength area, local stress concentrations may develop due to internal mismatches (Guedes and Xavier, 2013). An important element influencing the properties of composites is the connection of the matrix with the filler. The place where the matrix is in contact with the filler is exposed to stress concentration caused by external load. Such conditions can lead to detachment at the interface due to the weakening of the interaction between the filler and the matrix. One of the ways to modify composites is to improve these interactions; they can be carried out by improving or creating physical and chemical bonds or mechanical keying. Van der Waals forces play a significant role in the physical nature of the interphase bond; their increase is associated with an increase in the wettability of the filler by the matrix, which, on the other hand, can be improved by modifying the surface of the fillers. The surface can be chemically modified one can increase its roughness, leading to a strong mechanical keying, increasing the diffusion of the matrix (Radecka et al., 2016).

The definition of biocomposites covers a wide range of hybrid materials with no less than two phases, and both matrices and fillers are biodegradable, or at least one of them is (Figure 5.2; Kuciel and Rydarowski, 2012; Pegoretti et al., 2020).

The pursuit of a circular economy prompts us to focus on biodegradable composites containing a biodegradable matrix and a natural filler. Composites of biodegradable materials have been investigated for many years. It is mainly due to the reduction of the price of biodegradable polymers through the addition of filler – but also the modification of the properties of the obtained composites. It is significant to improve the efficiency of production of new materials, at the same time remembering about reducing the risk of disturbing the balance of the natural environment. Therefore, the research focuses not only on the development of new composites but also on the

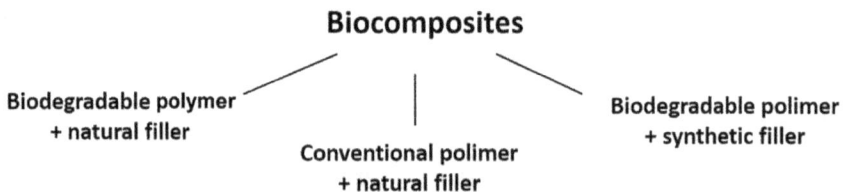

Biocomposites

Biodegradable polymer
+ natural filler

Conventional polimer
+ natural filler

Biodegradable polimer
+ synthetic filler

FIGURE 5.2 General classification of biocomposites.

processing technology, sources of raw materials acquisition and recycling of products made from composites. Appropriate selection of the matrix, filler or connecting materials as well as modification of the composition or production processes give a huge scale of possibilities to control the properties of biocomposites. In addition to economy, ecology is also becoming more and more important when choosing materials or finished products. As a result, biodegradable composites are increasingly appearing in subsequent industry sectors (Kuciel and Rydarowski, 2012).

Biodegradable polymers are becoming more and more popular. Their applications reach much further than the most popular areas such as medicine or the packaging industry. Due to their properties and price, they are often used in niche fields. The best known and used biodegradable polymers include: PHA, PLA, PCL, PVA, PBAT or TPS. These polymers are also among the most frequently chosen matrices in the production of composites. Most often in composites where a biodegradable polymer is used as a matrix, natural fibres are used as a filling. Apart from preserving the biodegradability of the composite obtained, natural fillers also have many positive features that add value to the final materials. However, the relatively low temperature at which natural fibres can be processed introduces a lot of limitations in the production of composites. Additionally, their high hydrophilicity causes problems in connection with the hydrophobic matrix, hence the aforementioned necessity to modify the fibres, described in detail in the previous work (Jurczyk et al., 2017).

Many reviews have already been written in this area (Girijappa et al., 2019; Obielodan et al., 2019; Aaliyah et al., 2021; Bari et al., 2021) so the following section will be limited to research over the past few years.

5.2.1 Preparation of Biodegradable Composites

The 3D printing method, which is the most popular method of obtaining personalised products, has also found its place in the production of objects from biodegradable composites. The main problem when using composites is to obtain an appropriate filament suitable for this method. PLA is one of the most popular biodegradable polymers, so it is not surprising that research is conducted also into its modifications in the field of producing filaments for 3D printing. PLA with organosolv lignin was mechanically mixed in order to obtain the homogeneous material, and then the extrusion was applied in filament production. Additionally, for comparison part of the composite was filled with the lignin which was treated with 3-aminopropyltriethoxysilane to modify its functional groups. This procedure was to enhance surface bonding and miscibility of composites. Fused filament fabrication (FFF) 3D printing applications were used to produce all tensile samples from obtained filaments. The results showed that, even with high filler loading, the composites were viable to melt extrusion. During the FFF-based 3D printing process high shear rate occurs, however, the developed composites can withstand it. Application of modified organosolv lignin does not significantly affect the processing of composites but improves the mechanical properties of the products obtained from them. Large possibilities of applications of the obtained composites result from the large number of available compositions (Singh et al., 2020).

Orthopaedic medicine is constantly looking for materials suitable for use as implants. Most calcium phosphate and calcium carbonate-based materials are too brittle and unsuitable for this type of application; they are used in 3D printed scaffolds, coatings or drug delivery. However, the composites with hydroxyapatite (HAP) use the filler's ability to provide osteoconductivity and the ability to bind bone. HAP is a material whose chemical composition is close to the natural apatite occurring in bone tissue. The possibility of using it in composites with HAP as a matrix for the PLA means the obtained materials possess suitable absorbability and easy processing. Here, achieving a high degree of homogeneity is essential but requires more complex processing methods. HAP nanoparticles (NP)s were used to obtain PLA composites for orthopaedic purposes. The method of wet chemical precipitation was utilised for preparation of HAP NPs from calcium oxide, orthophosphoric acid solution and water. Appropriate granules were obtained by the method of spray drying. Composites preparation required dissolving PLA in dichloromethane and suspension of HAP NPs in polymer solution. Then high-pressure impregnation, during which the PLA solution was infused into porous HAP NPs granules, was apply. Obtained composites were next formed by high pressure in different ways, by high-pressure pressing without warming, isostatic pressing with increase of temperature and by the mixing of those two methods. The two-step method, preparation of PLA/HAP NPs composite granules followed by consolidation of these granules, has been successfully applied to the receipt of homogeneous hybrid composite, with a high degree of densification. Among the composite forming methods used, the best results were obtained when both methods were combined. Material with 80 wt% of HAP NPs possesses the homogeneous structure and between PLA and filler good adhesion was observed. The proposed solution can be successfully used for the production of composites with a high content of the ceramic phase at temperatures below 200°C (Pietrzykowska et al., 2020).

The production of composites from biodegradable synthetic polymers with silk fibroin powder can be considered a new source of ecological materials. However, this requires not only mixing materials but also appropriate modification. Initially, the PBAT/PLA blend was prepared by the twin-screw extrusion method, and then, in a micro-extruder/injection moulding machine, it was mixed with silk powder of various contents. Composites with 10% silk powder content were selected for further modification because they showed high notch toughness. Epoxy functional oligomeric acrylic resins (Joncryl® ADR-4368) were used as compatibilisers in samples without and with 10 wt% of silk powder. The presence of chain extender in the samples with silk powder did not lower agglomerate formation and only improved compatibility between PBAT and PLA polymers. The occurrence of the compatibility process was confirmed by the observed enhancement of viscosity and melt storage modulus (Nakayama et al., 2018).

One of the key elements determining the profitability of the production of composites is the availability and price of the filler, therefore in various regions of the world different types of fillers will gain popularity depending on their availability. However, the possibility of obtaining large amounts of biodegradable polymers definitely increases the attractiveness of the region for the production of new materials. In Brazil, the fruit of the babassu palm (*Orbignya speciosa*) is widely available as a natural source of many valuable products, as well as starch, which is used in the production of thermoplastic starch. In addition to starch from babassu coconut it can also be obtained from the fibres that are used as reinforcement in composites. In order to choose the best possible way to obtain composites that meet specific expectations, babassu fibres have been modified. Three types of fibres were obtained: washed fibres (WF), bleached fibres (BF) and alkaline fibres (AF). TPS was processed from babassu by addition of glycerol and by using the thermopressure method. For the production of composites, materials containing 90% of the matrix and 10% of fillers were used. The ready-made test samples were obtained with the use of a hydraulic press. For comparison purposes, samples without filler were also made. Natural fibres contain hemicellulose, lignin and waxes; removing them by chemical treatment allows access to the internal fibrous structure. Each of the fibre modifications used introduced significant changes; however, in the case of bleaching, a reduction in moisture absorption was observed. The surface of the samples made of the pure TPS matrix was smooth and uniform. However, in the case of composites, the voids between the fibre and the matrix were observed. However, for composites containing bleached fibre (TPSBF) and alkalinised fibre (TPSAF) better adhesion was achieved in comparison to composites with washed fibre (TPSWF). The improvement of the interfacial interactions between the matrix and the fibres as a result of the modification of the latter also influenced the mechanical properties of the obtained composites. For composites, compared to pure TPS, higher values of tensile strength and modulus of elasticity were obtained, which makes the composites more rigid. As TPSBF composites contain fibres that have been significantly changed as a result of modification, the mechanical properties of these composites are better than for TPSAF or TPSWF. Composites obtained from TPS and babassu fibres can be widely used due to their properties and the possibility of obtaining both the matrix and fibres from one plant (Moura et al., 2021).

The enzymatic treatment can be used to modify natural fibres; for this, palm trunk waste was used, which, after modification, was then introduced as a filling for a composite with PBS as a matrix. During the development of the tested materials, several methods of fibre modification were used to improve the properties of the composites. The fibres themselves were preliminarily assessed after various types of modifications, compared with the raw ones. Then, composites with raw fibres as well as those showing the best properties after treating were made. The fibres selected for processing were chemically treated and treated with an enzyme mixture (xylanase and pectinase). The composites were prepared using an extruder; then they were cut into granules, and standard test specimens were obtained by injection moulding. The optimisation of the fibre modification process has led to a filling with better properties, and the procedure has been further simplified and shortened. The developed fibre materials after the modification showed an improvement in mechanical properties (Khlif et al., 2022).

In the aviation industry, composite laminates play an important role in the production of lightweight structures. This is mainly due to the high strength-to-weight ratio. Observation of the cross-section of the composites indicates a random distribution of the fibres. Cross-ply laminates are a type of material in which there are any number of layers, each of which has a fibre orientation of 0° or 90°

(Guedes and Xavier, 2013). Cross-ply layup of [0/90/0/90/0/90] was used in the PLA-based composites. The preparation of this type of composites required several steps. Initially, the PLA was mixed with the nano-hydroxyapatite in an internal mixer. Materials with various nano-HAP contents from 0 to 40% were prepared. A hot press machine was used to process these materials into films which served as the matrix for the composites. After hot pressing, the films were cold pressed. A conventional film-stacking method was used to produce the laminates from obtained films and unidirectional flax fabrics, adding them alternately. For consolidation the film and fabric were hot pressed and then prepared composite plates were cold pressed. The samples were characterised and the flame behaviours were investigated in accordance with the currently withdrawn ISO 1210 standard (ISO 1210:1992). For the obtained composites with an increased amount of nano-HAP, a reduction in the burning rate and prevention of flame development were observed. However, the presence of nano-HAP molecules in the tested composites caused the reduction of the mechanical properties, which may result from the agglomeration of nanoparticles, especially in materials with a higher content of them. For composites, an increase in water absorption was also observed, resulting from poor adhesion between the flax fibres and the matrix. The presence of nano-HAP also contributes to the increase in this property. The optimisation of the composition of the tested composites must be directly related to the expectations regarding these materials. It is necessary to find a compromise between fire resistance and mechanical properties or moisture absorption (Khalili et al., 2019).

The selection of appropriate processing conditions also has a significant impact on the obtained materials, therefore conducting comparative tests of materials made with the use of various methods is of great importance. For this purpose, two methods of obtaining PLA and PBAT composites containing a copy paper as a filler were applied. The use of this type of paper as a reinforcement for composites results from the need to solve the problems arising during the recycling of paper waste and to reduce its amount in landfills. Optimising the processing process can be of great importance in this case. In the first case, a miniature press with cutting edges was used. A dried semi-finished material, wrapped in Teflon foil, was placed between the work plates; it was heated and then the appropriate pressure was applied. The resulting composite was cooled between two aluminium plates. For the second method, a hot press was used where the layers of matrix and filler were stacked and placed in a preheated steel mould. Form with filling was introduced to the hot press and then without the pressure was molten for an established time. Then adequate pressure was applied for a specified period of time followed by the final step of consolidation. After the entire process was completed; the mould was placed in a cooling press. Both methods produced composites with different combinations of matrices as well as with different contents of copy paper. The diversification of the methods of obtaining composites mainly influenced the compaction of the material. At low compaction, a decrease in tensile properties was observed. However, in the case of using the PBAT matrix, an increase in impact strength was observed for composites with low density, but it did not improve this parameter for composites with PLA. The results of the conducted research indicate the possibilities of using paper waste in the production of composites for various applications (Graupnera et al., 2019).

Obtaining the film by pouring the prepared solution on the Petri dish and then peeling the film off after drying is one of the fairly simple methods of obtaining biodegradable composites. A simple method increases the possibilities of application and at the same time gives great opportunities for easy modification of the obtained materials. Composites containing PVA, carboxymethyl cellulose (CMC) and L-alanine surface modified CuO nanorods were made using this method. To obtain CuO nanorods, an aqueous extract of ash from mango skins was used, which was then mixed with copper nitrate trihydrate ($Cu(NO_3)_2 \cdot 3H_2O$). The resulting black precipitate indicates the formation of CuO nanorods. For the modification of CuO nanorods L-alanine amino acid and microwave strategy were utilised. The preparation of composites consisted of mixing PVA and CMC solutions and then adding the modified CuO nanorods dispersed in water. The whole was then sonicated in order to obtain a homogeneous solution. After that the solution was cast and let to dry. For comparison, films without the addition of CMC and CuO nanorods were made using the same method. During the characterisation of the films, a homogeneous dispersion of CuO nanorods in composites was demonstrated. The developed composites showed excellent antimicrobial and antifungal properties in relation to the tested microorganisms and fungi (*Escherichia coli, Staphylococcus aureus, Candida tropicalis* and *Candida albicans*). MTT assay was used to test the cytotoxicity of the investigated material. The obtained results indicated that composites showed a non-toxicity effect. Antimicrobial and antifungal properties along with the improvement of mechanical properties indicate the possibility of using the obtained materials as food packaging but also in medicine as wound dressings (Amaregouda et al., 2022).

5.2.2 Application of Biodegradable Composites

Due to the great possibilities of using various polymer matrices and different types of fillers, the possibilities of application of biodegradable composites are very wide. The 3D printing method has found a special place in fields related to medicine as an instrument for the production of biomedical tools and devices. The developed composite made of PLA and kenaf cellulose fibres is a material that meets the requirements for the production of certain medical devices. It can also be used in personalised prostheses and in the textile industry. Kenaf fibres before mixing with PLA were exposed to tetraethyl orthosilicate to enhance compatibility between them and the polymer. Processability for PLA/kenaf composites was improved by the incorporation of poly(ethylene glycol) as a plasticiser. The conducted modifications led to obtaining composites with better mechanical properties compared to pure PLA. The PLA/kenaf composite was successfully processed into a filament, which was then used in 3D printing (Aumnate et al., 2021).

The fillers present in composites may have specific properties that enable the use of these materials in niche applications. Addition of biochar to the polymer matrix can have the influence on the composite electrostatic properties. This feature can be used in the production of materials for personal protective equipment in the places where the explosion mixture can be present. Also, in the precise measurement and assembly laboratories, limitation of electrostatic discharges lead to less damage-sensitive electronic parts and improving the electrical measurements. In that kind of laboratory this type of materials can be applied as antistatic flooring. Electrostatic discharges can also lead to serious product quality problems and are therefore important for packaging materials. Antistatic packaging reduces the sticking of dust, which makes the packaging more inviting for consumers. Composites of Ecovio® F Mulch C2311 (PBAT/PLA) with biochar in different amounts were prepared in order to check the influence of the amount of filler on the electrostatic properties of the materials obtained. The results indicated increase of the surface resistivity reduction with the increase of biochar content. However, the obtained values did not enable these materials to be used in the places where the explosion of the mixture may occur. This opens up the possibility of applications in other areas as well as the search for optimal matrices for these composites (Musioł et al., 2022).

Biodiesel-residual glycerine and cassava peels were mixed with bentonite or zeolite to obtain the composites. Glycerine was used as a plasticiser to obtain the thermoplastic starch from cassava peels. The aim of these studies was to receive the materials for treating the soil contaminated by gasoline. Composites prepared by the extrusion contain 5 and 10% of fillers. Addition of the fillers was intended to increase the surface roughness. This feature is responsible for decreasing values of contact angle and support for the growth of microorganisms. Materials having this ability can be successfully used in soil bioremediation techniques, in which the microorganism's degrading pollutant is multiplicate. The use of biodegradable polymers as matrices in such composites is an added value, additionally being a source of nutrients for multiplying microorganisms. TPS/bentonite and TPS/zeolite composites show the growth of surface roughness in comparison to neat TPS and thus an increase in the amount of microorganisms in the soil to be cleaned of toxic substances. The cultivation of BTEX (benzene, toluene, ethylbenzene and xylene) degrading microorganisms with the use of the developed composites is an interesting alternative in soil bioremediation techniques. These composites are prepared from industrial waste like cassava peels and biodiesel-residual glycerine which significantly increases the value of the results obtained (Silva et al., 2022).

Environmental protection requires a sensible approach to natural resources but also to waste generated in the course of human activity. One of the biggest problems is still various types of packaging or disposable products, which are produced in huge amounts and, unfortunately, become waste very quickly. The use of biodegradable polymers for the production of packaging and disposable products, along with the use of an appropriate disposal method, can significantly contribute to improving the current situation in landfills. Additives for biodegradable polymers used in the development of new packaging materials may, in addition to lowering the price, introduce additional properties to composites. Introduction of titanium dioxide (TiO$_2$) nanoparticles to the PBAT/TPS blend obtains composites with enhanced ethylene-scavenging activity. The composites were prepared in two stages; at first the TPS was mixed with TiO$_2$ by a twin-screw extruder, and then extrudates in the form of pellets were mixed with PBAT pellets. The obtained mixture was processed also by twin-screw extruder, and then the materials were blown by a single-screw blown-film extruder. The effect of the presence of the additive, depending on its amount, on the mechanical properties of composites was noted. One percent of TiO$_2$ made it possible to obtain the homogenous matrices while increase in nanofiller contend caused the formation of a non-homogeneous mixture. This is indicated, inter alia, by a marked increase of tensile strength for the composite with 1% of TiO_2 and then its lowering with further nanofiller increase. The films obtained from the tested composites were used as packaging for bananas. By examining of these packages over time, it was found that the fruit darkening time was delayed in the packages containing the addition of TiO_2. The aforementioned research shows that these composites can be used in the packaging industry to a large extent (Phothisarattana et al., 2021).

Most often, the subject of biodegradable composites is associated with the packaging industry. This is mainly due to the huge amounts of packaging produced, the lifetime of which is usually very short. This has a particular impact on the development of the single-use packaging industry. This type of packaging should meet the biodegradability requirements but also allow for the use of waste as filling and thus reduce the amount of waste in landfills. Scientists went in this direction by developing composites from denim waste with corn starch as a matrix. The idea to use denim resulted from the richness of cellulose it contained and the possibility of its reuse. Different sizes of scraps of denim have been utilised due to the use of waste to make the composites. A smoother surface and higher tensile strength were observed for starch/denim composites containing smaller chunks of filler. The composite prepared from semi-gelatinised corn starch showed good structural compatibility and a nonwoven-like, completely elastic structure. The moisture and mechanical properties of the 50/50 starch/denim composites may indicate their application in single-use packaging (Haque and Naebe, 2022).

The utilisation of waste coconut shells as filling for biodegradable composites can also be used in the production of packaging for short-lived products or disposable tableware. By applying mechanical, chemical and physical methods, i.e. ball milling, acid hydrolysis and ultrasonication, respectively, cellulose nanofibres were obtained from coconut shells. After removing the non-cellulosic material from the fibres, they were bonded to a matrix which was PVA. In order to increase the application possibilities, linseed oil and lemon oil were added to the developed composites to improve the antioxidant properties. The obtained material can also find its place in the field of active packaging due to its antimicrobial properties, which it exhibits in relation to food-borne pathogens. In addition, the introduction of the filler led to an improvement in thermal and mechanical properties in relation to the pure PVA film, and the film also gained a hydrophobic character. These improvements meant that the obtained material could be used as an alternative to conventional packaging materials, significantly reducing the amount of packaging waste in landfills (Arun et al., 2022).

The researchers also confirmed the possibility of using composites in the packaging industry in which the matrices will be chitosan. Composites were processed by the casting method. Chitosan solution was prepared and then the slurry of micro ramie fibre and micro lignin were added with various amounts. The resulting mixtures were then poured off and the solvent contained therein evaporated. Developed films were evaluated as far as mechanical and chemical properties. An increase in tensile strength was observed after the addition of 20 wt% ramie fibres, while the addition of 20 wt% lignin caused an increase in water resistance and anti-oxidative activity. On the other hand, both fillers increased the thermal stability of the obtained composites. The tested composites were used as packaging for chicken breast and cherry tomatoes. In the case of packing meat, a smaller growth of microorganisms was observed than in the case of polyethylene film. Composites containing 10 wt% ramie fibres and 10 wt% lignin showed the best properties for packaging applications (Ji et al., 2022).

Biodegradable composites with the ability to conduct electricity can be used as micropatterns in interconnects. To obtain composites meeting the requirements in this area, PCL was used as a matrix and microparticles of iron (Fe) as a filling. Fe materials used at 40 vol% for biodegradable electrical connections have been successfully micropatterned in daisy-chain structures. The conducted degradation studies have shown the possibility of using in implantable electrical systems the micropatterned interconnects made from Fe-PCL (Zhang et al., 2019).

More and more often, in the area of medical care, information technology is included; this connection appears most often in the field of personalised electronics in the form of disposable foils or patches on the skin. However, due to their nature, these products will become waste quickly and in large quantities, hence the interest in developing conductive pastes based on biodegradable polymers. Biodegradable medical devices with controlled dissolution kinetics in bioliquids are an interesting research path in the field of biodegradable electronics, because such devices will not have to be removed from the body during surgical procedures. Composites of biodegradable polymers with metal microparticles allow for a wide range of materials for the production of conductive pastes. A paste was developed in which the matrix is PBAT and the filler is molybdenum (Mo) microparticles. To obtain a suitable material, PBAT was dissolved in chloroform to obtain a viscous solution (mass ratio of PBAT/chloroform = 1/3.75). Then, after adding the Mo particles, they were mechanically mixed. For the production of electronic devices, it was also necessary to add tetraglycol which improves the conductivity by acting as a lubricant and increases the dispersion of Mo particles in the entire paste volume. Final products made of the tested pastes can be made using the following methods: laser cutting or direct injection printing. The utilisation of biodegradable PBAT/Mo composites showed that perfect conductivity and extensibility were achieved. The introduction of tetraglycol resulted in additional improvement of these properties (Kim et al., 2022).

The most spectacular event in the field of biodegradable composites was the presentation of a car made of this type of material. An environmentally friendly electric car was presented at the Shell Eco-marathon in May 2017. A car called "Lina" was developed by TU/Ecomotive (student team based at the Eindhoven University of Technology). Plant-based materials were used for preparation of the car body. In these composites, sugar-beet-based PLA plastic was utilised as a matrix and flax as a filling. The capabilities of EconCore technology resulted in honeycomb PLA structure placed between two sheets of flax fibre composite. Owing to such a structure, the obtained composite exhibits properties similar to glass fibre in terms of the strength-to-weight ratio. "Lina", having an electric drive, can reach a maximum speed of 80 km/h; additionally it is equipped with technologically advanced functions increasing the usability of this car (Livekindly; Moore, 2017).

5.2.3 DEGRADATION OF BIODEGRADABLE COMPOSITES

The progress of degradation of biodegradable materials is influenced by many factors, from material composition to external conditions. In order for a material to receive a certificate of suitability for disposal under organic recycling, it must undergo a series of tests. Each newly developed material must pass these tests, but the thickness of the final product is also taken into account and tests must be carried out for each of the possible use thicknesses. In the case of composites, a different amount of filler as well as the types or amount of compatibilisers used may have a significant impact on the rate of their degradation.

Plant-based filler obtained from pinecone particles was used in composites where PCL was used as matrix and graphene as a compatibiliser. Materials with different filler and compatibiliser content were used for the tests. Degradation was conducted according to the ISO 20200 standard in the commercialised compost. The conducted research shows that an increase in the filler content causes an increase in the rate of composite disintegration. Materials with a decreasing amount of graphene show a similar direction of increase in disintegration. According to the obtained results, this type of filler together with the compatibiliser used in various mass ratios can be applied for products that are subject to organic recycling after use (ISO 20200:2015; Jha et al., 2021).

Influence on the biochar content on the degradation profile of biodegradable composites was investigated in abiotic and biotic environments under laboratory conditions. Commercial biodegradable PBAT/PLA blend (Ecovio®) was used as a matrix. Composites with different content of biochar were obtained by the extrusion and injection moulding preceded by mixing aimed to obtain the homogenous dry-blend. The degradation test under composting conditions was conducted in a Micro-Oxymax respirometer with the compost from industrial facilities. The abiotic degradation was carried out in accordance with International Standard: ISO 13781 in demineralised water at $70°C$ ($\pm$ $0.5°C$) (ISO 13781:2017). The biochar content had an influence only on degradation in water; after composting no difference in the degradation rate between various compositions was observed. During incubation in water the surface erosion was less visible for neat matrix than for composites. Also, the thermal properties differed from sample to sample, pointing to faster degradation of PLA in the neat PBAT/PLA blend determined by the disappearance of PLA T_g value. The obtained results indicate no significant influence of the filler presence on the degradation progress (Musioł et al., 2022).

Cork is a very interesting versatile natural origins material. Its properties make it an interesting alternative as a composites filler. Not only is it of natural origin, but also it possesses buoyant ability, low mass, density, impermeability to gases and liquids and good thermal, electrical properties. Poly(3-hydroxybutyrate-co-4-hydroxybutyrate) (P3HB4HB)/cork composites were investigated in term of degradation rate in different conditions. Degradation under industrial composting conditions was produced in two composting systems: BIODEGMA and KNERR. These systems are among the popular methods of composting organic waste. The main feature of the BIODEGMA system is that one of compostable fractions is an organic component of mixed municipal waste. Comparative studies were carried out in an abiotic environment and under laboratory composting conditions. In addition to the impact on the degradation progress of the obtained materials, a significant impact on the properties of the developed composites was also noted. The presence of the filler slowed down the degradation progress of the composites in relation to the pure matrix. The specific properties of cork limit the penetration of water into the composites, which reduces the share of hydrolysis in the degradation process. Using this type of filler in different amounts can lead to products with a certain life expectancy (Jurczyk et al., 2019).

A buffer medium with the diastase enzyme constituted a degradation study environment for composites made of PLA and coconut shell. To improve the composites properties the filler was modified with various methods of chemical treatment by salination with 3-aminopropyltriethoxysilane and by the maleic acid treated. Modified and unmodified coconut shell was used to obtain composites in which the matrix was PLA in order to evaluate the influence of filler in various forms on the degradation process. Increase in the interfacial adhesion between the polymer and filler observed for composites with modified filler caused the decrease of its biodegradation rate regardless of the modification method. Presence of the lactic acid in the PLA/coconut shell composites after degradation was confirmed by FTIR analysis and ultraviolet-visible spectroscopy analysis. The obtained results show a clear influence of filler modification on the degradation progress of composites (Tanjung et al., 2018).

Fillers in the form of fibres for composites in which the matrix is PLA, among others, are designed to strengthen the obtained materials. However, any interference with the material can have a clear impact on the degradation process. The sensitivity of fibres to degradation factors, especially water absorption, is of great importance in biodegradation processes involving microorganisms but also in the case of ordinary hydrolysis. PLA-based composites were filled with agave and coir waste fibres by compression moulding with (glycidyl methacrylate)-g-PLA as compatibiliser. After accelerated weathering the samples were subjected to abiotic and biotic degradation. The presence of fibres had a significant impact on the degradation progress, especially due to the water absorption of the filling, which led to a slower degradation (Campo et al., 2021).

Soil degradation tests conducted on biodegradable composites show the possibility of their use in agriculture but also inform about the behaviour of these materials during disposal in soil conditions. The samples of composites containing the millet husk fibre as filler and PLA as matrix were prepared in the aim to study degradation in soil. The millet husk fibres were milled but treated first with NaOH or by mercerisation and after those fibres were blended with PLA via Brabender. Obtained composites with different compositions were buried in soil and placed in special containers. The presence of filler greatly accelerates degradation in the soil. Increasing the amount of millet husk fibre increases the degradation progress under these conditions. The obtained test results indicate the possibility of using millet husks as a filler in composites with PLA (Hammajam et al., 2019).

The creation of composites is one of the forms of obtaining new biodegradable materials, in which the fibres can be modified to increase adhesion. However, the modification of the known and commonly used biodegradable polymer applied as a matrix significantly increases the application possibilities of biocomposites. For these newly obtained materials, it is necessary to conduct degradation studies under different conditions in order to assess the possibility of their use as well as subjecting them to various forms of disposal. Conducting soil degradation studies according to specific standards is commonly used in assessing material degradation under certain conditions. PLA resin thermoset was obtained by chemical reaction. *L*-Lactide was received by dehydration of *L*-lactic acid and then reacted with pentaerythritol which led to the star-shaped PLA with four arms. At the end the four-arm star-shaped PLA was capped with methacrylic anhydride which was used as a terminal functional reagent. The obtained four-arm star-shaped PLA after blending with di-*tert*-butyl peroxide was evenly applied to the ramie fabrics surface. The samples were then processed by hot-pressing machine. Both materials with and without ramie fabrics were subjected to the degradation test in the controlled soil conditions according to the GB/T19277. 1-2011 Chinese standard adapted from ISO 14855-1:2005, currently ISO 14855-1:2012 (GB/T19277.1-2011, 2011; ISO 14855-1:2012, 2012). Preliminary studies showed that thermoset PLA possesses better thermal stability than linear PLA, which follows from the crosslinking structure. Incubation in the soil of composite and pure matrices shows the decrease of degradation rate for thermoset PLA/ramie in comparison to thermoset PLA. During the degradation test in soil the CO_2 emissions and soil pH value were evaluated. Obtained results indicated that degradation products are not harmful to the environment. Composites of thermoset PLA/ramie possess good biodegradability and can find application in a wide range of industrial areas (He et al., 2021).

From the point of view of new materials introduced to the market, solutions generalising the degradation tests or extrapolating the results on the basis of preliminary test results seem interesting. The development of such a methodology requires taking into account a number of parameters having an impact on the degradation process. Taking into account the initial morphological and chemical features of composites, a model can be developed which allows one to see the final extent but also the triggering of degradation pathways. Composites were prepared with PLA as a matrix and *Posidonia Oceanica* flour as filler in various proportions. The internal batch mixer was used to produce the composites by melt compounding. In the next step the materials after cooling and being ground into pellets were compression moulded. Two different amounts and two different granulometries of filler were used. Research on the degradation of PLA/*Posidonia Oceanica* composites has led to the development of the morpho-chemical parameter as a simple indicator of chemical stability. Influence examination of PLA/*Posidonia Oceanica* intraphase and formulation on the behaviour during hydrolytic degradation was investigated. The tests were carried out at two different pHs (pH = 7.4 and pH = 10). During the investigation the residual mass degree, absorption of the solution, morphology and molar mass changes of the samples were evaluated. The obtained test results made it possible to develop the morphochemical parameter (*MCP*, Equation 2).

$$MCP = M_v \times intraphase \qquad \text{(Eq. 2)}$$

Where: M_v – viscosimetric molecular mass [kg/mol], *intraphase* – dimensionless parameter (value from 0 to 1) dependent on the ability of macromolecules to enter the empty channels of fillers.

For all samples, the time with a 5% mass loss was plotted as a function of their morphochemical parameters. A satisfactory linear fit was obtained in both buffers. This parameter can be treated as an indicator of the chemical stability of the systems: the higher it is, the longer the time to start losing mass. It can also be used to describe the final state of degradation in the systems tested; however, in this case it is less accurate. The type of filler used has a porous structure which, when creating composites, is filled with PLA macromolecules; such a structure allows the solution to be absorbed inside the composite. During the degradation tests, the mass loss of the samples after specific degradation times was assessed and correlated with the key characteristics of the materials, i.e., the initial molar mass of the polymer and the porosity of the samples. Both factors are related to the size and content of the filler. The conducted research gives a picture of the behaviour of PLA/*Posidonia Oceanica* composites during degradation in buffers, and the developed parameter may facilitate the assessment of degradation of similar materials under these conditions (Scaffaro et al., 2021).

5.3   CONCLUSION

Biodegradable polymers are an alternative to conventional materials to an increasing extent. Research related to the production of new biodegradable materials is carried out on an ongoing basis. Scientists indicate the possibilities of simple methods of obtaining new materials using the mixing process. Modifications of the process itself as well as the use of various additives significantly increase the application possibilities of the blends obtained, introducing significant changes in their properties. The changes that occur in the materials as a result of the formation of the blend also affect their biodegradability, changing the rate of its progress. That is why it is so important that each modified material is subjected to degradation tests under various conditions, despite the fact that each polymer has this property separately. The same is the case with composites where, in addition to the polymer matrix, various fillers are used, which can also significantly affect the properties of the final materials. The main goal of developing composites was to reduce the price and improve the mechanical properties. Large resources of waste natural fibres that can fill composites cause a significant reduction in the price of the final product. However, obtaining the appropriate properties requires a number of tests and modifications. Also, the introduction of new materials to the market, whether blends or composites, requires a number of activities and large-scale research.

REFERENCES

Aaliyah, B., Sunooj, K.V., and Lacknerhttps, M. 2021. Biopolymer composites: A review. *Int. J. Biobased Plast.* 3: 40–84.

Amaregouda, Y., Kamanna, K., Gasti, T., and Kumbar, V. 2022. Enhanced functional properties of biodegradable polyvinyl alcohol/carboxymethyl cellulose (PVA/CMC) composite films reinforced with *L*-alanine surface modified CuO nanorods. *J. Polym. Environ.* 30: 2559–2578.

Andersen, B. 2004. *Investigations on Environmental Stress Cracking Resistance of LDPE/EVA Blends.* Martin-Luther-Universität Halle-Wittenberg dissertation. https://sundoc.bibliothek.uni-halle.de/diss-online/04/04H140 (accessed August 17, 2021).

Arun, R., Shruthy, R., Preetha, R., and Sreejit, V. 2022. Biodegradable nano composite reinforced with cellulose nano fiber from coconut industry waste for replacing synthetic plastic food packaging. *Chemosphere* 291: 132786.

ASTM D5338-15(2021). 2021. *Standard Test Method for Determining Aerobic Biodegradation of Plastic Materials under Controlled Composting Conditions, Incorporating Thermophilic Temperatures*. New York: ASTM International.

Aumnate, Ch., Soatthiyanon, N., Makmoon, T., and Potiyaraj, P. 2021. Polylactic acid/kenaf cellulose biocomposite filaments for melt extrusion based-3D printing. *Cellulose* 28: 8509–8525.

Bari, E., Sistani, A., Morrell, J.J., Pizzi, A., Akbari, M.R., and Ribera, J. 2021. Current strategies for the production of sustainable biopolymer composites. *Polymers* 13: 2878.

Bhattacharya, S., Hailstone, R., and Lewis, C.L. 2020. Thermoplastic blend exhibiting shape memory-assisted self- healing functionality. *ACS Appl. Mater. Interfaces* 12: 46733–46742.

Campo, A.S.M., Robledo-Ortíz, J.R., Arellano, M., and Perez-Fonseca, A.A. 2021. Influence of agro-industrial wastes over the abiotic and composting degradation of polylactic acid biocomposites. *J. Compos. Mater.* 56: 43–56.

Chen, W., Qi, Ch., Li, Y., and Tao, H. 2021. The degradation investigation of biodegradable PLA/PBAT blend: Thermal stability, mechanical properties and PALS analysis. *Radiat. Phys. Chem.* 180: 109239.

Claro, P.I.C., Neto, A.R.S., Bibbo, A.C.C., Mattoso, L.H.C., Bastos, M.S.R., and Marconcini, J.M. 2016. Biodegradable blends with potential use in packaging: A comparison of PLA/chitosan and PLA/cellulose acetate films. *J. Polym. Environ.* 24: 363–371.

Corrêa, A.C., de Campos, A., Claro, P.I.C., Guimarães, G.G.F., Mattoso, L.H.C., and Marconcini, J.M. 2022. Biodegradability and nutrients release of thermoplastic starch and poly(ε-caprolactone) blends for agricultural uses. *Carbohydr. Polym.* 282: 119058.

Corrêa, A.C., Carmona, V.C., Simão, J.A., Mattoso, L.H.C., and Marconcini, J.M. 2017. Biodegradable blends of urea plasticized thermoplastic starch (UTPS) and poly(ε-caprolactone) (PCL): Morphological, rheological, thermal and mechanical properties. *Carbohydr. Polym.* 167: 177–184.

Favvas, E.P., Mitropoulos, A.Ch. 2008. What is spinodal decomposition? *J. Eng. Sci. Technol.* 1: 25–27.

Ferri, J.M., Garcia-Garcia, D., Rayón, E., Samper, M.D., and Balart, R. 2020. Compatibilization and characterization of polylactide and biopolyethylene binary blends by non-reactive and reactive compatibilization approaches. *Polymers* 12: 1344.

GB/T19277.1-2011. 2011. *Determination of the Ultimate Aerobic Biodegradability of Plastic Materials Under Controlled Composting Conditions. Method by Analysis of Evolved Carbon dioxide. Part 1: General Method*. Beijing: Standardization Administration of China. National Bio-based Materials and Biodegradable Products Standardization Technical Committee.

Girijappa, Y.G.T., Rangappa, S.M., Parameswaranpillai, J., and Siengchin, S. 2019. Natural fibers as sustainable and renewable resource for development of eco-friendly composites: A comprehensive review. *Front. Mater.* 6: 1–14.

Goonoo, N., Bhaw-Luximon, A., and Jhurry, D. 2015. Biodegradable polymer blends: Miscibility, physicochemical properties and biological response scaffolds. *Polym. Int.* 64: 1289–1302.

Graupnera, N., Prambauerb, M., Fröhlkinga, T., et al. 2019. Copy paper as a source of reinforcement for biodegradable composites – Influence of fibre loading, processing method and layer arrangement – An overview. *Compos. Part A Appl. Sci. Manuf.* 120: 161–171.

Guedes, R.M., and Xavier, J. 2013. Understanding and predicting stiffness in advanced fibre-reinforced polymer (FRP) composites for structural applications. In *Woodhead Publishing Series in Civil and Structural Engineering, Advanced Fibre-reinforced Polymer (FRP) Composites for Structural Applications*, ed. Bai, J., pp. 298–360. Sawston: Woodhead Publishing.

Guo, Y., Ma, J., Lv, Z., Zhao, N., Wang, L., and Li, Q. 2018. The effect of plasticizer on the shape memory properties of poly(lactide acid)/poly(ethylene glycol) blends. *J. Mater. Res.* 33: 4101–4112.

Haider, T.B., Völker, C., Kramm, J., Landfester, K., and Wurm, F.R. 2019. Plastics of the future? The impact of biodegradable polymers on the environment and on society. *Angew. Chem. Int. Ed.* 58: 50–62.

Hammajam, A.A., El-Jummah, A.M., and Ismarrubie, Z.N. 2019. The green composites: Millet husk fiber (MHF) filled poly lactic acid (PLA) and degradability effects on environment. *Open J. Compos. Mater.* 9(3): 300–311.

Haque, A.N.M.A., and Naebe, M. 2022. Sustainable biodegradable denim waste composites for potential single-use packaging. *Sci. Total Environ.* 809: 152239.

Hashimoto, K., Kurokawa, N., and Hotta, A. 2021. Controlling the switching temperature of biodegradable shape memory polymers composed of stereocomplex polylactide/poly(*D,L*-lactide-*co*-ε-caprolactone) blends. *Polymer* 233: 124190.

He, J., Yu, T., Chen, S., and Li, Y. 2021. Soil degradation behavior of ramie/thermoset poly(-lactic acid) Composites. *J. Polym. Res.* 28: 379.

Hernández-García, E., Vargas, M., Chiralt, A., and González-Martínez, C. 2022. Biodegradation of PLA-PHBV blend films as affected by the incorporation of different phenolic acids. *Foods* 11(2): 243.

Imre, B., and Pukánszky, B. 2013. Compatibilization in bio-based and biodegradable polymer blends. *Eur. Polym. J.* 49: 1215–1233.

Inoue, T. 2003. Morphology of Polymer Blends. In *Polymer Blends Handbook*, ed. Utracki, L. A., pp. 547–576. Dordrecht: Springer.

ISO 1210:1992. 1992. *Plastics – Determination of the Burning Behaviour of Horizontal and Vertical Specimens in Contact with a Small-flame Ignition Source*. Geneva: International Organization for Standardization. Technical Committee: ISO/TC 61/SC 4 Burning Behaviour

ISO 13781:2017. 2017. *Implants for Surgery – Homopolymers, Copolymers and Blends on Poly(lactide) – In vitro Degradation Testing*. Geneva: International Organization for Standardization. Technical Committee ISO/TC 150/SC 1 Materials.

ISO 14855-1:2012. 2012. *Determination of the Ultimate Aerobic Biodegradability of Plastic Materials Under Controlled Composting Conditions – Method by Analysis of Evolved Carbon dioxide – Part 1: General Method*. Geneva: International Organization for Standardization. Technical Committee ISO/TC 61/SC 14 Environmental Aspects.

ISO 20200:2015. 2015. *Plastics – Determination of the Degree of Disintegration of Plastic Materials Under Simulated Composting Conditions in a Laboratory-scale Test*. Geneva: International Organization for Standardization. Technical Committee: ISO/TC 61/SC 14 Environmental Aspects.

Jha, K., Tyagi, Y.K., Kumar, R., et al. 2021. Assessment of dimensional stability, biodegradability, and fracture energy of bio-composites reinforced with novel pine cone. *Polymers* 13: 3260.

Ji, M., Li, J., Li, F., et al. 2022. A biodegradable chitosan-based composite film reinforced by ramie fibre and lignin for food packaging. *Carbohydr. Polym.* 281: 119078.

Jurczyk, S., Kurcok, P., and Musioł, M. 2017. Multifunctional composite ecomaterials and their impact on sustainability. In *Handbook of Ecomaterials*, eds. Martínez, L.M.T., Kharissova, O.V., and Kharisov, B.I., pp. 1–31. Cham: Springer.

Jurczyk, S., Musioł, M., Sobota, M., et al. 2019. (Bio)degradable polymeric materials for sustainable future – Part 2: Degradation studies of P(3HB-*co*-4HB)/cork composites in different environments. *Polymers* 11: 547.

Kammer, H.W., and Kummerloewe, C. 1990. Phase separation in poly(*p*-phenylene terephthalamide)/polyamide 6 blends. *Acta Polym.* 41: 269–273.

Kataoka, T., Hiramoto, K., and Kurihara, H. 2014. Effects of melt annealing on the miscibility and crystallization of poly(butylene succinate)/poly(ethylene succinate) blends. *Polym. J.* 46: 405–411.

Khalifeh, S. 2020. 1 – Introduction to polymers for electronic engineers. In *Polymers in Organic Electronics,* ed. Khalifeh, S., pp. 1–31. Toronto: ChemTec Publishing.

Khalili, P., Liu, X., Zhao, Z., and Blinzler, B. 2019. Fully biodegradable composites: Thermal, flammability, moisture absorption and mechanical properties of natural fibre-reinforced composites with nano-hydroxyapatite. *Materials* 12: 1145.

Khlif, M., Chaari, R., and Bradai, C. 2022. Physico-mechanical characterization of poly(butylene succinate) and date palm fiber-based biodegradable composites. *Polym. Polym. Compos.* 30: 1–10.

Kim, K.-S., Yoo, J., Shim, J.-S., et al. 2022. Biodegradable molybdenum/polybutylene adipate terephthalate conductive paste for flexible and stretchable transient electronics. *Adv. Mater. Technol.* 7(2): 2001297.

Korycki, A., Garnier, C., Abadie, A., Nassiet, V., Sultan, C.T., and Chabert, F. 2021. Poly(etheretherketone)/poly(ethersulfone) blends with phenolphthalein: Miscibility, thermomechanical properties, crystallization and morphology. *Polymers* 13: 1466.

Kuciel, S., and Rydarowski, H. 2012. Wprowadzenie. In *Biokompozyty z Surowców Odnawialnych,* eds. Kuciel, S., and Rydarowski, H., pp. 9–11. Kraków: Collegium Columbium (In Polish).

Kultravut, K., Kuboyama, K., and Ougizawa, T. 2019. Effect of blending procedure on tensile and degradation properties of toughened biodegradable poly(lactic acid) blend with poly(trimethylene terephthalate) and reactive compatibilizer. *Macromol. Mater. Eng.* 304(11): 1900323.

Lewicka, K., Rychter, P., Pastusiak, M., Janeczek, H., and Dobrzynski, P. 2020. Biodegradable blends of grafted dextrin with PLGA-block-PEG copolymer as a carrier for controlled release of herbicides into soil. *Materials* 13: 832.

Livekindly. *Introducing 'Lina', The Eco-friendly Electric Car Made from Plants.* www.livekindly.co/eco-friendly-electric-vehicle-plants (accessed December 17, 2021).

Ma, P., Cai, X., Zhang, Y., et al. 2014. In-situ compatibilization of poly(lactic acid) and poly(butylene adipate-*co*-terephthalate) blends by using dicumyl peroxide as a free-radical initiator. *Polym. Degrad. Stab.* 102: 145–151.

Moore, S. 2017. World's first car made from bio composites makes global debut. *Plastics Today.* www.plasticstoday.com/automotive-and-mobility/worlds-first-car-made-bio-composites-makes-global-debut (accessed November 22, 2021).

Moura, C.V.R., Cruz Sousa, D., Moura, E.M., Araújo, E.C.E., and Sittolin, I.M. 2021. New biodegradable composites from starch and fibers of the babassu coconut. *Polímeros* 31: 2021007.

Musioł, M., Rydz, J., Janeczek, H., et al. 2022. (Bio)degradable biochar composites – Studies on degradation and electrostatic properties. *Mater. Sci. Eng. B.* 275: 115515.

Musioł, M., Sikorska, W., Janeczek, H., et al. 2018. (Bio)degradable polymeric materials for a sustainable future – Part 1. Organic recycling of PLA/PBAT blends in the form of prototype packages with long shelf-life. *Waste Manage.* 77: 447–454.

Muthuraj, R., Misra, M., and Mohanty, A.K. 2018. Biodegradable compatibilized polymer blends for packaging application: A literature review. *J. Appl. Polym. Sci.* 135: 45726.

Nakayama, D., Wu, F., Mohanty, A.K., Hirai, S., and Misra, M. 2018. Biodegradable composites developed from PBAT/PLA binary blends and silk powder: Compatibilization and performance evaluation. *ACS Omega* 3: 12412–12421.

Navarro-Baena, I., Sessini, V., Dominici, F., Torre, L., Kenny, J.M., and Peponi, L. 2016. Design of biodegradable blends based on PLA and PCL: From morphological, thermal and mechanical studies to shape memory behavior. *Polym. Degrad. Stab.* 132: 97–108.

Nevoralová, M., Koutný, M., Ujcic, A., et al. 2020. Structure characterization and biodegradation rate of poly(ε-caprolactone)/starch blends. *Front. Mater.* 7: 141.

Nunes, E.C.D., Souza, A.G., and Rosa, D.S. 2019. Effect of the joncryl® ADR compatibilizing agent in blends of poly(butylene adipate-*co*-terephthalate)/poly(lactic acid). *Macromol. Symp.* 383: 1800035.

Obielodan, J., Vergenz, K., Aqil, D., Wu, J., and Mc Ellistrem, L. 2019. *Characterization of PLA/lignin Biocomposites for 3D Printing. Solid Freeform Fabrication 2019: Proceedings of the 30th Annual International Solid Freeform Fabrication Symposium – An Additive Manufacturing Conference: Reviewed Paper.* http://utw10945.utweb.utexas.edu/sites/default/files/2019/087%20Characterization%20of%20PLA_Lignin%20Biocomposites%20for%203.pdf.

Pang, M.Z., Qiao, J.J., Jiao, J., Wang, S.J., Xiao, M., and Meng, Y.Z. 2008. Miscibility and properties of completely biodegradable blends of poly(propylene carbonate) and poly(butylene succinate). *J. Appl. Polym. Sci.* 107: 2854–2860.

Pegoretti, A., Dong, Y., and Slouf, M. 2020. Editorial: Biodegradable matrices and composites. *Front. Mater.* 7: 265.

Phothisarattana, D., Wongphan, P., Promhuad, K., Promsorn, J., and Harnkarnsujarit, N. 2021. Biodegradable poly(butylene adipate-*co*-terephthalate) and thermoplastic starch-blended TiO$_2$ nanocomposite blown films as functional active packaging of fresh fruit. *Polymers* 13: 4192.

Pietrzykowska, E., Romelczyk-Baishya, B., Wojnarowicz, J., et al. 2020. Preparation of a ceramic matrix composite made of hydroxyapatite nanoparticles and polylactic acid by consolidation of composite granules. *Nanomaterials* 10: 1060.

Qiu, Z., Ikehara, T., and Nishi, T. 2003. Miscibility and crystallization behaviour of biodegradable blends of two aliphatic polyesters. Poly(3-hydroxybutyrate-*co*-hydroxyvalerate) and poly(butylene succinate) blends. *Polymer* 44: 7519–7527.

Radecka, I.K., Jiang, G., Hill, D.J., and Kowalczuk, M.M. 2016. Poly(hydroxyalkanoates) composites and their applications. In *Green polymer composites technology*, ed. Inamuddin, pp. 163–175. Boca Raton: CRC Press.

Rahman, M.O., Zhu, F.C., and Yu, B. 2022. Improving the compatibility of biodegradable poly(lactic acid) toughening with thermoplastic polyurethane (TPU) and compatibilized meltblown nonwoven. *Open J. Compos. Mater.* 12: 1–15.

Rydz, J., Sikorska, W., Musioł, M., et al. 2019. 3D-printed polyester-based prototypes for cosmetic applications – Future directions at the forensic engineering of advanced polymeric materials. *Materials* 12: 994.

Sabzi, F., and Boushehri, A. 2006. Compatibility of polymer blends. I. Copolymers with organic solvents. *J. Appl. Polym. Sci.* 101: 492–498.

Sangroniz, A., Gonzalez, A., Martin, L., Irusta, L., Iriarte, M., and Etxeberria, A. 2018. Miscibility and degradation of polymer blends based on biodegradable poly(butylene adipate-*co*-terephthalate). *Polym. Degrad. Stab.* 151: 25–35.

Sawyer, L.C., Grubb, D.T., and Meyers, G.F. 1987. *Polymer Microscopy.* 3rd ed., pp. 1–25. New York: Springer.

Scaffaro, R., Maio, A., and Gulino, E.F. 2021. Hydrolytic degradation of PLA/*Posidonia oceanica* green composites: A simple model based on starting morpho-chemical properties. *Compos. Sci. Technol.* 213: 108930.

Shen, J., Wang, K., Ma, Z., Xu, N., Pang, S., and Pan, L. 2021. Biodegradable blends of poly(butylene adipate-*co*-terephthalate) and polyglycolic acid with enhanced mechanical, rheological and barrier performances. *J. Appl. Polym. Sci.* 138: 51285.

Sikorska, W., Rydz, J., Wolna-Stypka, K., et al. 2017. Forensic engineering of advanced polymeric materials – Part V: Prediction studies of aliphatic-aromatic copolyester and polylactide commercial blends in view of potential applications as compostable cosmetic packages. *Polymers* 9: 257.

Silva, L.C.S., Camani, P.H., de Lima, E.C., and Rosa, D.S. 2022. Biodegradable composites made by cassava peels, residual glycerin, bentonite, and zeolite: The contribution to the treatment of BTEX in gasoline-contaminated soils. *Waste Biomass Valor.* 13: 1965–1980.

Silva, M., Ferreira, F.N., Alves, N.M., and Paiva, M.C. 2020. Biodegradable polymer nanocomposites for ligament/tendon tissue engineering. *J. Nanobiotechnol.* 18. https://jnanobiotechnology.biomedcentral.com/track/pdf/10.1186/s12951-019-0556-1.pdf.

Singh, S., Singh, G., Prakash, Ch., Ramakrishna, S., Lamberti, L., and Pruncu, C.I. 2020. 3D printed biodegradable composites: An insight into mechanical properties of PLA/chitosan scaffold. *Polym. Test.* 89: 106722.

Şirin, K., Seziş, U.G., and Ay, E. 2021. Preparation and characterization of dialdehyde cellulose/polylactic acid blends. *El-Cezerî J. Sci. Eng.* 8: 1158–1169.

Smola-Dmochowska, A., Śmigiel-Gac, N., Kaczmarczyk, B., et al. 2020. Triple-shape memory behavior of modified lactide/glycolide copolymers. *Polymers* 12: 2984.

Tabasi, R.Y., and Ajji, A. 2015. Selective degradation of biodegradable blends in simulated laboratory composting. *Polym. Degrad. Stab.* 120: 435–442.

Tanjung, F.A., Arifin, Y., and Husseinsyah, S. 2018. Enzymatic degradation of coconut shell powder–reinforced polylactic acid biocomposites. *J. Thermoplast. Compos. Mater.* 33: 800–816.

Van Der Vegt, A.K., and Elmendorp, J.J. 1986. Blending of incompatible polymers. In *Integration of Fundamental Polymer Science and Technology*, eds. Kleintjens, L.A., and Lemstra, P.J., pp. 381–389. Dordrecht: Springer.

Wang, R., Sun, X., Chen, X., and Liang, W. 2021. Morphological and mechanical properties of biodegradable poly(glycolic acid)/poly(butylene adipate-*co*-terephthalate) blends with *in situ* compatibilization. *RSC Adv.* 11: 1241–1249.

Wang, X., Peng, S., Chen, H., Yu, X., and Zhao, X. 2019. Mechanical properties, rheological behaviors, and phase morphologies of high-toughness PLA/PBAT blends by *in-situ* reactive compatibilization. *Compos. B. Eng.* 173: 107028.

Wang, Y., Wei, Z., and Li, Y. 2016. Highly toughened polylactide/epoxidized poly(styrene-*b*-butadiene-*b*-styrene) blends with excellent tensile performance. *Eur. Polym. J.* 85: 92–104.

Yua, D., Xu, Ch., Chen, Z., and Chen, Y. 2014. Crosslinked bicontinuous biobased polylactide/natural rubber materials: Super toughness, "net-like"-structure of NR phase and excellent interfacial adhesion. *Polym. Test.* 38: 73–80.

Zanela, J., Casagrande, M., Shirai, M.A., Lima, V.A., and Yamashita, F. 2016. Biodegradable blends of starch/polyvinyl alcohol/glycerol: Multivariate analysis of the mechanical properties. *Polímeros* 26: 193–196.

Zhang, T., Tsang, M., Du, L., Kim, M., and Allen, M.G. 2019. Electrical interconnects fabricated from biodegradable conductive polymer composites. *IEEE Trans. Compon. Packag. Manuf. Technol.* 9: 822–829.

Zhang, Y., Jia, S., Pan, H., et al. 2021a. Preparation, characterization and properties of biodegradable poly(butylene adipate-*co*-butylene terephthalate)/thermoplastic poly(propylene carbonate) polyurethane blend films. *Polym. Adv. Technol.* 32: 613–629.

Zhang, Z., He, F., Wang, B., et al. 2021b. Biodegradable PGA/PBAT blends for 3D printing: Material performance and periodic minimal surface structures. *Polymers* 13: 3757.

Zhao, H., and Zhao, G. 2016. Mechanical and thermal properties of conventional and microcellular injection molded poly(lactic acid)/poly(ε-caprolactone) blends. *J. Mech. Behav. Biomed. Mater.* 53: 59–67.

Zhao, Y., Li, Y., Xie, D., and Chen, J. 2021. Effect of chain extender on the compatibility, mechanical and gas barrier properties of poly(butylene adipate-*co*-terephthalate)/poly(propylene carbonate) bio-composites. *J. Appl. Polym. Sci.* 138: 50487.

6 New Research Strategy in Forecasting Directions of (Bio)degradable Polyester Applications

Joanna Rydz

CONTENTS

6.1 RELATIONSHIP BETWEEN STRUCTURE AND PROPERTIES OF (BIO)DEGRADABLE POLYESTERS

The studies of the relationship between the structure of (bio)degradable polymers and their properties, the analysis of (bio)degradation processes in various environments as well as assessing the possibility of their use in various areas, including (bio)degradable packaging for products with a long shelf life, are important elements of the research strategy in forecasting directions of environmentally friendly polymers applications, in particular (bio)degradable polyesters (Rydz et al., 2015b; Filiciotto and Rothenberg, 2021).

Conventional plastics, as one of the basic engineering materials, have become ubiquitous in almost every area of human life and replace those effectively used for centuries such as wood, glass or metals. The use of conventional plastics in various fields, especially as food and cosmetics packaging, causes particularly a significant increase in waste, which results in an increase in environmental pollution because,

DOI: 10.1201/9780429352799-7

due to their chemical structure, such polymers do not decompose under natural conditions. In addition, difficulties related to their recycling, resulting, among other things, from impurities that are difficult to remove, such as oily substances from foodstuffs or cosmetics, mean that typical disposal options (depending on the properties of plastics: material recycling or monomer recovery) are not always possible and only energy recovery remains (Millet et al., 2019).

In the natural ecosystem, carbon is part of a biological life cycle, and disruption of this cycle leads to irreversible changes in the environment. Therefore, it is so important that carbon-based polymers do not interfere with its global carbon cycle. The conversion rate of fossil raw materials (petroleum-based) by conventional plastics obtained from them is characterised by a total imbalance with the rate of their consumption and renewal; while using natural resources (e.g. biomass) as raw materials for the production of polymers (as well as synthetic (bio)degradable polymers), the carbon dioxide recovery rate is balanced by its consumption, which leads to environmental sustainability (Figure 6.1).

Environmental issues, as well as the gradual depletion of global oil resources, lead scientists to look for alternative sources of raw materials, and therefore polymers from renewable raw materials are playing an increasingly important role. The development of environmentally friendly polymers obtained in accordance with the idea of sustainable development (both (bio)degradable and/or from renewable raw materials with a minimised carbon footprint) is justified from an economic and, above all, ecological point of view (Narayan, 2012; Urbaniec et al., 2017; Kyulavska et al., 2019; Sikorska et al., 2019).

(Bio)degradable polyesters such as polylactide (PLA), polyhydroxyalkanoates (PHA) or copolyesters with aliphatic units are a group of compounds that exhibit

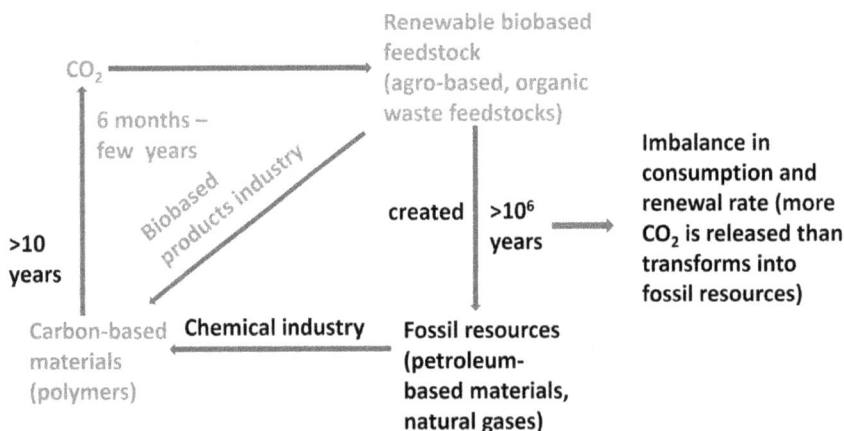

FIGURE 6.1 Global carbon life cycle.

Source: reprinted with permission from Rydz et al., Sustainable future alternative: (bio) degradable polymers for the environment. In *Encyclopaedia of renewable and sustainable materials*, eds Hashmi, S., and Choudhury, I.A. (Oxford: Elsevier, Academic Press, 2020), 274–284

similar properties and mechanisms of reactions and phenomena in which they participate. As a result, finished products from these polymers provide specific, time-limited, environmentally friendly applications and are suitable for organic recycling. Most (bio)degradable polyesters have great potential for packaging applications because they have thermoplastic properties and similar processing parameters. Their properties depend on the chemical structure of the main chain as well as the structure and size of the side chains. Aliphatic polyesters with short side chains exhibit properties similar to polypropylene. As the side chains elongate, the polymer acquires elastomer-like properties (Naser et al., 2021).

Increasing the share of (bio)degradable polymers in the plastics market creates new opportunities but also risks. The rate of degradation and mechanical properties are key factors in many applications of (bio)degradable polymers, especially those with a long service life. Therefore, it is extremely important to design packaging made of (bio)degradable plastics that would be safe for human health and the environment and at the same time indicate in a responsible manner and in accordance with the idea of sustainable development new areas where their unique properties can be exploited. It is also extremely important to determine whether and how the contents of the packaging may interact with the (bio)degradable polymer, in particular for packaging of products with a long shelf life (Moshood et al., 2022).

Compostable packaging plastics can minimise the increase in currently generated onerous packaging waste from conventional plastics. On the other hand, (bio)degradation processes can also occur when using packaging made of (bio)degradable polymers. The physicochemical changes caused by these processes affect the thermal and mechanical properties of the plastic, causing its morphological and structural transformations, which entails deterioration of its quality, which in turn may be a key factor in many applications of (bio)degradable polymers. The main challenge is therefore to determine the conditions in which the use of (bio)degradable plastics is beneficial, as well as the criteria for their use. The new research strategy in the field of (bio)degradable polymers should minimise any possible failures related to the future use of such polymers. Designing packaging from "plastics of the future" should enable the implementation of the new strategy on plastics and significantly help implement the principles of the circular economy (Huda et al., 2002; Prieto, 2016).

The use of (bio)degradable polymers as packaging for products with a long shelf life, especially as cosmetic packaging, is a new trend in the production and management of solid waste. Therefore, the development of this area of research is particularly important. Knowledge of environmental conditions of use and degradation mechanisms occurring in the plastic enables proper forecasting of its applications. This is especially necessary for (bio)degradable polymers, including compostable ones. The aging tests of (bio)degradable plastics in situ conditions of use, as well as in simulated conditions, not only provide knowledge about the "life span" of plastics but also enable the development of ways to modify properties both at the stage of polymerisation and processing so as to adapt this time to specific applications. Understanding and assessing the relationship between the structure, properties and behaviour of (bio)degradable polyesters before, during and after practical applications will allow one to precisely define and predict the possibilities and limitations both at the stage of designing the finished product as well as during its exploitation and recycling through (bio)degradation (Shaikh et al., 2021).

6.1.1 STRUCTURE AND PROPERTIES STUDIES OF (BIO)DEGRADABLE POLYMERS IN CONTACT WITH COSMETIC SIMULANTS – RESEARCH METHODOLOGY

Cosmetics represent a group of products containing a chemical substance or their mixture intended to come into contact with the human body and to perform specific functions, which, however, do not interfere with the structure and metabolic processes taking place in the body. They must, as well as the packaging, meet acceptable safety standards during use and exhibit physicochemical stability during storage or transport. It is important to maintain a barrier not only against microbial contamination but also against chemical contamination resulting from overheating or other decomposition reactions during processing or from improper storage. Also, interactions between cosmetics and packaging components can lead to the formation of new chemical structures or harmful degradation products of polymeric containers. One of the main requirements for polymer packaging is to limit the migration of low molar-mass ingredients to cosmetic formulations. It is therefore important to identify in advance the interactions between the packaging of (bio)degradable polymers and the cosmetic formulation. In this case, the complete replacement of conventional plastics with environmentally friendly packaging is impossible to achieve without recognising and assessing the possible risks (Gilbert, 1975; Lewis, 1998; Bucci et al., 2005; Rieger, 2001).

In the case of (bio)degradable polymers for packaging applications, environmental conditions that can lead to degradation of the product during storage should be avoided, especially for liquid-state products. In addition, compounds originated from cosmetic formulations, such as lipids or aromatic compounds, may interact with the packaging and cause modification and deterioration of their gas and aroma barrier properties (Colomines et al., 2008).

The degradation processes of polyesters in various environments such as distilled and deionised water (Musioł et al., 2011; Musioł et al., 2022); buffers (Rydz et al., 2013a); humidity environment (Thor et al., 2021); sea (Rutkowska et al., 2008); river and lake water (He et al., 2000; Volova, 2015); soil (Rychter et al., 2006); domestic, laboratory and industrial compost (Endres et al., 2011; Jurczyk et al., 2019; Musioł et al., 2022); compost with activated or anaerobic sludge (Rutkowska et al., 2008); weathering tests (Rydz et al., 2019) as well as in a mineral media in the presence of microorganisms (Hakkarainen et al., 2000) have been extensively studied. In contrast, degradation in selected model liquids as selected ingredients used in the preparation of cosmetics is much less reported and is based on standards for medical, pharmaceutical and food packaging. The cosmetics industry does not use specific guidelines concerning packaging/content interaction testing. However, under the European Union (EU) Cosmetic Regulation 1223/2009, there is a requirement to report packaging material characteristics such as stability, impurities and traces as part of the product safety assessment. Skin exposure to undesirable substances from packaging must be at a level which is safe for human health (2013/674/EU, Connolly et al., 2019).

Initial research into the suitability of polymers as cosmetic packaging has involved the development of a methodology as there are no standardised testing procedures for cosmetic product packaging to predict behaviour before, during and after use (Marx, 2004). It is not easy to directly measure the migration into a cosmetic formulation. The migration behaviour of the plastic packaging should be easily measurable using stability model tests, in media such as ethanol, distilled water, acetic acid or alkanes, such as in the case of food packaging (CEN EN 1186-1:2002; Avella et al., 2005). The correct choice of simulant for a packaging-product degradation experiment depends on several factors, including the material being tested and the cosmetic products that the material interacts with. The pH of cosmetic formulations ranges from 2 for anti-aging creams to 10 for shaving creams. Paraffin, glycerine, propylene glycol, ethyl alcohol and water play an important role in cosmetology and are ingredients commonly found in cosmetic formulations. Considering the factors influencing the rate of the (bio)degradation process both during use and after use (composting of used packaging, improper waste management – illegal dumping), an innovative methodology has been developed – tests determining the stability of packaging in various conditions corresponding to the actual use of the product, i.e. suitable cosmetic simulants, were selected such as distilled water, liquid paraffin, anhydrous glycerine, propylene glycol, ethyl alcohol as well as pH = 4.00 and pH = 10.00 buffer solutions (Winter, 2009; Rydz et al., 2013a).

First, PLA film degradation in selected cosmetic simulants (degradation media) was examined (Rydz et al., 2013b). The polyesters chain cleavage processes induce morphological and mechanical changes that affect the disintegration process of polyester film and cause its degradation (Santonja- Blasco et al., 2010). Thermal analysis techniques, such as dynamic mechanical thermal analysis (DMTA), differential scanning calorimetry (DSC) and thermogravimetric analysis (TGA) have been successfully used to study the effect of degradation on the physicochemical properties of (bio)degradable polymers, including PLA (Rydz et al., 2015a). An evaluation of the rate of degradation using thermoanalytical techniques provides valuable information related to the optimal processing conditions and predicting the shelf life of final products. The films after degradation in selected media (paraffin, glycerine and propylene glycol) were subjected to comprehensive thermal analysis in order to determine the impact of the tested cosmetic simulants as degradation media on the polymer properties (thermal properties, crystallinity, thermal stability and changes in the kinetics of decomposition; Rydz et al., 2013b).

The macro- and microscopic evaluation of the surface of test items in the form of PLA film strips showed surface erosion in all tested media, also at 37°C, although much slower than at 70°C, as the rate of degradation increases at temperatures above glass transition temperature (T_g) of the polymer tested (Andersson et al., 2010). At the beginning of degradation, the surface erosion depended on the solubility of degradation products in the studied simulants. This was confirmed by the results of gel permeation chromatography (GPC) analysis and molar-mass distribution curves, which remained unimodal in propylene glycol, because the low molar-mass products diffused into the degradation medium, making the film surface more diverse. It was observed that degradation in paraffin occurred much faster than in glycerine. Paraffin is a non-polar and chemically unreactive medium in which degradation

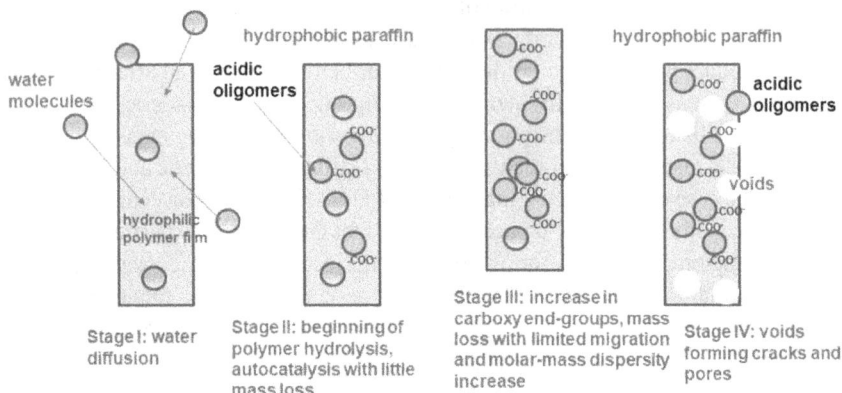

FIGURE 6.2   Schematic representation of hydrolytic degradation of polymer in hydrophobic medium.

should not occur. It is also considered one of the most hydrophobic substances and water-repellent agents (Zbik et al., 2006). Thus, the explanation for this unexpected degradation in paraffin may be the residual moisture content, which can generate an autocatalytic effect. For PLA film, an absorption effect is observed, during which the residual moisture content in paraffin (0.016% according to the Karl Fischer method) penetrates into the polymer matrix, starting the autocatalysis process (Rydz et al., 2013a). The scheme of PLA hydrolytic degradation in a hydrophobic environment is presented in Figure 6.2.

Confirmation of this statement may be the results of multistage mass spectrometry with electrospray ionisation (ESI-MS[n]) analysis of degradation products. The results of this analysis for the residue of PLA film after degradation indicate that even a small amount of moisture contained in the cosmetic formulation can initiate the process of hydrolytic degradation of PLA film. The presence in the spectrum of low molar-mass lactic acid oligomers containing hydroxy and carboxy end groups indicates a significant decrease in the molar mass of the PLA film due to the random hydrolysis of ester bonds in the studied polyester, which characterise hydrolytic degradation occurring in water. In addition, lactic acid and its dimer molecules were identified. The presence of lactic acid in the remaining PLA film after degradation confirmed that the degradation process due to limited migration was also associated with the acid-catalysed internal degradation process taking place inside the polymer matrix (Rydz et al., 2013a). Faster internal degradation of PLA is considered a general phenomenon. Hydrolytic chain cleavage causes an increase in the concentration of carboxy end groups with degradation time, which accelerates internal degradation (autocatalytic effect) and enhances the degradation differentiation on the surface and in the polymer matrix (Siparsky et al., 1988). Due to the limited migration, low-molar-mass degradation products, insoluble in paraffin, remained in a hydrophilic environment (PLA film), and the molar-mass dispersity increased rapidly, and bimodal GPC curves were also observed as a result of the autocatalytic mechanism and due to the presence of two populations of macromolecules from degradation occurring at different rates on the surface and in the PLA matrix (Li et al., 1990; Li and Vert, 2002). In contrast, the molar-mass distribution during degradation in propylene glycol remained unimodal (continuous function for which a maximum of one local extreme exists in a given range) due to the diffusion of low molar-mass products into the degradation medium (Rydz et al., 2013a).

Analysis of TGA thermograms showed that the thermal decomposition of PLA film is a single stage, first order decomposition process. The almost constant value of activation energy (E_a) during degradation experiments with increasing incubation time indicates that the same degradation reaction pathway was involved throughout the degradation process. Changes in thermal properties, crystallinity and thermal stability of PLA films during degradation indicates morphological and structural transformations that cause deterioration of the properties of the film in all tested media (Rydz et al., 2013b).

The results of the conducted research demonstrated that the selected media affect the deterioration of the performance of PLA films, which limits the future use of PLA for the packaging of products with a long shelf-life application. It can be concluded that PLA is an imperfect alternative to classic polymers in these applications, and its degradation depends largely on the degradation environment. So, degradation of PLA film occurs not only in the presence of polar solvents (ethyl alcohol, glycerine, propylene glycol) but also in the presence of paraffin (hydrophobic and chemically inert medium).

An aliphatic-aromatic poly(1,4-butylene adipate-*co*-1,4-butylene terephthalate) (PBAT) copolyester or poly(3-hydroxybutyrate) (PHB) can be a more attractive alternative to conventional plastics. Degradation studies at 70°C in paraffin were carried out on specimens obtained from a blend of 76 mole% PBAT (containing 47 mole% aromatic segments), more resistant to hydrolytic degradation, with 24 mole% PLA and compared with degradation in water, as well as specimens obtained from a blend of PLA with 3, 9, 12 and 15 mole% of poly[(*R,S*)-3-hydroxybutyrate] ((*R,S*)-PHB) and for comparison with PLA alone and with unprocessed synthetic (*R,S*)-PHB ((*R,S*)-PHB does not form films). Polymer-paraffin interactions (surface erosion, molar mass changes, degradation products, thermal properties and polymer crystallinity) were monitored during degradation experiments (Rydz et al., 2015c; Sikorska et al., 2017).

Macro- and microscopic evaluation of the surface of PLA/(*R,S*)-PHB films with different PLA content after degradation in paraffin showed erosion of the materials' surface. It has been found that higher levels of the (*R,S*)-PHB component result in a prolonged disintegration time in immiscible blends, whereas the visual assessment of the surface of the tested PBAT/PLA dumbbell-shaped bars after incubation in paraffin showed no change in their surface or disintegration. No significant microscopic surface changes were found. It was only during degradation in water (used as a reference medium) that the dumbbell-shaped bar structure cracked and then disintegrated (Rydz et al., 2015c; Sikorska et al., 2017).

A decrease in transparency of the examined specimen surfaces was observed in both PLA-based films from the beginning of the degradation process in paraffin and for PBAT/PLA dumbbell-shaped bars degraded in water after longer incubation times (caused by molecular reorganisation or an increase in structural irregularity resulting from the formation of new spherulites; Cam et al., 1995). For thicker rigid PLA film, this process was much faster than for thin PLA film used in preliminary degradation experiments. Films on the microscale are degraded in paraffin homogeneously and more slowly (erosion is more limited to the surface) than films with a greater thickness (Grizzi et al., 1995). Because thickness can affect the properties of the final product, this feature may be important from the perspective of the use of (bio)degradable polyesters in cosmetics packaging (Rydz et al., 2013a; Sikorska et al., 2017).

The degradation process in paraffin (in an environment with a residual moisture content) caused a continuous decrease in the molar mass of all rigid PLA/(R,S)-PHB films from the beginning of the experiment. During the degradation of unprocessed (R,S)-PHB, a decrease in molar mass was also observed, although it occurred more slowly than with PLA. The poorly hydrophilic (R,S)-PHB degrades more slowly than the more hydrophilic PLA (Rydz et al., 2015c). However, despite the lower hydrophilicity, the hydrolysis process takes place from the beginning of the experiment because moisture is trapped in the polymer matrix (the drying process is more difficult for amorphous areas of the polymer) and the amount of moisture absorbed depends on the crystallinity of the polymer (amorphous areas of the polymer absorb moisture faster than crystalline ones; Freier et al., 2002; Reliance Industries Limited. 2003).

The results of the GPC analysis showed that the addition of the (R,S)-PHB component at the initial step of degradation reduces the degree of degradation in paraffin in the case of rigid films only in blends with good miscibility (97PLA/3(R,S)-PHB). In blends, miscibility is one of the most important factors influencing the final properties of the material. The degradation progress of rigid films with PLA/(R,S)-PHB depended on the miscibility of the blends. The degree of film heterogeneity before degradation increased with increasing amounts of the (R,S)-PHB component in the blends. Only in the blend with good miscibility (97PLA/3(R,S)-PHB film) was the surface roughness low. As the degradation process progressed, the heterogeneity of the film (surface roughness) increased significantly for the immiscible films (Rydz et al., 2015c).

The results of the ESI-MS[n] analysis for the residual specimens (PLA, its blends and unprocessed (R,S)-PHB) confirm that, after the degradation in paraffin, the obtained oligomers had mainly hydroxy and carboxy end groups due to the hydrolytic degradation caused by the residual moisture content of the paraffin films (Rydz et al., 2015c).

During degradation of PBAT/PLA dumbbell-shaped bars, a significant shift of GPC curves towards lower values of molar masses was found for specimens incubated in water, while for incubated in paraffin, only a slight shift towards lower values of molar masses was found, which did not lead to loss of integrity of tested dumbbell-shaped bars. It was noticed that the presence of the PBAT component significantly improved the stability of the PBAT/PLA blend in contact with paraffin (Sikorska et al., 2017).

The effect of PBAT/PLA incubation time during the degradation process in paraffin and distilled water on changes in the thermal properties of selected dumbbell-shaped bars was characterised by DSC. During the first calorimetric trace at a rate of 10°C/min (first heating run), after incubation in water, due to complete degradation of the PLA component, the residue contained only the PBAT component. Both melting point (T_m) and T_g of the PLA component were not observed. However, after 52 weeks of incubation in paraffin, slight changes in the thermal properties of the PLA component were observed. Due to entrapped of the PLA in the polymer matrix with a predominance of PBAT (76 mole%), it appears to be "protected" from access by residual moisture content in paraffin (Sikorska et al., 2017).

During the cooling run at a rate of 10°C/min, the dumbbell-shaped bars incubated in water showed significant changes in the crystallisation region due to the fact that, during degradation of a polymer matrix, mainly degrades the PLA component and the residue contained only the PBAT component. Comparing DSC thermograms of specimens before degradation and incubated in paraffin it was found that the PBAT/PLA blend in paraffin showed only some differences within the crystallisation region after 52 days of incubation (Sikorska et al., 2017).

The ESI-MS[n] analysis of PBAT/PLA specimens after degradation in the paraffin was impossible because their molar mass was too high. To identify emerging degradation products, ESI-MS[n] analysis of water-soluble hydrolytic degradation products from degradation media was performed. It was interesting that during the ESI-MS[n] analysis of PBAT/PLA degradation products in the aqueous medium no signals from lactic acid oligomers were observed (Sikorska et al., 2017). Also, the slow rate of PLA hydrolytic degradation in the blend may indicate a chain-end scission mechanism. In general, polyesters undergo hydrolysis by random-chain scission. The mechanism of chain-end scission may result from the short distance between the carbonyl group and the PLA alkoxy group entrapped in the polymer matrix and/or increasing acidity, creating a good environment for the occurrence of hydrolysis from the end of the chain in a privileged way (Gleadall et al., 2014; Rydz et al., 2017).

The compatibility and migration of components from PLA/organoclay nanocomposites in simulants (vegetable oil) and two different cosmetic formulations was also investigated in accordance with the guidelines for materials intended for contact with food. It was observed that migration was controlled both by the type of polymer and the processing method and simulant used. The overall migration of extracts was greater in more lipophilic simulants, but it was within acceptable levels (< 10 mg/dm^2). No toxicity was also observed for any of the tested nanocomposites and in vitro migrating components. The obtained PLA nanocomposites can be potentially used in cosmetic packaging (No 10/2011, 2011; Connolly et al., 2019).

6.2 PREDICTING THE BEHAVIOUR OF TEST ITEMS RECEIVED BY THE THREE-DIMENSIONAL (3D) PRINTING METHOD IN VARIABLE (BIO)DEGRADATION CONDITIONS

Due to the fact that additive manufacturing is increasingly used for personalised consumer goods and such processing may affect mechanical and thermal properties, especially in the case of (bio)degradable polymers, the behaviour of dumbbell-shaped bars made using three-dimensional (3D) printer was analysed depending on the geometry of the element (printing orientation: the orientation of the product with respect to the printer build platform, filament arrangement according to the algorithm used to process the specimens – horizontal and vertical build directions) and then prototypes of packaging obtained by 3D printing (material extrusion) were used in cosmetic applications and the impact of various environmental conditions on the stability of the prototypes have been investigated (Rydz et al., 2017; Sikorska et al., 2017; Gonzalez Ausejo et al., 2018a; Gonzalez Ausejo et al., 2018b; Rydz et al., 2019; Rydz et al., 2020).

3D printing is a rapidly developing additive manufacturing process, relatively simple to produce almost any 3D object of any shape, with relatively high resolution and low cost using digital computer-aided design (CAD). Additive manufacturing technologies are dynamically developing in many industries in prototyping processes but also in the production of highly complex components, small-lot production and in the area of innovative solutions to problems and limitations of traditional technologies but also increasingly in the case of personalised consumer products, such as packaging or everyday items. The 3D printing process consists creating items based on adding building material, usually layer by layer. The rapid development of additive manufacturing technologies has enabled the development of the market of (bio)degradable polymer "inks" in the form of filaments, beads, powders, solutions and gels for the production of specific products. Thermoplastics are the only group of materials that can be processed by extrusion and injection moulding. Above a certain temperature limit, thermoplastics pass into a plasticised state in which they exhibit the ability to undergo large deformations. This enables pressure forming as well as additive manufacturing. The combination of 3D printing process with (bio)degradable polymers and/or renewable raw materials gives almost unlimited application possibilities (Włodarczyk et al., 2017).

Processing may affect mechanical and thermal properties, especially for (bio) degradable polymers. The extrusion process often results in a decrease in viscosity and a reduction in the average molar mass, which worsens the mechanical properties. Mixing time, temperature and drying also affect the degradation of PLA-based plastics (Sikorska et al., 2012). That is why it is so important to determine the impact of printing conditions and orientation on the polymer properties and hydrolytic degradation process and the relationship between processing conditions and 3D processing build directions (orientation of the test item on the printer build platform) based on molar mass and its dispersity, chemical structure, thermal and mechanical properties of PLA and PLA/PHA polyester filament currently used most often except the popular PLA. Furthermore, studies on the hydrolytic degradation of dumbbell-shaped bars at 50°C and 70°C were performed to assess the effect of incubation temperature on the course of hydrolytic degradation by monitoring changes in mechanical and thermal properties, molar mass and its dispersity, chemical structure, water absorption and changes in pH of solution (Gonzalez Ausejo et al., 2018a; Gonzalez Ausejo et al., 2018b).

To assess the impact of layer orientation, specimens with two different processing build directions were printed: horizontally (crisscross pattern, contact with the printer build platform 15 min) and vertical (transverse pattern, contact with the printer build platform 40 min from one end of the specimens). While for containers, the bottom and the top (lid) were obtained with a flat processing build direction (cross pattern). The wall of the container was obtained by printing individual layers, consisting of two concentric adjacent rings (Gonzalez Ausejo et al., 2018a).

6.2.1 PLA/PHA FILAMENT COMPOSITION VERIFICATION

Commercially available (bio)degradable filaments, often patented, do not contain information about their composition, and therefore their response to adverse environmental factors (abiotic and biotic) is unspecified. That is why it is so important to know the composition of the filament made of (bio)degradable polymers and to predict the properties of printed items to be able to accurately understand the impact of environmental conditions on the properties of final goods and match the right raw materials for specific applications. Thus, the molecular structure of the PLA/PHA filament (88 wt% PLA based on TGA) was determined to learn about the impact of processing conditions (contact time and surface size of the test item in contact with the printer build platform during printing) and printing orientation (filament arrangement according to the algorithm used – in the horizontal and vertical build direction) on the hydrolytic degradation profile at 50°C and 70°C of the tested dumbbell-shaped bars and the prototype of packaging (container) obtained by 3D printing from PLA/PHA filament and for comparison with PLA filament. Ex ante examination of plastics to define and minimise potential failures of new products from (bio)degradable polymers before they arise is badly needed (Gonzalez Ausejo et al., 2018a; Gonzalez Ausejo et al., 2018b; Rydz et al., 2019; Rydz et al., 2020).

Investigating the behaviour of polymers in various environments requires a thorough understanding of their composition and molecular structure. The composition of PLA/PHA filament was determined on the basis of proton nuclear magnetic resonance (^{1}H NMR) and ESI-MS^n spectra. ^{1}H NMR analysis showed the presence of two main components of the blend: PLA and PHB, as well as azelaic acid (AZA). Signals from azelaic acid were also seen in the ^{1}H NMR spectrum of the PLA filament (Gonzalez Ausejo et al., 2018a). Azelaic acid and its derivatives are often added to polyesters as a plasticiser or antibacterial agent (Asrar and D'haene, 1999; Corma et al., 2007).

Accurate quantitative analysis using NMR was not possible because the PLA/PHA filament is not completely soluble in chloroform due to the high molar mass of the PHA component. Therefore, the detailed molecular characteristic of the high molar mass PLA/PHA blend were carried out by means of ESI-MS^n. The ESI-MS^n technique allows the determination of the sequence distribution of monomer units in PHA, ranging from dimer to oligomers with a molar mass up to about $M_n = 2000$ g/mol. In order to verify the chemical structure of the PLA/PHA filament, controlled partial depolymerisation of the blend in a neutral medium at 70°C to the chloroform-soluble PHA oligomers was performed. ESI-MS^n spectra of PLA/PHA filament degradation products showed the presence of signals from lactic acid, 3-hydroxybutyric acid and their oligomers, AZA as well as poly(3-hydroxybutyrate-*co*-3-hydroxyvalerate) (PHBV). The random distribution of hydroxyvalerate (HV) constitutional repeating units along the PHBV chain was confirmed by fragmentation in positive-ion mode. The PHA component of the PLA/PHA blend consists of a small number of HV repeating units randomly distributed along the PHB chain. Thus, analysis of the commercially available PLA/PHA filament showed that the PHA component in the blend mainly contains hydroxybutyrate (HB) units and a small amount of HV units (Gonzalez Ausejo et al., 2018a).

6.2.2 INFLUENCE OF 3D PRINTING ORIENTATION ON THE PROPERTIES OF OBTAINED TEST ITEMS

The printing orientation determines not only the time and volume of the raw material used and the quality of the resulting items but also its properties. 3D printing in both vertical and horizontal directions had a significant impact on the structure and morphology of the obtained dumbbell-shaped bars as well as their thermal and mechanical properties (Gonzalez Ausejo et al., 2018a; Gonzalez Ausejo et al., 2018b; Rydz et al., 2020).

For the PLA filament only T_g was observed, suggesting that the filament is amorphous. During printing, tensile forces act upon the filament that cause tension-induced crystallisation and orientation of the layers in the printed items (Anderson et al., 2011). In the first heating run, the PLA/PHA filament showed the crystallisation phenomenon associated with the presence of the PHA component (nucleation). An increase in melting enthalpy (ΔH_m) was also observed with respect to the PLA filament. Contact with the printer build platform maintained at a constant temperature further affects the thermal properties of the dumbbell-shaped bar in contact with the platform. When the dumbbell-shaped bars were printed in a horizontal direction, the upper and underside layers had surfaces with different characteristics. This is due to the fact that the first layer was in direct contact with the printer build platform. In the case of fused deposition modelling, the heated thermoplastic polymer (filament) is extruded through the nozzle tip moving in the XY plane, where it is heated to the melting temperature and then deposited on the platform, forming the underside layer first. The extrusion head then deposits layer by layer in the Z-axis. Consequently, the upper layer is the farthest from the printer build platform. The printer platform is kept at a constant, not very high temperature (approx. 65°C) for the thermoplastic polymer to harden quickly. As a result, the polymer is permanently maintained at a constant temperature during printing for some time (the printing time of the dumbbell-shaped bar: 15 min for horizontal printing, 40 min for vertical printing), which affects the subsequent properties of the printed items. Printing in the vertical direction does not affect the surface structure of the dumbbell-shaped bars, because the contact with the printer platform concerned only a small area at the interface of the bar base and the printer build platform, and there it could only affect the thermal properties of the polymer. ΔH_m for the dumbbell-shaped bars printed in the horizontal direction had a higher value indicating an increase in the crystalline phase during printing in that direction. ΔH_m of dumbbell-shaped bars printed vertically from the platform side also increased slightly compared to the other side and the filament. In addition, the central section of the dumbbell-shaped bars (narrower than the ends), when printing in the horizontal direction at the same temperature of the printer build platform, accumulated more heat because it has a smaller surface, which resulted in a greater ordering in this place (higher ΔH_m and cold crystallisation enthalpy (ΔH_{cc}) and a lower T_m and cold crystallisation temperature (T_{cc})) which was not found for dumbbell-shaped bars printed in a vertical direction. On the other hand, for vertically printed dumbbell-shaped bars, the upper part is farther from the printer build platform and thus the lowest ordering was observed there due to a lower temperature than at the end of the dumbbell-shaped bars which are in contact with the printer build platform. Due to the temperature gradient, the ordering should be different in different sections of the dumbbell-shaped bars – highest in the underside layer where it meets the printer platform and lowest at the opposite end. Therefore, processing (3D printing) causes not only an increase in the crystalline phase of the polymer after printing compared to the original filament but also an increase related to the contact with the heated printer build platform. After printing, a slight increase in T_g was also observed as a result of an increase in the crystal domains, which may also increase the stiffness of the polymer. Not only the does contact time with the 3D printer platform lead to an increase in the crystalline phase during printing, but also parts of dumbbell-shaped bars with a smaller surface area accumulate more heat, which causes an increase in ordering (Gonzalez Ausejo et al., 2018a; Gonzalez Ausejo et al., 2018b; Rydz et al., 2020).

Similar relationships were observed in the case of printing a prototype of a cosmetic package (container) from PLA and PLA/PHA. During 3D printing, its particular parts had different contact times with the printer build platform. The bottom of the container and the lid top had a longer contact time (15–18 min) and during this time they were exposed to an elevated temperature causing the crystalline phase to grow as opposed to the wall vertically to the platform (ΔH_m and ΔH_{cc} were lower for the container walls). Therefore, the processing conditions, in particular the orientation of the printing of the various parts of the container, influenced its properties, which can then affect the life time and degradation process of the entire container (Rydz et al., 2019).

Dumbbell-shaped bars printed vertically with layers oriented perpendicular to the load direction had better mechanical properties (Young's modulus, tensile strength and elongation were higher) than horizontally printed dumbbell-shaped bars with layers oriented parallel to the load direction. This is probably due to the higher uniformity and packing density of the polymer matrix, as well as to deformation-induced crystallisation. The orientation results in more crystallisation, while the chains and hence the entire crystal domain are directed towards the stretching, which improves tensile and impact strength, stiffness, clarity and durability and leads to higher values of the elastic modulus of the polymers. On the other hand, orientation can have an adverse effect on elongation at break (Zhao et al., 1999).

Biodegradable polymers and their additives, depending on the application, should be environmentally friendly, non-toxic and create biocompatible systems in these applications. For these reasons, toxicity studies of PLA/PHA dumbbell-shaped bars were conducted to evaluate the potential hazards (Shaw et al., 2002; Thomas and Smart, 2005). As long as the substrate is replaced regularly, both cell lines (WI-38 and HEK293) used can continue to grow on PLA/PHA dumbbell-shaped bars without any toxicity effect, and this growth also depends on the build direction of printing. Thus, the dumbbell-shaped bars obtained by 3D printing showed no effect on the viability of the cells, which also underlines the potential suitability for medical applications (Gonzalez Ausejo et al., 2018a).

Taking into account the fact that, during the 3D printing process, the influence of the contact time of the tested shapes on their properties (increase in the crystalline phase and structure order) was observed, the prototypes of cosmetic packaging were also examined in terms of determining differences in the properties of its individual parts – the bottom of the container or the lid, as well as the walls. In order to assess changes in thermal properties of the tested packaging prototypes as a result of thermal history during processing (3D printing), DSC analysis was performed (Rydz et al., 2019).

The T_m and cold crystallisation enthalpies for the bottom of the PLA container were higher than for the container wall, which indicates that the processing, especially the contact time with the printer platform, causes an increase in the crystalline phase of these parts of the container. In the case of the PLA/PHA container, these effects were slightly lower than in the case of the PLA container. The results of the conducted research showed that the packaging prototypes also crystallise during warming when the item is printing (Rydz et al., 2019).

6.2.3 Effect of 3D Printing Build Directions
on the Degradation Profile of the Obtained Test Items

The progress of hydrolytic degradation under laboratory conditions of PLA/PHA dumbbell-shaped bars and for comparison bars of PLA at 50°C and 70°C was carried out for a period of 70 days. The influence of the printing build direction on the degradation of the dumbbell-shaped bars is more visible when the degradation occurs at a lower temperature (50°C). PLA/PHA dumbbell-shaped bars printed horizontally after degradation showed erosion only in the form of cracks, while bars printed vertically underwent significant disintegration. Thus, printing in the horizontal direction caused an increase in the disintegration time. The faster disintegration of vertically printed bars is due to the weaker cohesion between printed layers, which also resulted in higher mass loss. The results also showed a reduction in mass loss in the first degradation step at both temperatures, greater for PLA dumbbell-shaped bars. It is known that in the first step the presence of PHA, particularly PHB, reduces the rate of degradation due to its more hydrophobic nature as well as the increased crystallinity of PLA/PHA blends (Gonzalez Ausejo et al., 2018a; Gonzalez Ausejo et al., 2018b).

PLA dumbbell-shaped bars printed in the horizontal direction showed significant deformation during degradation as a result of the printing algorithm used (crisscross pattern). The bars printed in the vertical direction (transverse pattern) showed slightly less deformation during degradation. It is known that the main disadvantage of PLA is its deformation at relatively low temperatures, especially above T_g (Liang et al., 2012). The PHA component in the PLA/PHA blend resulted in a significant reduction in deformation during the degradation of printed dumbbell-shaped bars in both directions, as well as increased thermal stability during degradation, therefore these bars were also stiffer and stronger (Young's modulus and tensile strength were higher). Young's modulus was lower at the beginning of the hydrolytic degradation of the dumbbell-shaped bars as it causes the polymer to be soften due to the plasticisation effect of oligomeric degradation products. In general, the bars become softer but less strong as the elongation at break has decreased for all bars. After a longer period of degradation, especially at a temperature of 70°C, the degree of crystallinity and therefore the brittleness of the polymer increases and it becomes stiffer and less strong (Gonzalez Ausejo et al., 2018a; Gonzalez Ausejo et al., 2018b).

TG analysis of PLA and PLA/PHA dumbbell-shaped bars showed that no significant changes in the thermal stability of PLA were observed during the degradation at 50°C. In the case of the thermal decomposition of PLA/PHA dumbbell-shaped bars, which is a two-step mass loss, a slight shift towards a higher temperature of maximum rate of mass loss (T_{max}) value of PHA was noticed, which indicates an increase in the thermal stability of the polymer, in particular the PHA component, as a result of improved miscibility of components due to the reduction of molar mass and the plasticisation effect of oligomeric products formed during the degradation of the PLA/PHA blend (Lai et al., 2017). This effect was more pronounced for dumbbell-shaped bars degraded at 70°C.

Due to the fact that, during hydrolytic degradation, both the crystallinity and the weaker cohesion between the printed layers of PLA/PHA and PLA dumbbell-shaped bars caused changes during degradation, prototypes of cosmetic packaging were also tested to determine differences in degradation behaviour of individual parts – container bottom or lid and walls. For the purposes of the planned research, a number of tests (procedure) were developed to check the course of (bio)degradation of polymers. This procedure involved virtual tests under various environmental conditions, including an accelerated aging test, packaging/cosmetic formulation compatibility test with real cosmetic formulations, composting under laboratory and industrial conditions as well as natural weathering test. The influence of cosmetic contamination on the course of (bio)degradation in selected environments was also investigated (Gonzalez Ausejo et al., 2018a; Gonzalez Ausejo et al., 2018b).

The effects of aging of the polymer and therefore the long-term storage up to the product expiry date (3 months, 6 months, one year, 3 years, etc.) of the packaging/product system can be simulated by accelerated aging tests. Accelerating the changes taking place in a normal environment makes it possible to determine the shelf life of the product under real conditions (2013/674/EU). The progress of aging of the test items was estimated on the basis of failure analysis (macroscopic observations of the surface), changes in molar mass and thermal properties. A preliminary accelerated aging experiment at 55°C of PLA packaging prototypes in the presence of selected media (paraffin, ethanol and deionised water) and an empty container (blank test) over 37 days (equivalent to real-time aging of one year) showed that the cosmetic containers were deformed from the beginning of the experiment and confirmed their unsuitability for long-term applications (Rydz et al., 2019).

The higher degree of crystallinity of the PLA/PHA container increased its stability at elevated temperatures. The molar mass loss during aging with paraffin and blank test occurred in the following order: PLA with paraffin > PLA/PHA blank test > PLA blank test > PLA/PHA with paraffin. The residual moisture content, as in the case of the films, resulted in faster degradation of PLA containers filled with paraffin. Paraffin also has a heat buffering capacity and cooling effect (Bremerkamp et al., 2012; Cao, 2016), therefore it slowed the degradation of PHA in the blend and thus the degradation of the entire container (Rydz et al., 2019).

The research carried out for the PLA/PHA packaging prototypes showed that the degradation of the PLA/PHA blend caused significant changes in properties only after 111 days of aging with paraffin (which corresponds to real-time aging of three years), which is beneficial from the point of view of applications with a long shelf life use (Rydz et al., 2019).

The packaging/cosmetic formulation compatibility test with real cosmetic formulations consists in checking the interaction of the cosmetic mass with the packaging in the surrounding environment in order to verify their fit. The packaging should be compatible with the product, which means that all the ingredients of the cosmetic formulation do not adversely affect the packaging, and, vice versa, the components of the packaging do not react with the cosmetic masses. The evaluation criteria taken into account for both the packaging and the cosmetic formulation are appearance, colour and odour. In the case of PLA containers, the packaging/cosmetic formulation compatibility test was negative. In the case of PLA/PHA containers, during the packaging/cosmetic formulation compatibility test after 12 weeks at ambient temperature, regardless of the type of cosmetic formulation, the containers opened and the mass loss was 50–70% for moisturising masses and 30–40% for oily cosmetic masses. In the case of containers with cosmetic formulations incubated at 45°C, a significant mass loss of cosmetic masses was observed, amounting to 75–80% for moisturising and 60–75% for oily masses with a slight change in colour (darkening), as well as in the odour (unpleasant, pungent). No odour change occurred when tested in conventional polypropylene cosmetic containers. During the packaging/cosmetic formulation compatibility test, the deformation of the cosmetic containers was in the range of 1–5%. There were no cracks and the colour of the containers did not change. Due to the PHA component of the blend, the PLA/PHA containers have better compatibility with cosmetic formulations but are still unsatisfactory (Rydz et al., 2019).

In order to investigate the influence of contamination with cosmetic formulations on the course of degradation under the laboratory and industrial composting conditions, as well as natural weathering conditions, a comparative degradation test of empty cosmetic containers (blank test) and cosmetic containers with a small amount of paraffin was carried out. Biodegradation tests were also carried out under laboratory conditions simulating aerobic composting in industrial composting plants. The carbon dioxide released during the biodegradation of the containers was measured according to the test procedure at constant process parameters (Rydz et al., 2019).

Biodegradation of empty PLA containers during laboratory composting was slower than PLA/PHA containers, because the PLA component of the blend, although commonly considered a biodegradable polymer, as previously proven decomposes in the compost by simple chemical hydrolysis taking place relatively quickly under these conditions, while the presence of natural PHA that degrades mainly enzymatically accelerates the entire process. The degradation process of containers with paraffin contamination in both laboratory and industrial compost was higher compared to the empty prototypes. Under the influence of an environment rich in enzymes produced by bacteria and fungi, paraffin accelerated (bio)degradation as an additional source of carbon and energy. On the other hand, under the influence of natural weathering conditions in which temperature and humidity were important, paraffin slowed down the degradation. Macroscopic visual evaluation of PLA/PHA and PLA containers after degradation under laboratory and industrial composting conditions showed erosion by fragmentation, especially at the point of contact between the wall and the bottom of the container, where the cohesion between printed layers was weaker. The

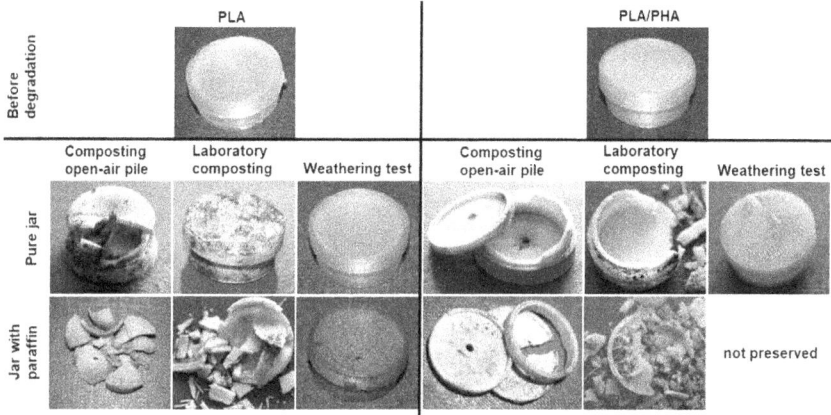

FIGURE 6.3 Macrographs of PLA packaging prototypes before degradation and after 84 days of composting under laboratory and industrial conditions as well as after 365 days of degradation under natural weathering conditions.

Source: reprinted with permission from Rydz et al., 3D-printed polyester-based prototypes for cosmetic applications-future directions at the forensic engineering of advanced polymeric materials, *Materials* no 12(6) (2019):994. For more details, see the CC BY 4.0

parts with a greater structural ordering of the polymer (contact with the printer build platform), the bottom and the lid of the container degrade more slowly compared to the wall of the container. In the case of cosmetic containers with the addition of paraffin, the disintegration was much faster because the paraffin contamination accelerated the degradation (Figure 6.3) (Rydz et al., 2019).

6.2.4 Plastic Defects and Their Impact on Degradation Behaviour

The use of (bio)degradable polymers, especially in medical applications, requires a proper understanding of their properties and behaviour in various environments. Structural elements made of such polymers may be exposed to changing environmental conditions, which may cause defects. During standard hydrolytic degradation study, the PLA and PLA/PHA dumbbell-shaped bars obtained by 3D printing showed an unexpected shrinkage phenomenon, representing about 50% of the bar length, regardless of the printing direction. Usually polymers such as PLA break down already in the initial step of hydrolytic degradation (at 70°C after 7 days). However, in this case, no significant disintegration occurred after 70 days of hydrolytic degradation at 70°C (Figure 6.4; Rydz et al., 2020).

Due to the fact that this is a rather significant defect, research was undertaken on the impact of a specific processing method, i.e. processing parameters and conditions, on the properties expressed by the change in molar mass and thermal properties of dumbbell-shaped bars obtained by the 3D printing from PLA and PLA/PHA filaments in order to identify the causes of this phenomenon and to study its impact on the degradation process (Rydz et al., 2020).

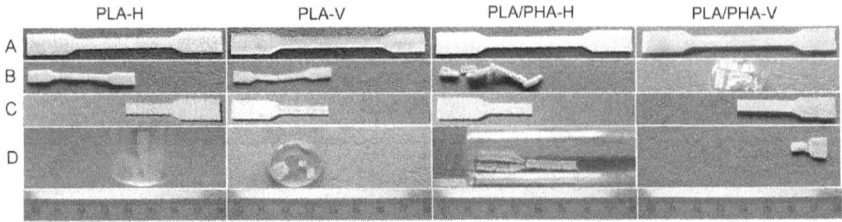

FIGURE 6.4   Macrographs of PLA and PLA/PHA dumbbell-shaped bars obtained by 3D printing in the horizontal (H) and vertical (V) directions before (A and C) and after 70 days of hydrolytic degradation at 70°C (B – whole bars, D – bars cut in half).

Source: reprinted with permission from Rydz et al., Three-dimensional printed PLA and PLA/PHA dumbbell-shaped specimens: Material defects and their impact on degradation behavior. *Materials* no 13(8) (2020):2005. For more details, see the CC BY 4.0

During processing, polymers may have a tendency not only to orient their structure but also to shrink. The shrinkage (longitudinally and/or transversely) begins at the molecular level as the polymer first melts and then cools (Lampman, 2003). The shrinkage effect can be considered a shape memory property, because after being permanently deformed under the influence of mechanical loads, the polymer can recover its original shape only under the action of temperature. In the case of semicrystalline polymers, the shape memory property is manifested by the thermal retraction of the polymer when the molecular structure has been oriented by mechanical load. After the imposed strain has been applied, the shape remains immovable until the temperature is equal to or lower than the deformation temperature. If the temperature rises, a retraction occurs or shrinkage stress arises (Scalet et al., 2015).

Microscopic observation is a useful tool in detecting 3D printing defects. The scanning electron microscopy (SEM) images clearly show that the width of the filament has decreased in the range from 41% to 66% (Figure 6.5) according to the following order: PLA-H > PLA/PHA-H > PLA-V > PLA/PHA-V for whole bars and PLA/PHA-V > PLA/PHA-H for bars cut in half (taking into account the average of both layers in the horizontal direction of printing; Rydz et al., 2020).

Interestingly, in the case of whole bars, the highest shrinkage was observed for PLA obtained by 3D printing in the vertical direction (51% of the total length, 40% for the vertical direction). Shrinkage is the result of thermal and mechanical treatment. Treatment during the printing process heated the filament to a temperature above the T_m of PLA and PHA; then the polymers were cooled, leading to shrinkage caused by three possible causes. First, as the polymer phase changes from molten to solid, a change in contraction rate may occur. Second, the crystalline regions have a lower expansion coefficient than the amorphous regions. PLA and PHA can crystallise below the T_m and their volume changes after crystallisation. If the ambient temperature (the temperature of the degradation medium or conditioning environment) has been higher than the T_g, unexpected warping and/or shrinkage may occur. Third, normal thermal contraction may also occur during cooling due to normal thermal expansion (Rydz et al., 2020 and references cited therein).

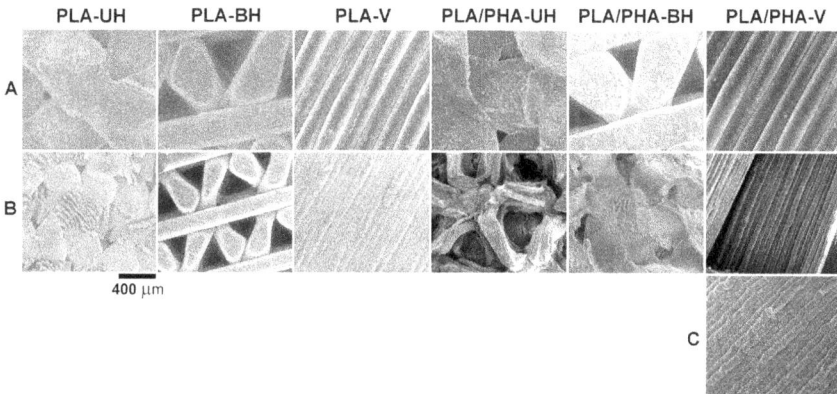

FIGURE 6.5 Selected SEM images of the upper (UH) and underside (BH) layers of PLA and PLA/PHA dumbbell-shaped bars obtained by 3D printing in the horizontal (H) and vertical (V) directions before (A) and after 70 days of hydrolytic degradation at 70 °C (B – whole bars, C – bars cut in half).

Source: reprinted with permission from Rydz et al., Three-dimensional printed PLA and PLA/PHA dumbbell-shaped specimens: Material defects and their impact on degradation behavior, *Materials* no 13(8) (2020):2005. For more details, see the CC BY 4.0

The morphology of the plastics, especially in the case of blends, plays a significant role on the shrinkage effect. Mixing different polymers modifies the phase morphology, also by nucleation, which to some extent affects the shrinkage effect. The shrinkage is related to the molar mass of the dispersed phase and the viscosity ratio between the components of the polymer matrix, respectively. Nucleation usually increases the rate of crystallisation of polymers. This reduces the size of the crystals and causes an increase in their number and better dispersion. In this case, the lower shrinkage of PLA/PHA dumbbell-shaped bars could have been affected by the nucleation effect due to the presence of the PHA component of the blend (acting as a nucleating agent; Rydz et al., 2020).

There were no significant differences during the hydrolytic degradation in the values of the molar mass loss between the dumbbell-shaped bars with different printing directions within each polymer, PLA and PLA/PHA blend. However, when comparing PLA and PLA/PHA, a greater shift towards low molar masses was observed in the case of PLA dumbbell-shaped bars. The lower molar mass reduction of PLA/PHA dumbbell-shaped bars was due to the different degree of shrinkage of the individual components of the blend, causing the warping effect of the filament itself forming the dumbbell-shaped bar and reducing the space between the individual filaments, which limited the penetration of water into the polymer matrix and the removal of low-molar-mass degradation products. Additional stress caused by cutting the bars in half resulted in an increase in the molar-mass dispersion, which suggests that cutting both PLA and PLA/PHA dumbbell-shaped bars obtained by the 3D printing in the vertical direction (V) led to a greater shrinkage compared to the uncut, which further limited water penetration (Rydz et al., 2020).

Comparing the DSC thermograms of the PLA and PLA/PHA dumbbell-shaped bars before and after 70 days of hydrolytic degradation at 70°C during the first heating run, a melting effect was observed, followed by an evaporation effect, which suggests evaporation of part of the bars, which confirmed the mass reduction of the analysed samples. In subsequent heating runs, further slow evaporation took place, which resulted in a shift of the T_g of the PLA component (also in the blend) towards higher values. This suggests that the dumbbell-shaped bars are degraded to amorphous oligomers and hydroxy acids, which, due to remaining in the polymer matrix, evaporated during analysis, causing an increase in the T_g of the PLA component. Usually, during hydrolytic degradation, the amorphous phase decomposes faster and the T_m increases, however, in this case a large amount of amorphous low-molar-mass degradation products have been trapped in the polymer matrix due to shrinkage and less water penetration into the matrix, acting as a plasticiser, which resulted in a decrease in T_m (Rydz et al., 2020).

The factors that may influence the shrinkage of the final product are the structure and properties of the polymer, processing parameters (printing time and speed, temperature and cooling speed), element geometry and its thickness. Non-uniform cooling of the printer build platform can cause stresses between different areas of the printed item and can also cause distortion. The shrinkage phenomenon can also be influenced by various external factors, for example, the action of forces on the filament during printing, the temperature of the printer's build platform, partial closure of the nozzle (giving too little polymer), the pH of the degradation medium, external temperature or sunlight falling on the polymer during processing. Therefore, the conditions during the printing of dumbbell-shaped bars were analysed. However, there were no clear changes that could be indicative of shrinkage, either in the case of a change in printing time and speed – which could cause additional force on the filament – or in the case of changes in ambient temperatures and cooling speed. The nozzle fed the polymer in the same amount (the same mass of the bars). The phenomenon of shrinkage after heat treatment occurred both in unconditioned and conditioned dumbbell-shaped bars, while it was observed that the shrinkage was influenced by the element geometry and the temperature of the printer's build platform and the nozzle. Unexpected shrinkage phenomenon was observed after heat treatment at 110°C but only for bars with a specific geometry (Rydz et al., 2020).

In order to eliminate the shrinkage effect at 110°C, the printing conditions of the dumbbell-shaped bars were changed; in particular, different temperatures of the build plate and the nozzle were used. When the temperature of the printer build platform and the nozzles increased from 70°C and 220°C to 120°C and 240°C, respectively, shrinkage and deformation decreased but were not completely eliminated. It is also worth mentioning that the shape memory property, manifested by the return of the shape of the polymer due to temperature, may play a significant role in the defects of PLA-containing plastics (Figure 6.6) (Rydz et al., 2020).

Shrinkage phenomena during degradation resulted in amorphous oligomers and hydroxy acids trapped in the polymer matrix. Low-molar-mass degradation products, due to lower water penetration into the matrix, remained therein, increasing the molar mass dispersity, while causing less disintegration of the dumbbell-shaped bars. The additional stress caused by cutting the bars in half disrupted the order

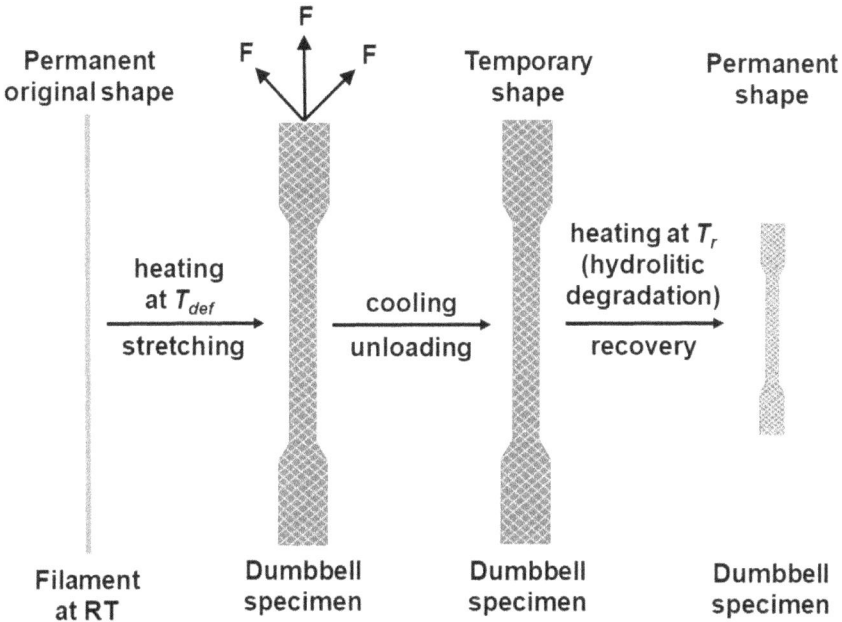

FIGURE 6.6 Shrinkage effect considered as a shape memory property; T_{def} – deformation temperature, T_r – temperature of retraction process.

Source: reprinted with permission from Rydz et al., Three-dimensional printed PLA and PLA/PHA dumbbell-shaped specimens: Material defects and their impact on degradation behavior, *Materials* no 13(8) (2020):2005. For more details, see the CC BY 4.0

of the polymer structure and led to a further increase in the molar mass dispersity, which suggests that cutting both PLA bars (both directions of printing) as well as PLA/PHA obtained by 3D printing in the vertical direction led to a greater shrinkage compared to uncut bars, which further limited water penetration (Rydz et al., 2020).

6.3 CONCLUSION

Determining the time of safe use of products made of (bio)degradable polymers, optimisation and understanding of physicochemical changes in their structure are crucial for their numerous applications. The continual development of new plastics that are stronger, lighter or more versatile than the previous ones must not only improve safety but also reduce environmental concerns as the complexity of recovering the value of the used product increases. Current product design challenges are leading to the development of plastics that are stable in use and at the same time susceptible to microbial attack during organic recycling. For each application of polymers, understanding which ones are optimal for the desirability of their targeted applications allows for accurate prediction of behaviour and quality assessment throughout their life cycle, under real conditions. Potential failures can be avoided if all factors are taken into account early in the process of developing new products.

The degradation process of PLA-based material in an environment with a residual moisture content, such as paraffin, takes place and indicates the occurrence of morphological and structural changes in the tested materials. Blends with a higher content of the (R,S)-PHB component (12–15%) in the polymer matrix extend the time of material disintegration, and blends with good miscibility reduce the degree of degradation in the first weeks of the process. The aliphatic-aromatic PBAT/PLA copolyester blend is less prone to degradation in paraffin than the PLA/(R,S)-PHB blend. Studies have shown good stability of the PBAT/PLA blend during degradation in paraffin, which means that it can be a suitable material for the production of compostable cosmetics packaging. In addition, hydrolysis from chain-end of the PLA is observed during incubation in water, which slows down its degradation in the blend. It was also found that the degree of erosion of the rigid PLA/(R,S)-PHB films depended on the composition of the blend and the miscibility of both components. Knowledge of the degradation and damage phenomena as well as the deterioration of performance properties under operating conditions of (bio)degradable polymers – and therefore the prediction of their behaviour and selection of appropriate test procedures – determines the scope of applications, possibilities and limitations of the finished product, as well as its durability. The main challenge was the development of a new methodology and procedures, i.e. a strategy for the comprehensive characterisation of (bio)degradable polymers, including the study of their behaviour during disposal (organic recycling, storage, simulating packaging contamination) of waste. Therefore packaging prototypes were tested under simulated and real conditions corresponding to their life cycle. The performed study showed that: (1) it has been observed that the degradation of the tested aliphatic polyesters occurs not only in such cosmetic simulants as ethyl alcohol, glycerine and propylene glycol but also in the presence of an inert medium such as paraffin. The degradation of PLA in paraffin is due to the residual moisture content which causes the autocatalytic effect. The oligomers are not released into the hydrophobic medium and the erosion takes place in the polymer matrix. A blend of the aliphatic-aromatic copolyester PBAT and PLA is less prone to degradation in paraffin than PLA itself or a blend of PLA with (R,S)-PHB; (2) in polymer blends containing PLA, a change in the degradation mechanism was observed from the random hydrolysis of the polymer chain (privileged) to chain-end scission, which can significantly change the degradation rate and extend the shelf life of the polymer; (3) the influence of processing conditions and printing parameters of test items on their surface structure, properties and abiotic degradation were observed. Not only the contact time but also the size of the surface in contact with the printer's build platform had a significant impact on the growth of the crystalline phase and the ordering of the structure of the test bars during the 3D printing process. The build direction and the orientation of layers turned out to be a more important parameter determining degradation than the hydrophobicity of polymers; (4) the unexpected phenomenon of shrinkage was also observed in the case of dumbbell-shaped bars made of PLA or with its addition, which is quite a significant defect that can be of fundamental importance in specialised applications.

REFERENCES

2013/674/EU. *Commission Implementing Decision of 25 November 2013 on Guidelines on Annex I to Regulation (EC) No 1223/2009 of the European Parliament and of the Council on cosmetic products*. Brussels.

Anderson, K.A., Randall, J.R., and Kolstad, J.J. 2011. *Polylactide Molding Compositions and Molding Process*. Nature Word Patent WO 2011085058 A1.

Andersson, S.R., Hakkarainen, M., Inkinen, S., Södergård, A., and Albertsson, A.-C. 2010. Polylactide stereocomplexation leads to higher hydrolytic stability but more acidic hydrolysis product pattern. *Biomacromolecules* 11: 1067–73.

Asrar, J., and D'haene, P. 1999. *Modified Polyhydroxyalkanoates for Production of Coatings and Films*. International Patent Application. PCT/US1998/019461.

Avella, M., De Vlieger, J.J., Errico, M.E., Fischer, S., Vacca, P., and Volpe, M.G. 2005. Biodegradable starch/clay nanocomposite films for food packaging applications. *Food Chem.* 93: 467–474.

Bremerkamp, F., Seehase, D., Nowottnick, M. 2012. Heat protection coatings for high temperature electronics. In *Proceedings of the 35th International Spring Seminar on Electronics Technology*. Bad Aussee, Austria, 9–13 May, 9–14.

Bucci, D.Z., Tavares, L.B.B., and Sell, I. 2005. PHB packaging for the storage of food products. *Polym. Test* 24(5): 564–571.

Cam, D., Suong-Hyu, H., and Ikada, Y. 1995. Degradation of high molecular weight poly(L-lactide) in alkaline medium. *Biomaterials* 16: 833.

Cao, H. 2016. Smart coatings for protective clothing. In *Active Coatings for Smart Textiles*, ed. Hu, J., pp. 375–389. Cambridge: Woodhead Publishing Series in Textile, Elsevier.

CEN EN 1186-1:2002. 2002. *Materials and Articles in Contact with Foodstuffs – Plastics – Part 1: Guide to the Selection of Conditions and Test Methods for Overall Migration*. Brussels: European Committee for Standardization.

Colomines, G., Domenek, S., Ducruet, V., and Guinault, A. 2008. Influences of the crystallisation rate on thermal and barrier properties of polylactide acid (PLA) food packaging films. *Int. J. Mater. Form.* 1: 607–610.

Connolly, M., Zhang, Y., Brown, D.M., Ortuño, N., Jordá-Beneyto, M., Stone, V., Fernandes, T.F., and Johnston, H.J. 2019. Novel polylactic acid (PLA)-organoclay nanocomposite bio-packaging for the cosmetic industry; migration studies and *in vitro* assessment of the dermal toxicity of migration extracts. *Polym. Degrad. Stab.* 168: 1089382.

Corma, A., Iborra, S., and Velty, A. 2007. Chemical routes for the transformation of biomass into chemicals. *Chem. Rev.* 107: 2411–2502.

Endres, H.-J., and Siebert-Raths, A. 2011. End-of-life options for biopolymers. In *Engineering Biopolymers. Markets, Manufacturing, Properties and Applications*, pp. 225–243. Munich, CT: Hanser Publishers.

Filiciotto, L., and Rothenberg, G. 2021. Biodegradable plastics: Standards, policies, and impacts. *ChemSusChem*, 14(1): 56–72.

Freier, T., Kunze, C., Nischan, C., et al. 2002. *In vitro* and *in vivo* degradation studies for development of a biodegradable patch based on poly(3-hydroxybutyrate). *Biomaterials* 23: 2649–2657.

Gilbert, S.G. 1975. Low molecular weight components of polymers used in packaging. *Environ. Health Perspect.* 11: 47–52.

Gleadall, A., Pan, J., Kruft, M.-A., Kellomäki, M. 2014. Degradation mechanisms of bioresorbable polyesters. Part 1. Effects of random scission, end scission and autocatalysis. *Acta Biomater.* 10: 2223–2232.

Gonzalez Ausejo, J., Rydz, J., Musioł, et al. 2018a. A comparative study of three-dimensional printing directions: The degradation and toxicological profile of a PLA/PHA blend. *Polym. Degrad. Stab.* 152: 191–207.

Gonzalez Ausejo, J., Rydz, J., Musioł, et al. 2018b. Three-dimensional printing of PLA and PLA/PHA dumbbell-shaped specimens of crisscross and transverse patterns as promising materials in emerging application areas: Prediction study. *Polym. Degrad. Stab.* 156: 100–110.

Grizzi, I., Garreau, H., Li, S., and Vert, M. 1995. Hydrolytic degradation of devices based on poly(*D,L*-lactic acid) size-dependence. *Biomaterials* 16: 305–311.

Hakkarainen, M., Karlsson, S., and Albertsson, A.C. 2000. Rapid (bio)degradation of polylactide by mixed culture of compost microorganisms – Low molecular weight products and matrix changes. *Polymer* 41: 2331–2338.

He, Y., Asakawa, N., and Inoue, Y. 2000. Blends of poly(3-hydroxybutyrate)/4,4′-thiodiphenol and poly(3-hydroxybutyrate-*co*-3-hydroxyvalerate)/4,4′-thiodiphenol: Specific interaction and properties. *J. Polym. Sci. B Polym. Phys.* 38: 2891–2900.

Huda, M.S., Yasui, M., Mohri, N., Fujimura, T., and Kimura, Y. 2002. Dynamic mechanical properties of solution-cast poly(*L*-lactide) films. *Mater. Sci. Eng. A.* 333: 98–105.

Jurczyk, S., Musioł, M., Sobota, M., Klim, M., Hercog, A., Kurcok, P., Janeczek, H., and Rydz J. 2019. (Bio)degradable polymeric materials for sustainable future – Part 2: Degradation studies of P(3HB-*co*-4HB)/cork composites in different environments. *Polymers* 11(3): 547.

Kyulavska, M., Toncheva-Moncheva, N., and Rydz, J. 2019. Biobased polyamide ecomaterials and their susceptibility to biodegradation. In *Handbook of Ecomaterials*, eds. Martínez, L.M.T., Kharissova, O.V., and Kharisov, B.I., pp. 2901–2934. Cham: Springer International Publishing AG.

Lampman, S. 2003. *Characterization and Failure Analysis of Plastics*. Materials Park: ASM International.

Lai, S.M., Liu, Y.H., Huang, C.T., and Don, T.-M. 2017. Miscibility and toughness improvement of poly(lactic acid)/poly(3-hydroxybutyrate) blends using a meltinduced degradation approach. *J. Polym. Res.* 24(102): 12.

Lewis, C. 1998. Clearing up cosmetic confusion. *FDA Consum.* 32(3): 6–11.

Li, S.M., Garreau, H., and Vert, M. 1990. Structure-property relationships in the case of the degradation of massive aliphatic poly-(α-hydroxy acids) in aqueous media. *J. Mater. Sci. Mater. Med.* 1: 123–130.

Li, S.M., and Vert, M. 2002. *Degradable Polymers. Principles and Applications*. Dordrecht: Kluwer Academic Publishers.

Liang, F., Todd, B.L., and Saini, R.K. 2012. *Reinforcing Amorphous PLA with Solid Particles for Downhole Applications*. European Patent EP 2764068 A1.

Marx, S. 2004. Guidelines on stability testing of cosmetics product. *Cosmetics Europe – The Personal Care Association, Colipa.* www.packagingconsultancy.com/pdf/cosmeticscolipa-testing-guidelines.pdf (accessed February 10, 2022).

Millet, H., Vangheluwe, P., Block, Ch., Sevenster, A., Garcia L., and Antonopoulos R. 2019. The nature of plastics and their societal usage, in plastics and the environment. In *Plastics and the Environment*, eds. Harrison, R.M., and Hester, R.E., pp. 1–20. Cambridge: The Royal Society of Chemistry.

Moshood, T.D., Nawanir, G., Mahmud, F., Mohamad, F., Hanafiah Ahmad, M., and AbdulGhani, A. 2022. Biodegradable plastic applications towards sustainability: A recent innovations in the green product. *Clean. Eng. Technol.* 6: 100404.

Musioł, M., Rydz, J., Janeczek, H., et al. 2022. (Bio)degradable biochar composites – Studies on degradation and electrostatic properties. *Mater. Sci. Eng. B.* 275: 115515.

Musioł, M.T., Rydz, J., Sikorska, W.J., Rychter, P.R., and Kowalczuk, M.M. 2011. A preliminary study of the degradation of selected commercial packaging materials in compost and aqueous environments. *Pol. J. Chem. Tech.* 13: 55–57.

Narayan, R. 2012. Biobased & Biodegradable plastics: Rationale, Drivers, and technology exemplars. In *Degradable Polymers and Materials: Principles and Practice*, pp. 13–31 (ACS Symposium Series). Washington, DC: American Chemical Society.

Naser, A.Z., Deiab, I., and Darras, B.M. 2021. Poly(lactic acid) (PLA) and polyhydroxyalkanoates (PHAs), green alternatives to petroleum-based plastics: A review. *RSC Adv.* 28: 17151–17196.

No 10/2011. 2011. *Commission Regulation (EU) No 10/2011 of 14 January 2011 on Plastic Materials and Articles Intended to Come into Contact with Food*. Brussels: European Commission regulation (EC).

Prieto, A. 2016. To be, or not to be biodegradable . . . that is the question for the bio-based plastics. *Microb. Biotechnol.* 9(5): 652–657.

Reliance Industries Limited. 2003. *Drying RELPET*. Mumbai: Marketing Technical Services.

Rieger, M.M. 2001. Cosmetics. In *Kirk-Othmer Encyclopedia of Chemical Technology*, pp. 820–865. New York: John Wiley & Sons.

Rutkowska, M., Krasowska, K., Heimowska, A., et al. 2008. Environmental degradation of blends of atactic poly[(*R,S*)-3-hydroxybutyrate] with natural PHBV in Baltic sea water and compost with activated sludge. *J. Polym. Environ.* 16: 183–191.

Rychter, P., Biczak, R., Herman, B., et al. 2006. Environmental degradation of polyester blends containing atactic poly(3-hydroxybutyrate). Biodegradation in soil and ecotoxicological impact. *Biomacromolecules* 7: 3125–3131

Rydz, J., Adamus, G., Wolna-Stypka, K., Marcinkowski, A., Misiurska-Marczak, M., and Kowalczuk, M. 2013a. Degradation of polylactide in paraffin and selected protic media. *Polym. Degrad. Stab.* 98(1): 316–324.

Rydz, J., Musioł, M., and Janeczek, H. 2015a. Thermal analysis in the study of polymer (bio) degradation. In *Reactions and Mechanisms in Thermal Analysis of Materials*, eds. Tiwari, A., and Raj, B., pp. 103–126. Beverly: Wiley-Scrivener Publishing LLC.

Rydz, J., Musioł, M., and Kowalczuk, M. 2017. Polymers tailored for controlled (bio)degradation through end-group and in-chain functionalization. *Curr. Org. Chem.* 14(6): 768–777.

Rydz, J., Sikorska, W., Kyulavska, M., and Christova, D. 2015b. Polyester-based (bio)degradable polymers as environmentally friendly materials for sustainable development. *Int. J. Mol. Sci.* 16(1): 564–596.

Rydz, J., Sikorska, W., Musioł, M., et al. 2019. 3D-printed polyester-based prototypes for cosmetic applications-future directions at the forensic engineering of advanced polymeric materials. *Materials* 12(6): 994.

Rydz, J., Włodarczyk, J., Gonzalez Ausejo, J., et al. 2020. Three-dimensional printed PLA and PLA/PHA dumbbell-shaped specimens: Material defects and their impact on degradation behavior. *Materials*, 13(8): 2005.

Rydz, J., Wolna-Stypka, K., Adamus, G., Janeczek, H., Musioł, M., Sobota, M., Marcinkowski, A., Krzan, A., and Kowalczuk, M. 2015c. Forensic engineering of advanced polymeric materials. Part 1 – Degradation studies of polylactide blends with atactic poly[(*R,S*)-3-hydroxybutyrate] in paraffin. *Chem. Biochem. Eng. Q.* 29(2): 247–259.

Rydz, J., Wolna-Stypka, K., Musioł, M., Szeluga, U., Janeczek, H., and Kowalczuk, M. 2013b. Further evidence of polylactide degradation in paraffin and in selected protic media. A thermal analysis of eroded polylactide films, *Polym. Degrad. Stab.* 98(8): 1450–1457.

Santonja-Blasco, L., Moriana, R., Badía, J.D., and Ribes-Greus, A. 2010. Thermal analysis applied to the characterization of degradation in soil of polylactide: I. Calorimetric and viscoelastic analyses, *Polym. Degrad. Stab.* 95: 2185–2191.

Scalet, G., Auricchio, F., Bonetti, E., et al. 2015. An experimental, theoretical and numerical investigation of shape memory polymers. *Int. J. Plasticity* 67: 127–147.

Shaikh, S., Yaqoob, M., and Aggarwal, P. 2021. An overview of biodegradable packaging in food industry. *Curr. Res. Food Sci.* 4: 503–520.

Shaw, G., Morse, S., Ararat, M., and Graham, F.L. 2002. Preferential transformation of human neuronal cells by human adenoviruses and the origin of HEK 293 cells. *FASEB J.* 6: 869–871.

Sikorska, W., Musioł, M., Zawidlak-Węgrzyńska, B., and Rydz, J. 2019. Compostable polymeric ecomaterials: Environment-friendly waste management alternative to landfills. In *Handbook of Ecomaterials*, eds. Martínez, L.M.T., Kharissova, O.V., and Kharisov, B.I., pp. 2733–2764. Cham: Springer International Publishing AG.

Sikorska, W., Richert, J., Rydz, J., et al. 2012. Degradability studies of poly(*L*-lactide) after multireprocessing experiments in extruder. *Polym. Degrad. Stabil.* 97: 1891–1897.

Sikorska, W., Rydz, J., Wolna-Stypka, K., et al. 2017. Forensic engineering of advanced polymeric materials – Part V: Prediction studies of aliphatic-aromatic copolyester and polylactide commercial blends in view of potential applications as compostable cosmetic packages. *Polymers* 9(7): 257.

Siparsky, G.L., Voorhees, K.J., and Miao, F. 1988. Hydrolysis of polylactic acid (PLA) and polycaprolactone (PCL) in aqueous acetonitrile solutions: autocatalysis. *J. Polym. Environ.* 6: 31–41.

Thomas, P., and Smart, T.G. 2005. HEK293 cell line: A vehicle for the expression of recombinant proteins. *J. Pharmacol. Toxicol. Meth.* 51(3): 187–200.

Thor, P., Jolly, M., Montgomery, J., et al. 2021. Humidity as a use condition for accelerated aging of polymers. *Medtronic,* February 19. www.mddionline.com/testing/humidity-use-condition-accelerated-aging-polymers (accessed February 11, 2022).

Urbaniec, K., Mikulcic, H., Rosen, M.A., and Duic, N. 2017. A holistic approach to sustainable development of energy, water and environment systems. *J. Clean. Prod.* 155: 1–11.

Winter, R.A. 2009. *Consumer's Dictionary of Cosmetic Ingredients.* New York: Three Rivers Press.

Włodarczyk, J., Sikorska, W., Rydz, J., et al. 2017.3D processing of PHA containing (bio) degradable materials. In *Current Advances in Biopolymer Processing & Characterization*, ed. Koller, M., pp. 121–68. New York: Nova Science Publishers.

Volova, T., Boyandin, A.N., Prudnikova, S.V., et al. 2015. Biodegradation of polyhydroxyalkanoates in natural water environments. *J. Sib. Fed. Univ. Biol.* 2(8): 168–186.

Zbik, M., Horn, R.G., and Shaw, N. 2006. AFM study of paraffin wax surfaces. *Colloids Surf A* 287: 139–146.

Zhao, Y., Keroack, D., and Prud¢homme, R. 1999. Crystallization under strain and resultant orientation of poly(ε-caprolactone) in miscible blends. *Macromolecules* 32: 1218–1225.

7 (Bio)degradable Polymers of Biological and Biomedical Interest

Barbara Zawidlak-Węgrzyńska

CONTENTS

7.1 INTRODUCTION

In the last few decades, polymeric materials based on biodegradable polymers have been increasingly used in various medical fields. Initial acceptance of these materials was somewhat sceptical and slow, but as natural or synthetic biodegradable polymers have proven their value, demand has increased. However, it must be determined whether the characteristics of a material match its application and therefore all polymeric biomaterials for specific applications must be assessed for biocompatibility, mechanical properties and biodegradability. Biodegradation describes the process by which a material breaks down in nature. However, in the case of biomaterials for medical applications, biodegradation focuses on the biological processes occurring in the body that lead to the gradual breakdown of the material. A very important aspect is also degradation, which affects the ability of biomaterials to function in a specific application for a specific period of time. Biodegradation occurs through the action of enzymes and/or chemical decomposition associated with living organisms. This process takes place in two stages. The first stage involves the fragmentation of polymers into a low-molar-mass form through abiotic reactions, i.e. oxidation, photodegradation hydrolysis or biotic reactions, i.e. degradation by microorganisms. In the second stage, the bioassimilation of polymer fragments by microorganisms and their mineralisation occurs. The biodegradability is influenced not only by the origin of the polymer but also by its chemical structure and degradation conditions in the environment. The fact that biodegradable polymers are easily absorbed by the human body influences both the range of applications and the extent to which they are used in medicine, pharmacy, dentistry and tissue engineering (Kalirajan et al.,

DOI: 10.1201/9780429352799-8

2021). Examples of applications include: various medical and dental devices or components, surgical sutures and staples or biodegradable components used in ligament repair. Natural and synthetic biodegradable polymers have also found wide application in the manufacture of instruments used during various surgical procedures, as well as in the production of devices that are permanently implanted. Orthopaedics, cardiac procedures and intestinal surgery depend on the availability of instruments, tools and implantable components made from biodegradable polymers. The availability of film-coated or time-release drugs requires the availability of carefully selected and formulated polymers to release specific doses of the drug evenly over time. Biodegradable polymers are widely used as drug delivery systems and also as site-specific drug delivery systems (see Figure 7.1).

7.2 NATURAL POLYMERS

7.2.1 POLYSACCHARIDES AND THEIR DERIVATIVES

Chitin and chitosan are non-toxic, biodegradable and biocompatible natural polymers (Singh, et al., 2017). Chitin is one of the most common polymers in nature, found in the exoskeletons of crustaceans and molluscs, cuticles of insects and fungi.

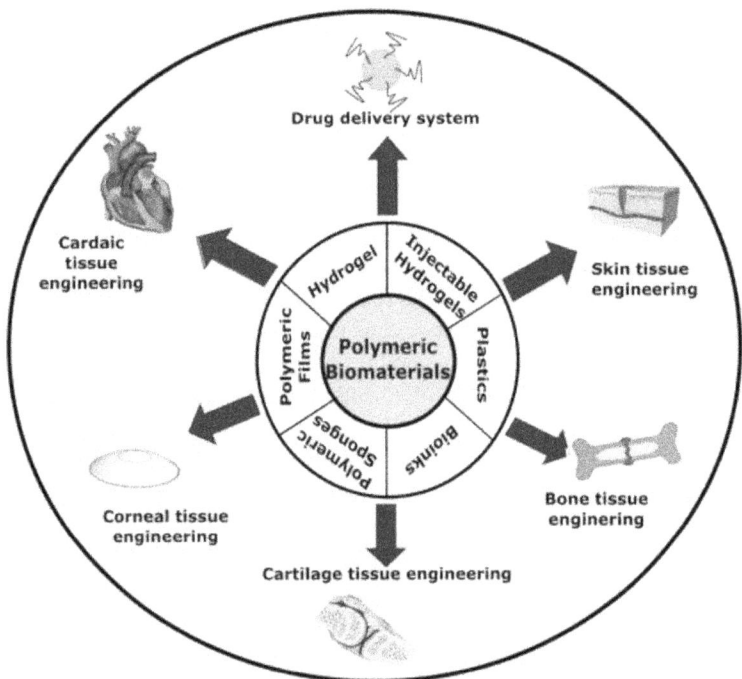

FIGURE 7.1   Polymeric biomaterials for medical applications.

Source: reprinted with permission from Kalirajan et al., Critical review on polymeric biomaterials for biomedical applications, *Polymers* no 17 (2021):3015. For more details see the CC BY 4.0

Chitosan, on the other hand, is a derivative of chitin and is obtained by deacetylation of chitin. Chitosan possesses antitumour, antioxidant and antimicrobial properties. Among commercially available, large-scale production of natural polymers, chitosan is the only polymer showing antimicrobial activity. Its spectrum of activity includes fungi, yeasts, algae, some viruses and bacteria, with greater activity against Gram-positive bacteria than against Gram-negative (Yan et al., 2021). The polymer chains have a positive charge which results in chitosan solutions having bactericidal and bacteriological properties. This is because the positive charge on the polymer chain adheres to the bacterial surface, which causes changes in the permeability of the membrane wall and thus prevents microbial growth (Goy et al., 2009). Chitosan can be conjugated with organic materials as well as biomolecules. It can be used in controlled drug delivery and in the preparation of nanoparticles that contain biologically active ingredients, for example, DNA, proteins, anticancer drugs and insulin or tissue engineering, gene therapy systems and wound healing. The potential for pharmaceutical applications of chitosan was initially mainly concerned with its potential use as excipient in drug formulations and delivery systems (Badwan et al., 2015). As drug delivery systems these materials can be used in the form of solutions, tablets, capsules, nanoparticles, gels, gummies, hydrogels, conjugate, bars, films and sponges (Akbar and Shakeel, 2018). Therefore, they can be used for oral, ocular, nasal, vaginal, buccal, parenteral, intravesical and transdermal administration, as well as drug delivery implants. Microsphere-based drug delivery systems offer the ability to control the release profile of the drug and determine its target site. Chitosan can be used as a promising candidate for a polymeric carrier of anticancer agents such as doxorubicin (DOX), paclitaxel (PTX), docetaxel (DTX), methotrexate (MTX) and curcumin (Lin et al., 2015; Yadav et al., 2022). The anticancer drug DOX was chosen as a model drug. The effect of pH and initial drug concentration content on the drug release profile was investigated. The study showed that the samples exhibited higher release of the model drug at pH = 5. Chitosan is also used for the preparation of nanoparticles, which can be used as drug, protein or gene carriers (Huang et al., 2004). Chitosan microspheres crosslinked with glutaraldehyde were used as a biodegradable device for long-term delivery of mitoxantrone. The study showed that the rate of release of this anticancer drug depends on the degree of crosslinking of the microspheres and for highly crosslinked microspheres it was 25% in 36 days (Jameela et al., 1995). Magnetic microspheres of chitosan and Fe_3O_4 magnetic nanoparticles were also prepared for use in protein drug delivery systems using microemulsion polymerisation (Li et al., 2020). Bovine serum albumin (BSA) was chosen as the protein drug model in the experiments. The study showed that the magnetic chitosan microspheres exhibited a high drug release rate of about 87.8%. Magnetic microspheres in the study were shown as promising drug carriers for biomedical applications. Chitosan, due to its low cost and biocompatibility of hydrophilic gel-forming polymers, is also used as a matrix for oral dosing of extended-release drugs. Chitosan tablets can be prepared using wet granulation or direct compression methods. Mucoadhesive vaginal tablets containing metronidazole were made by direct compression of the natural cationic polymer chitosan, loosely crosslinked with glutaraldehyde, sodium alginate with or without the addition of sodium carboxymethyl cellulose (CMC). Drug release studies conducted in pH = 4.8 buffer and distilled water and tablet swelling showed that a formulation containing 6% chitosan, 24% sodium alginate, 30% sodium CMC and 20% microcrystalline cellulose (MCC) showed adequate drug release properties (El-Kamel et al., 2002). Sustained release tablets based on chitosan and anionic polymers such as xanthan gum, carrageenan, sodium CMC and sodium alginate have also been developed (Sachan et al., 2015; Shao et al., 2015). They chose sodium valproate and valproic acid as model drugs. In drug release studies under simulated gastrointestinal conditions, they found that chitosan/xanthan gum tablets exhibited the best sustained release profile for the selected model drugs. Another application of chitosan is the possibility of making hydrogels. These are three-dimensional (3D) networks of crosslinked polymers with the ability to absorb large quantities of water. Chitosan was used to obtain multifunctional hydrogels using oxidised succinoglycan. The resulting chitosan/oxidised succinoglycan hydrogel exhibited 90% antibacterial efficacy, which can be used for pH-controlled drug release. The study showed that changing the pH from 7.4 to 2.0 had an effect on accelerating the release of 5-fluorouracil from 60 to 90%. The materials also showed very high cell viability and proliferation after 7 days of testing and therefore these hydrogels could potentially be used in wound healing, tissue engineering and drug release systems (Kim et al., 2022). Chitosan hydrogels modified with Ser-Ile-Lys-Val-Ala-Val (SIKVAV) peptide have been developed and their positive effect on the healing process of skin wounds has been demonstrated (Chen et al., 2017). The U.S. Food and Drug Administration (FDA) has approved for use as a hemostatic dressing made of chitosan for the rapid staunching of fatal bleeding, such as Chitoseal®, Celox®, Chitoflex® and HemCon® (Hu et al., 2018). A nanocomposite hydrogel based on sodium alginate/chitosan/hydroxyapatite was also obtained using gamma radiation as a crosslinking agent (Abou Taleb et al., 2015). Lipid-coated chitosan-DNA nanoparticles were also obtained, which showed optimal properties for gene carriers for in vivo applications (Baghdan et al., 2018). Chitosan nanoparticles for intranasal delivery of genistein were made by ionic gelation technique using sodium hexametaphosphate. The study showed that small (200–300 nm diameter) and homogeneous nanoparticles had the effect of improving the penetration of genistein through the nasal mucosa compared to pure genistein (Rassu et al., 2019). In regenerative medicine, chitosan is used as a scaffolding material, an analogue or extracellular matrix that acts as a support for the regeneration of damaged tissues. Chitosan can be easily reweighed and shaped into a host tissue or tissue interface biomaterial. It can be used to produce scaffolds with porous structures. This structure enables cell settlement, proliferation, migration and nutrient exchange. Furthermore, the controlled porosity of scaffolds made from this material promotes angiogenesis, which is essential for the regenerative function of soft tissues (Ko et al., 2010). This material elicits only a very weak foreign body response in vivo. Furthermore, scaffolds made of this material have shown biocompatibility in in vivo studies and no cytotoxicity reaction in in vitro studies. Chitosan can be mixed with other natural polymers such as collagen or fibronectin to produce a biomaterial with better cell affinity and also for use in the skeletal system (Huang et al., 2009). Sponges obtained by lyophilisation of frozen chitosan solution can also be used in bone tissue engineering. Cell density, alkaline phosphatase activity and calcium deposition were monitored in vitro for 56 days on chitosan sponges with a porous structure and histological studies showed that bone formation occurs in the sponges (Seol et al., 2004). On the other hand, the tested chitosan/collagen sponges (with ratio of 1/1 and 1/2) showed that the 1/1 composite sponge had the highest compressive strength. Furthermore, this composite also had better mechanical and physical properties than other sponges. It was also found that these composites promoted osteogenic differentiation of rat bone marrow stromal cells (Arpornmaeklong and Suwatwirote, 2008).

Starch is a white, odourless, amorphous and tasteless powder. This natural polymer is insoluble in water and other common organic solvents. It belongs to a group of carbohydrates consisting of two similar molecules – linear and helical amylose and the branched amylopectin. Starch can be extracted from the leaves, stems, tubers, seeds and roots of plants, where it serves as an energy reserve. In the pharmaceutical industry, starch is used as a binder, diluent and disintegrant. Starch, like chitosan, has the potential to be used as a haemostatic material due to its special surface area, lack of immunogenicity and absorption by the human body. Starch is used for the production of tablets but can also be used in other drug delivery systems (intranasal, periodontal and film forms; Krogars et al., 2003). Starch and its derivatives are widely used in the pharmaceutical industry in recent years. Starch derivatives are used in the development of drug delivery systems and, depending on the modification, can be used in solid dosage forms, for encapsulation and drug release in pH-sensitive systems, for site-directed controlled release of drugs, for new smart solid dosage forms and for many other things. Starch can be chemically modified by oxidation, etherification and esterification. Commercially launched chemically modified starches on the pharmaceutical market are Contramid® (Labopharm), 400 L-NF, Glycosys® and Lycoat RS 720® (Roquette & Freres) Hyswell® (Maruti Chemicals), Expotab® (JRS Pharma), Amprac 01® (Rofarma), (JRS Pharma) and Primojel® (DEF Pharma), which are widely used in the pharmaceutical industry (Lemos et al., 2021). Superparamagnetic nanoparticles functionalised with Fe_3O_4 that can be used as potential carriers for targeted-release drugs were prepared using a starch hydrolysis product, with starch dialdehyde as the envelope and epichlorohydrin as the crosslinking agent (Lu et al., 2013). A biodegradable microgel system based on glycerol-1,3-diglycidyl ether crosslinked with 2,2,6,6-tetramethylpiperidinyloxy (TEMPO)-oxidised potato starch was prepared for controlled protein (lysozyme) uptake and release (Zhao et al., 2015). Dialdehyde starch nanoparticles that were directly linked to an anticancer drug for breast cancer with DOX may release the drug for longer and are more effective against cancer cells (Yu et al., 2007). Also, microparticles based on a polyelectrolyte complex of carboxymethyl starch and chitosan may be a promising strategy in controlling drug release that can be used in specific regions of the gastrointestinal tract. Bovine serum albumin was also incorporated into these microparticles. Experimental studies in simulated gastric fluid (pH = 1.2) and simulated intestinal fluid (pH = 6.8) showed that these microparticles exhibited a pH-dependent release profile (Quadrado and Fajardo, 2020). Mucoadhesive cheek films were obtained from two rice varieties differing in amylose content with potential applications in drug delivery systems. The rice powders were modified to produce a carboxyl derivative. The study showed that when glycerol was added as a plasticiser to the prepared films with higher amylose content higher transparency and better mucoadhesive properties were shown, while the addition of Tween® 20 has the effect of increasing the tensile strength and decreasing the elongation of the films

(Okonogi et al., 2014). This study showed that rice grains with high amylose content are a promising natural source of pharmaceutical film-forming agents. Starch can be used as a promising candidate for a polymeric carrier of anticancer agents such as DOX. A redox-sensitive hydroxyethyl starch-DOX conjugate was synthesised and in vitro studies confirmed the cytotoxic effect of this conjugate presented a prolonged plasma half-life with increased accumulation in the tumour compared to free DOX. Redox-sensitive conjugate may be a new DOX prodrug with the potential for clinical application. This prodrug could be used in chemotherapy with safe and effective delivery of DOX to tumour cells for destruction (Hu et al., 2016). Starch is also used in scaffolds in bone tissue engineering due to its biocompatibility, excellent mechanical properties and ability to regulate degradation time (Roslan et al., 2016). There are many varieties of starch derived from identical materials and with different properties, such as strength and water absorption. Scaffolds for bone regeneration can be obtained by many methods: gas foaming, solvent casting, electrospinning, phase separation and freeze-drying. Starch derived from brown rice was selected to produce composite scaffolds with hydroxyapatite. Composites with starch contents of 50, 60, 70 and 80% were used and it was shown that the porosity of the scaffolds increased with increasing starch content. However, the scaffold containing 60% of starch had the highest percentage of water absorption in the water absorption test and is moreover able to promote cell regeneration in tissue engineering as it shows the ability to absorb fluids (Mohd Nasir et al., 2018). The mesh scaffolds based on a mixture of starch and poly(ε-caprolactone) (PCL) have a positive effect on the proliferation and osteogenic differentiation of bone marrow stromal cells. A mixture of starch and PCL was used to produce a bilayer scaffold consisting of a membrane obtained by solvent casting and a mesh of wet-spun fibres. These fibres were in some cases functionalised with osteoconductive silanol groups (starch/PCL-Si). The usefulness of these scaffolds in jawbone regeneration was evaluated in comparison with a commercial collagen membrane. The study showed that starch/PCL and starch/PCL-Si scaffolds induced significantly higher levels of new bone formation compared to collagen membrane (Requicha et al., 2014). The reticulated fibre scaffolds also based on a mixture of starch and PCL with porosities of 50 and 75% showed that scaffolds with 75% porosity had a significant effect on bone marrow stromal cell proliferation. These results also indicate that scaffold porosity affects the sequential development of osteoblastic cells (Gomes et al., 2006). Due to its similarity to the natural cellular environment, starch can also be used to create scaffolds or wound dressings (Weinstein-Oppenheimer et al., 2010). A method of obtained 3D scaffolds by electrospinning was also developed. These scaffolds were based on starch and poly(vinyl alcohol) (PVA). In vitro cellular studies using mouse L929 fibroblast cells have shown that these materials are non-toxic and have the ability to promote cell proliferation (Waghmare et al., 2017).

The most common polysaccharide on earth is cellulose. It can be obtained from many different sources, including the cell walls of plants, some species of bacteria and algae. Due to the diversity of organic material sources from which cellulose is derived or due to specific biosynthetic and processing conditions, it occurs in many morphological fibre forms such as microfibrils/nanofibrils and micro/nanocrystalline cellulose. Cellulosic materials are biocompatible, biodegradable and have low cytotoxicity; for this reason, they can be successfully used in medical applications such as drug delivery systems (Gunduz et al., 2013), tissue engineering (Ninan et al., 2013) and wound care dressings (Solway et al., 2011). The origins of the use of cellulose in the pharmaceutical industry are primarily in the coating of tablets after mixing with various excipients for oral administration. Currently, research is ongoing into the use of this natural polymer and its derivatives in advanced drug release systems. One direction of research is to use the biopolymers or their derivatives to increase the dissolution rate of tablets as suitable excipients or prolonged drug release as novel drug carriers (Yan et al., 2019). The most significant application of cellulose derivatives is their use in matrix systems for solid oral dosage forms intended for oral, topical or parenteral administration in formulations such as transdermal, vaginal or ocular (Qiu et al., 2017). The control of drug release from such systems is based on its diffusion through the gel layers. This is due to the fact that cellulose derivatives swell in water (Grund et al., 2014). For example, ethyl cellulose or sodium CMC used in these systems allows the drug release to be sustained (Mehta et al., 2014). Cellulose powder and MCC are used as excipients in pharmacy. MCC is one of the most common excipients used in direct pressing as a dry binder, disintegrant, absorbent or tablet filler and capsule diluent. It is used to bind poorly tableted active pharmaceutical ingredients. MCC is also used as an ingredient in a carrier used for the preparation of oral suspension (Chaerunisaa et al., 2019). Cellulose ether derivatives include methyl cellulose, ethyl cellulose, hydroxyethyl cellulose, hydroxypropyl methyl cellulose and CMC. Methyl cellulose is not metabolised and digested in the body and is insoluble in most organic solvents. It is used in matrix tablets as thickeners, binders, emulsifiers and stabilisers. It can also be used in the manufacture of capsules in dietary supplements, where it successfully replaces gelatine. It is also important in the treatment of dry eye as a substitute for saliva or tears (Nasatto et al., 2015). Ethyl cellulose is an inert hydrophobic polymer, due to its lack of toxicity, stability during storage and good compressibility and most importantly its ability to swell in the presence of gastric juice, it can be used to produce modified-release drug formulations, providing a constant drug concentration (Wasilewska and Winnicka, 2019). Ethyl cellulose can be used to form hydrophobic coatings, fibres or support layers and also as a coating material in extended-release formulations (see Figure 7.2).

There are many commercial preparations containing ethyl cellulose on the pharmaceutical market. Among others is Micro-K® containing microcapsules made of ethyl cellulose with potassium chloride (KCl). The ethyl cellulose acts as a semi-permeable membrane in this formulation and provides a controlled release of K+ and Cl– ions. Another example is Theo-24®; this preparation contains theophylline. The drug is coated with a complex using ethyl cellulose and the release of this complex takes place within 24 h. Ethyl cellulose is also used in Metadate CD® capsules and many other such as Diffucaps, DiffCORE™, Geomatrix® and SODAS, as well as Cotempla® XR-ODT or Adzenys XR®-ODT for patients suffering from ADHD (Wasilewska and Winnicka, 2019). Hydroxyethyl cellulose is a cold- or hot-water soluble, odourless non-ionic polymer. In pharmacy it is mainly used as a thickener, protective colloid, adhesive, dispersant and stabiliser. It can be used for the production of films and as a material for the preparation of sustained release drug formulations. It is used in the production of topical drugs (administered in the form of cream, ointment and eye drops) and tablets and capsules. Hydroxyethyl

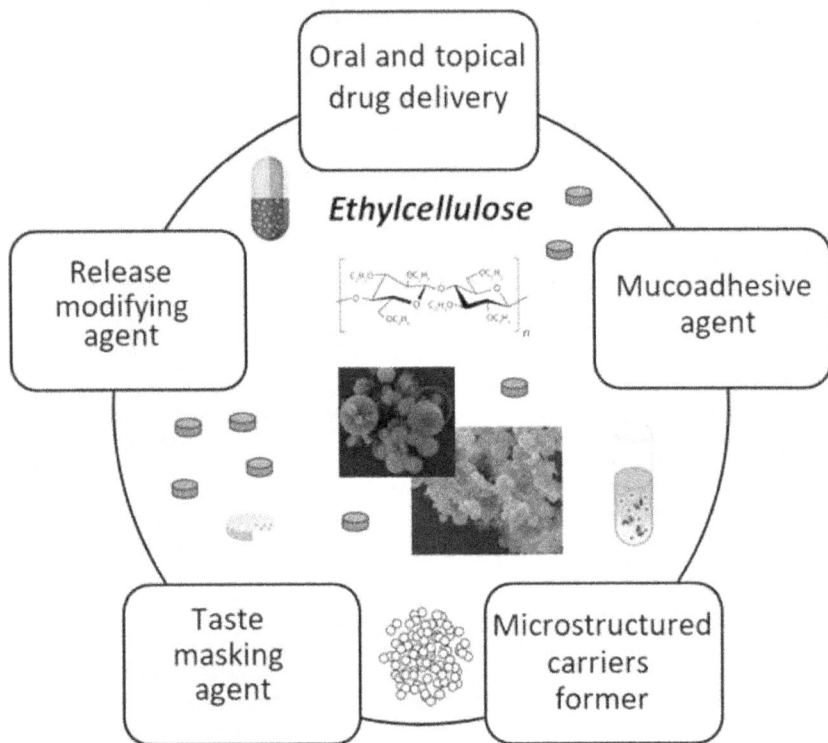

FIGURE 7.2 Ethyl cellulose in drug delivery applications.

Source: reprinted with permission from Wasilewska and Winnicka, Ethylcellulose – A pharmaceutical excipient with multidirectional application in drug dosage forms development, *Materials* no 12 (2019):3386. For more details see the CC BY 4.0

cellulose is included in the European as well as United States Pharmacopoeia. Hydroxyethyl cellulose was used to produce hydroxyethyl cellulose conjugates with ibuprofen (IBU). The study showed that such conjugates can provide slow release of IBU and higher bioavailability, which can also affect the reduction of drug dose. In addition, conjugates have the effect of reducing gastric exposure to the drug by decreasing the gastric side effects associated with the irritant effects of non-steroidal anti-inflammatory drugs (Abbasa et al., 2017). Hydroxypropyl cellulose is a derivative of cellulose, soluble in both water and organic solvents. Due to its very good solubility, it can be used as a tablet binder or thickening agent for emulsions or suspensions and as a topical protectant or lubricant in ophthalmology (Mesdour et al., 2008). Low molar mas hydroxypropyl cellulose can be used as a solubility enhancer for poorly soluble drugs (Martin-Pastor et al., 2021). Besides, it can be used to produce amorphous solid dispersions (Lübbert et al., 2021). Hydroxypropyl methyl cellulose is a widely used excipient in oral sustained release drug delivery matrix systems. Its widespread use in pharmacy is due to the fact that this polymer can have different molar masses, which influence the viscosity of the solution and its physicochemical properties. A topical antibiotic composite vancomycin hydrochloride release system based on hydroxypropyl methyl cellulose microparticles and thermosensitive chitosan/glycerophosphate hydrogel was developed for the treatment of osteomyelitis. The hydrogel reduces not only the release rate of the antibiotic but also affects the amount of drug released. It follows that the combination of macroparticles and hydrogel can be used for the topical administration of antibiotics (Mahmoudian and Ganji, 2017). Hydroxypropyl methyl cellulose was used to produce a floating gastroretentive drug for controlled release of theophylline (Swain et al., 2009). Cellulose ester derivatives such as cellulose acetated and sulphate also were used as materials for pharmaceutical used (Su et al., 2019). Electrospun cellulose acetate nanofibres have been used to deliver anti-inflammatory, anti-cancer, anti-microbial drugs, as well as vitamins and amino acids as patches through the skin (Wsoo et al., 2020). In tissue engineering, cellulose is used as an additive or basic material. The most commonly used cellulose derivatives in tissue engineering include bacterial cellulose, cellulose acetate, hydroxyethyl cellulose, cellulose acetate CMC, methyl cellulose, ethyl cellulose and hydroxypropyl cellulose (Vatankhah et al., 2014; Zennifer et al., 2021). A nanocomposite scaffold of bacterial cellulose was synthesised with magnetite (Fe_3O_4) and hydroxyapatite nanoparticles by ultrasound irradiation method. Studies have shown that the scaffold is non-toxic to mouse fibroblast L929 cells and can be used for bone tissue engineering (Dutta et al., 2019). Cellulose dressings are used to cover chronic wounds such as shin ulcers of various aetiologies, as well as bedsores and wounds resulting from metabolic diseases, e.g. diabetes. Due to their strong hygroscopic properties, these dressings are referred to as "eternally moist". The results of studies have shown that cellulose is capable of absorbing the volume of fluid even one hundred times higher than its mass (Almeida et al., 2011). This makes it possible to both moisten the wound and remove exudate from oozing defects (Czaja et al., 2006). This feature of bacterial cellulose means that dressings made from it can be successfully used where it is important to maintain the correct level of moisture in the wound. An additional advantage that is associated with the high absorbency of bacterial cellulose is the ability to saturate the dressing with therapeutic substances: antibiotics, antiseptics and anti-inflammatory drugs (Gupta et al., 2019; De Mattos et al., 2020; Gupta et al., 2020). Nanocellulose can also be successfully used as a dressing for skin grafts due to the fact that it adheres more tightly to the skin than traditional dressings and maintains hydration in the implant site at an optimal level for much longer, which speeds up the healing process (Zheng et al., 2020). Phosphoric acid derivatives of water-soluble cellulose have been obtained for use as a scaffold material for tissue engineering. The cytocompatibility of the films made by direct contact and also by indirect contact in human skin fibroblasts was investigated. The study showed that phosphorylated cellulose could be a good biomaterial for composite scaffolds due to its cytocompatibility and lack of toxicity (Petreus et al., 2014). Bacterial cellulose may also have potential use as a small diameter blood vessel (Zhang et al., 2015). An in vivo evaluation study of bacterial cellulose as a potential biomaterial for biosynthetic blood vessels was conducted on golden Syrian hamsters. The study demonstrated good biocompatibility of the biomaterial compared to materials commonly used in vascular surgery and thus bacterial cellulose could be a new biomaterial for tissue engineering applications of vascular grafts (Esguerra et al., 2010).

7.2.2 PROTEINS

Zein is a protein found in approximately 50% of the maize endosperm. It is a prolamine-rich protein, insoluble in water but soluble in alcohol. Due to its amino acid composition and its solubility, it can be divided into α-zein (19 and 22 kDa), β-zein (14 kDa), γ-zein (16 and 27 kDa) and δ-zein (10 kDa). Zein consists of the following amino acids: glutamine, leucine, proline and alanine but almost no lysine and tryptophan. It has been approved by the FDA as a safe excipient in the coating of pharmaceuticals. Zein in the form of microspheres, nanoparticles, nanocomplexes, fibres, membranes or films and gels, microspheres or scaffolds is widely used in drug delivery or bioactive molecules and tissue engineering applications (Guo et al., 2009; Corradini et al., 2014; Li et al., 2017). Much research has been devoted to zein-based nanoparticles and microparticles and to exploring their possible applications in various fields of medicine. For instance, zein was used to produce microparticles for oral drug delivery systems (Lau et al., 2012). The microparticles obtained using a phase separation method were loaded with prednisolone. In drug release studies of the fabricated microparticles under simulated gastric and small intestinal conditions; it was shown that a prednisolone release rate of almost 70% was achieved in the first case after 3 h and in the second case after 4 h. Furthermore, it was found that a clinically relevant dose of the drug could be reached in the body after administration of 100 mg of microparticles. Prednisolone-loaded microparticles may have applications in oral drug delivery systems. Zein/Tremella polysaccharide (from *Tremella Fuciformis*) nanoparticles were fabricated by antisolvent precipitation to evaluate the ability of the polysaccharide to stabilise the nanoparticles and the possibility of loading such nanoparticles with curcumin. In this study, it was shown that the chosen polysaccharide affects the hydrophilicity of zein and the stability of the obtained nanoparticles. Nanoparticles composed of zein and polysaccharide showed a greater ability to encapsulate curcumin (93.34%) than nanoparticles made only from zein (54.73%). The in vitro studies also showed an increase in the bioavailability of curcumin in nanoparticles with polysaccharides; thus the produced nanoparticles can be used as a potential carrier for the delivery of bioactive substances (Li et al., 2022). Novel hybrid drug-polymer with lipid nanoparticles containing zein and lipids and tamoxifen citrate as a model anticancer drug were fabricated by a modified coaxial electrospraying method. Studies have shown that such nanoparticles can be used to combine edible proteins and lipids to produce advanced nanomaterial-based drug delivery systems (Kang et al., 2021). Quercetin-containing zein nanoparticles by electrospinning were produced, which had the effect of increasing the bioavailability of quercetin. The obtained systems may have applications in pharmaceutics by increasing the bioavailability of quercetin in the bloodstream (Rodríguez-Félix et al., 2019). The protein also has major applications in tissue engineering (Pérez-Guzmán, et al., 2020). Zein was used to produce nanofibrous calcium phosphate ($Ca_3(PO_4)_2$) composite scaffolds. $Ca_3(PO_4)_2$ was used as a bone-like coating produced on a biodegradable polymer scaffold. In vitro studies with stem cells obtained from adipose tissue showed that mineralisation of zein fibres with $Ca_3(PO_4)_2$ could be used to produce nanofibrous scaffolds that could find applications in bone tissue engineering (Zhang et al., 2014).

Soy protein is a globular protein obtained from the soybean seed, with a spherical shape. Its main advantages are stability during long-term storage and biocompatibility. Soy protein as protein source can be found in soy flour, soy protein concentrate and soy protein isolate, and its content varies from 50% to 90% depending on the source. The most purified form of soy protein is soy protein isolate, containing 90% or more protein. It contains amino acids such as glutamate, aspartate, leucine arginine, glycine, aspartic acid and glutamine (Gorissen et al., 2018). Soy protein-based hydrogels, films, nanoparticles and scaffolds have also found applications in tissue engineering for scaffold or wound healing, as transdermal drug delivery dressings, drug carriers in controlled release systems and in the encapsulation of highly hydrophobic drugs (Teng et al., 2012). Microparticles of soy protein isolate and alginate with *Lactobacillus casei* lactic acid bacteria encapsulated inside have been obtained. The viability of the probiotic after 4 m of storage was 9.47 and 9.20 log CFU/g (CFU – colony-forming unit). Infrared spectroscopy stability studies confirmed the stability of the probiotic after microencapsulation. The obtained probiotic microparticles can be used in pharmaceutical products (Hadzieva et al., 2017). Studies of the inulinase enzyme from *Aspergillus niger* covalently immobilised on synthesised magnetic nanoparticles (Fe_3O_4) consisting of a soy protein isolate functionalised with bovine serum albumin nanoparticles with spherical morphology and an average size ranging from 80 to 90 nm showed that the degree of inulinase immobilisation on the nanoparticle surface was about 80% (Torabizadeh et al., 2017). Soy protein is used to prepare crosslinked scaffolds using a mixture of soy, chitosan and tetraethyl orthosilicate as the crosslinking agents combining a sol-gel process with the freeze-drying technique. Studies have shown that the addition of tetraethyl orthosilicate affects the mechanical stability and degradation rates, as well as the porosity of the scaffolds. Due to the presence of silanol groups, these hybrids were considered for cartilage tissue engineering applications (Silva et al., 2006). Also, a mixture of soy protein isolate and cellulose was used to produce porous membranes. The membranes were made by casting and coagulation from an aqueous solution of 5 wt% acetic acid or 5 wt% sulphuric acid (Luo et al., 2008). The addition of soy protein isolate resulted in increased mechanical properties and improved biocompatibility and biodegradability in vivo. Porous soy protein isolate/cellulose membranes were also used to obtain nerve guide tubes (Luo et al., 2015). In a study of sciatic nerve injury repair in rats, tubes made from these membranes in combination with a nerve growth factor (pyrroloquinoline quinone) reconstructed the damaged nerve and could potentially be used as nerve conduits in nerve tissue engineering. Tofu was used to fabricate porous soy scaffolds due to its porous structure and high soy protein content. One of the scaffolds was based on the traditional tofu manufacturing processes, while the other was modified using covalent crosslinking. The physicochemical properties of the scaffolds, their morphology and their biocompatibility in vitro and in vivo were investigated. The obtained scaffolds showed good cell proliferation and no inflammatory reaction in subcutaneous implantation tests. The results showed that these scaffolds could potentially be used in tissue engineering (Huang et al., 2018).

7.3   BIODEGRADABLE POLYESTERS

Among the polyesters, polyhydroxyalkanoates (PHA) are widely used in medicine to produce a wide range of medical devices, biomedical tissue engineering and drug delivery systems. Mainly polyesters obtained by chemical synthesis are used in medical devices and pharmaceuticals, but there is also growing interest in the use of polyesters obtained by biotechnological methods such as PHAs. Both synthetic and bio-based polymers in preclinical studies conducted according to the ISO 10993 standard show fairly good biocompatibility compared to many other materials and can be used as implants that come into contact with soft tissue, bone and blood (ISO 10993–1:2018). The biodegradation of poly(α-hydroxy acid)s such as polyglycolide (PGA), polylactide (PLA) or poly(lactide-*co*-glycolide) (PLGA) random copolymer has caused problems with their use in medicine. This is because the products that are formed during the hydrolytic degradation process of these polymers cause a drop in pH near the implant, which can cause chronic inflammation. To eliminate chronic inflammation, various anti-inflammatory drugs can be added to reduce the pro-inflammatory properties of implants, e.g. poly(*L*-lactide) (PLLA) fibres and spirals and simulating curcumin-coated stents have shown in vitro that curcumin impregnation can reduce the inflammatory response to bioresorbable PLLA fibres (Su et al., 2005). In contrast, degradation of PHAs does not result in acidification of the substrate as they are more resistant to hydrolysis in an aqueous environment. The predominant degradation product of poly(3-hydroxybutyrate) (PHB) is 3-hydroxybutyric acid, much weaker than lactic acid, which is the biodegradation product of PLA and PLGA (Bonartsev et al., 2019). The use of biodegradable polyesters in medicine dates back to the 1970s. The first application of biodegradable polymers was VICRYL® surgical threads based on PLGA. Today, these biodegradable polymers are widely used in medicine, ranging from dressings, surgical threads, various types of screws, pins, vascular grafts, tissue engineering of heart valves, nerve conduits, bone, cartilage, drug delivery systems that enable targeted delivery and prolonged action with reduced toxicity and increased stability of the drugs used. PHAs, due to their mechanical properties, biocompatibility and repair properties, are an excellent biomaterial that can be used in skin tissue repair. One of the applications of PHAs is dressings used to accelerate wound healing, especially in large areas of skin tissue loss. Hybrid wound dressings based on bacterial cellulose and a copolymer of 3-hydroxybutyric and 4-hydroxybutyric acids (poly(3-hydroxybutyrate-*co*-4-hydroxybutyrate) (P3HB4HB)) containing the drug actovegin were produced. In vivo studies showed that these dressings promoted healing of third-degree skin wounds better than the commercial VoskoPran dressing (Volova et al., 2019). Nanofibrous meshes consisting of poly(3-hydroxybutyrate-*co*-3-hydroxyvalerate) (PHBV) with different contents of 3-hydroxyvalerate (3HV) units were produced by haloarchaea. These meshes were used in the wound healing process to reduce excessive scarring. The study showed that wounds to which the manufactured mesh was applied as a dressing were softer and more flexible than wounds treated with film. PHBV nanofibre mesh can potentially be used as a wound dressing as it reduces excessive scarring (Kim et al., 2020).

   (Bio)degradable polymers such as PHB, PHBV, PGA, PLA, PLGA, PCL, poly(*p*-dioxanone), poly(glycolide-*b*-trimethylene carbonate) block copolymer and poly(glycolide-*co*-ε-caprolactone) can be used as suture materials. Suture selection must be tailored to the procedure, and ophthalmic, neurological and cardiovascular procedures require special care in suture selection. The risk-benefit analysis for each suture type is reviewed and investigated as new applications emerge. PHB and PHBV synthesised by the bacterium *Ralstonia eutropha* B 5786 were used to produce sutures and animal studies of those sutures showed no adverse effects on physiological, biochemical and functional parameters in the postoperative period. Moreover, the tissue reaction to the foreign body at the site of implantation was less intense than the reaction to sutures made of silk and catgut, which are the reference for the tested biomaterials (Volova et al., 2003). Two types of surgical sutures were also developed. The first fibre was made from poly(3-hydroxybutyrate-*co*-3-hydroxy-hexanoate) (PHBHHx), and the second fibre was made from a mixture of PHBV and PLA twisted together. The fibre made from PHBHHx showed very little effect on the surrounding tissues during implantation. Furthermore, even 96 weeks after implantation, the fibre retained half of its mechanical properties, whereas sutures made from a PHBV/PLA mixture degraded 36 weeks after implantation. Research has shown that these polymers can be used to produce degradable surgical sutures that degrade to non-toxic compounds (He et al., 2014).

The field of orthopaedics, due to its complexity and range of applications, continues to drive the need for the development of polymer devices to replace traditional titanium and other surgical steel devices. There is a continuous and growing demand for replacement joints, components used to repair fractures, bone fillers and bone cement and artificial orthopaedic structures used to replace those parts that have been irreparably damaged. Many different groups of biodegradable polymers have been developed for this purpose such as polyphosphazenes or polyurethanes asoly(α-hydroxy acid)s, which include PGA, PLA and their copolymer PLGA. PLGA allowed for making implantable orthopaedic devices. These polymers have also been used to product scaffolding that facilitates the formation of new cartilage material in the human body. These biodegradable polymers are particularly desirable for orthopaedic devices due to their excellent biocompatibility and possibilities to adjust physical properties and degradation properties to a specific application (Doyle et al., 1991; Shamsuria et al., 2004; Cool, 2007). Also accepted for use in orthopaedic devices are poly(*p*-dioxanone), PCL, PHB, polyhydroxyvalerate (PHV) and poly(aspartic acid). Implants are being developed using biopolymers such as chitosan, silk, starch, collagen, elastin or hyaluronic acid. The synthetic and natural polymers have been combined with nano-hydroxyapatite or β-tricalcium phosphate, which allows for the design of implants that have bone-like composition (Cichoń et al., 2019). These implants transfer stress to the damaged area, allowing the healing of the tissue and eliminating the need for surgery to remove metal implants. Benefits to the patient are obvious: shorter treatment time and no risk associated with follow-up surgery. In order to accelerate bone formation poly(aspartic acid) and its derivatives containing its composition $Ca_3(PO_4)_2$ can be used, which covers bone implants. Different examples of applications of those polymers are composites that simulate growth of bone tissue (Codreanu et al., 2022). During the mineralisation of collagen sponge it was observed that the use of the sodium salt of poly(aspartic acid) allowed one to get the material which was characterised by a much greater similarity in the construction of the fibre to the natural bone. Polymers like PCL and it composites, PGA, PLA, PLGA, PHB, PHBV, poly(propylene carbonate) and poly(butylene succinate) (PBS) can be used as a coating on medical devices or completely biodegradable devices such as anastomosis rings and tissue staples, ligating clips, tissue engineering scaffolds, nerve conduit tissue and vascular grafts and stents (Zhao et al., 2003; Grabow et al., 2007). For example, PHB has been used as a surgical implant or in surgery, as seam threads for the healing of wounds and blood vessels. PCL can be used as tissue engineering scaffolds for the regeneration of nerve, bone, skin and vascular tissues. A number of promising polymers are being investigated for potential use in medical applications that will result in a scaffold that mimics the 3D structure of host tissue. Renewable PHBHHx and synthetic PCL approved by the FDA were used to produce biodegradable stents for small-calibre blood vessels. In an in vitro study, the resulting stents showed good endothelial cell proliferation and excellent thromboresistance (Puppi et al., 2017). The properties of vascular patches electro-welded from a PCL and PHBV blend modified with Arg-Gly-Asp (RGD) peptides were evaluated. These vascular patches can significantly promote the formation of new vascular systems in vivo. RGD-modified PCL/PHBV patches induced milder platelet aggregation than the commercially available xeno-pericardial KemPeriplas-Neo patch (Sevostianova et al., 2020). A novel mixture of natural medium chain length PHA with synthetic PCL has been developed for the treatment of sciatic nerve injuries in rats. The material showed excellent neuroregenerative properties and a good bioresorption rate, and the porous tubes made from it, 10 mm long, had mechanical properties similar to rat sciatic nerves (Mendibil et al., 2021). Biodegradable polyesters can also be widely used as drug delivery systems as well as site-specific drug delivery systems (Prakash et al., 2022). Some of the most popular and widely used biodegradable polymers for drug delivery applications are PLA, PGA, PHB, poly(ester amide)s and especially PLGA. Poly(ester amide)s can be functionalised and conjugated with different drugs, peptides or molecules for cell signalling. Poly(ester amide)s had been used for microspheres, nanoparticles and hydrogel formation. A biodegradable polymeric drug delivery system such as PolyActive™ was made of poly(ether ester) multiblock copolymers, based on poly(ethylene glycol) (PEG) and poly(butylene terephthalate) (PBT). PolyActive™ can be used as an excellent tool for controlled release due to the fact that it has great linear drug release properties. This system has applications in pharmaceuticals and medical technology. Another example of a precise delivery system is OctoDEX™, one that allows the production of drug delivery microspheres in the range of 10–50 µm. These microspheres can be injected subcutaneously through a 25 g needle. One of the advantages of this method of drug delivery is the elimination of the burst effect: no high initial release of the encapsulated drug. There is complete control over the release profile and after the designated time the active ingredient will be fully released from OctoDEX™ matrix. Studies of PHB coated with iron oxide-based magnetic nanoparticles loaded with siRNA-etoposide have shown that magnetic nanoparticles were successfully taken up by cancer cells; moreover, these magnetic nanoparticles could be a potential targeted therapy agent to overcome drug resistance (Yalcin and Gündüz, 2021). PHB was used to encapsulate DOX and sorafenib and conjugated to PEG. The study showed that the drug release from the nanoparticles of both DOX and sorafenib was slowed down; moreover DOX was released in an acidic environment, i.e. in the tumour-specific environment (Babos et al., 2020).

7.4 CONCLUSION

There is a large variety of biodegradable polymers, both natural and synthetic, with applications in medicine including tissue engineering, drug delivery and wound healing. As these materials vary in composition quality, it is impossible to say which of them is most suitable for medical applications. Synthetic biomaterials show better mechanical properties than natural polymers because their structure and molar mass can be controlled during synthesis. Natural polymers, on the other hand, are characterised by greater biocompatibility. Many new methods of synthesis and modification of biodegradable polymers have been developed over the years, but it is still important to develop new technologies to obtain these polymers. Recent years have also seen an increase in research into hybrid materials, i.e. various combinations of polymers that can have a profile specific to a particular medical application.

REFERENCES

Abbasa, K., Amina, M., Hussain, M.A., Shera, M., Bukhari, S.N.A., and Edgar, K.J. 2017. Design, characterization and pharmaceutical/pharmacological applications of ibuprofen conjugates based on hydroxyethylcellulose. *RSC Adv.* 7: 50672–50679.

Abou Taleb, M.F., Alkahtani, A., and Mohamed, S.K. 2015. Radiation synthesis and characterization of sodium alginate/chitosan/hydroxyapatite nanocomposite hydrogels: A drug delivery system for liver cancer. *Polym. Bull.* 72: 725–742.

Akbar, A., and Shakeel, A. 2018. A review on chitosan and its nanocomposites in drug delivery. *Int. J. Biol. Macromol* 109: 273–286.

Almeida, J.F., Ferreira P., Lopes A., and Gil, M.H. 2011. Photocrosslinkable biodegradable responsive hydrogels as drug delivery systems. *Int. J. Biol. Macromol.* 49: 948–954.

Arpornmaeklong, P., and Suwatwirote, P.N. 2008. Properties of chitosan–collagen sponges and osteogenic differentiation of rat-bone-marrow stromal cells. *Int J Oral Maxillofac Surg.* 37: 357–366.

Babos, G., Rydz, J., Kawalec, M., et al. 2020. Poly(3-hydroxybutyrate)-based nanoparticles for sorafenib and doxorubicin anticancer (Drug Delivery). *Int. J. Mol. Sci.* 21: 7312.

Badwan, A.A., Rashid, I., Omari, M.M., and Daras, F.H. 2015. Chitin and chitosan as direct compression excipients in pharmaceutical applications. *Mar Drugs* 13: 1519–1547.

Baghdan, E., Pinnapireddy, S.R., Strehlow, B., Engelhardt, K.H., Schäfer, J., and Bakowsky, U. 2018. Lipid coated chitosan-DNA nanoparticles for enhanced gene delivery. *Int. J. Pharmaceut.* 535: 473–479.

Bonartsev, A.P., Bonartseva, G.A., Reshetov, I.V., Kirpichnikov, M.P., and Shaitan, K.V. 2019. Application of polyhydroxyalkanoates in medicine and the biological activity of natural poly(3-hydroxybutyrate). *Acta naturae.* 11(2): 4–16.

Chaerunisaa, A.Y., Sriwidodo, S., and Abdassah, M. 2019. Microcrystalline cellulose as pharmaceutical excipien. In *Pharmaceutical Formulation Design – Recent Practices*, eds. Ahmad, U., and Akhtar, J., pp. 1–21. London: IntechOpen.

Chen, X., Zhang, M., Chen, S., et al. 2017. Peptide-modified chitosan hydrogels accelerate skin wound healing by promoting fibroblast proliferation, migration, and secretion. *Cell Transplant.* 26(8): 1331–1340.

Cichoń, E., Haraźna, K., Skibiński, S., et al. 2019. Novel bioresorbable tricalcium phosphate/polyhydroxyoctanoate (TCP/PHO) composites as scaffolds for bone tissue engineering applications. *J. Mech. Behav. Biomed. Mater.* 98: 235–245.

Codreanu, A., Balta, C., Herman, H., et al. 2022. Bacterial cellulose-modified polyhydroxyalkanoates scaffolds promotes bone formation in critical size calvarial defects in mice. *Materials* 13(6): 1433.

Cool, S.M., Kenny, B., Wu, A., Nurcombe, V., Trau, M., Cassady, A.I., and Grøndahl, L. 2007. Poly(3-hydroxybutyrate-*co*-3-hydroxyvalerate) composite biomaterials for bone tissue regeneration: *In vitro* performance assessed by osteoblast proliferation, osteoclast adhesion and resorption, and macrophage proinflammatory response. *J. Biomed. Mater. Res. A.* 82: 599–610.

Corradini, E., Curti, P.S., Meniqueti, A.B., Martins, A.F., Rubira, A.F., and Muniz, E.C. 2014. Recent advances in food-packing, pharmaceutical and biomedical applications of zein and zein-based materials. *Int J Mol Sci.* 15(12): 22438–22470.

Czaja, W., Krystynowicz, A., Bielecki, S., and Brown, R.M. Jr. 2006. Microbial cellulose – The natural power to heal wounds. *Biomaterials* 27(2): 145–151.

De Mattos, I.B., Nischwitz, S.P., Tuca, A.C., et al. 2020. Delivery of antiseptic solutions by a bacterial cellulose wound dressing: Uptake, release and antibacterial efficacy of octenidine and povidone-iodine. *Burn.* 46: 918–927.

Doyle, C., Tanner, E.T., and Bonfield, W. 1991. *In vitro* and *in vivo* evaluation of polyhydroxybutyrate and of polyhydroxybutyrate reinforced with hydroxyapatite. *Biomaterials* 12: 841–847.

Dutta, S.D., Patel, D.K., and Lim, K.T. 2019. Functional cellulose-based hydrogels as extracellular matrices for tissue engineering. *J. Biol. Eng.* 13: 55.

El-Kamel, A., Sokar, M., Naggar, V., et al. 2002. Chitosan and sodium alginate – Based bioadhesive vaginal tablets. *AAPS J.* 4: 224–230.

Esguerra, M., Fink, H., Laschke, M.W., et al. 2010. Intravital fluorescent microscopic evaluation of bacterial cellulose as scaffold for vascular grafts. *J. Biomed. Mater. Res.* 93: 140–149.

Gomes, M.E., Holtorf, H.L., Reis, R.L., and Mikos, A.G. 2006. Influence of the porosity of starch-based fiber mesh scaffolds on the proliferation and osteogenic differentiation of bone marrow stromal cells cultured in a flow perfusion bioreactor. *Tissue Eng.* 12(4): 801–809.

Gorissen, S.H.M., Crombag, J.J.R., Senden, J.M.G., et al. 2018. Protein content and amino acid composition of commercially available plant-based protein isolates. *Amino Acids.* 50(12): 1685–1695.

Goy, R.C., de Britto, D., and Assis, O.B.G. 2009. A review of the antimicrobial activity of chitosan. *Polímeros* 19(3): 241–247.

Grabow, N., Bünger, C.M., Schultze, C., et al. 2007. A biodegradable slotted tube stent based on poly(*L*-lactide) and poly(4-hydroxybutyrate) for rapid balloon-expansion. *Ann. Biomed. Eng.* 35: 2031–2038.

Grund, J., Koerber, M., Walther, M., and Bodmeier R. 2014. The effect of polymer properties on direct compression and drug release from water-insoluble controlled release matrix tablets. *Int. J. Pharm.* 469: 94–101.

Gunduz, O., Ahmad, Z., Stride, E., and Edirisinghe, M. 2013. Continuous generation of ethyl cellulose drug delivery nanocarriers from microbubbles. *Pharm. Res.* 30: 225–237.

Guo, H.X., and Shi, Y.P. 2009. A novel zein-based dry coating tablet design for zero-order release. *Int. J. Pharm.* 370(1–2): 81–86.

Gupta, A., Briffa, S.M., Swingler, S., et al. 2020. Synthesis of silver nanoparticles using curcumin-cyclodextrins loaded into bacterial cellulose-based hydrogels for wound dressing applications. *Biomacromolecules* 21: 1802–1811.

Gupta, A., Keddie, D., Kannappan, V., et al. 2019. Production and characterisation of bacterial cellulose hydrogels loaded with curcumin encapsulated in cyclodextrins as wound dressings. *Eur. Polym. J.* 118: 437–450.

Hadzieva, J., Mladenovska, K., Crcarevska, M.S., et al. 2017. Lactobacillus casei encapsulated in soy protein isolate and alginate microparticles prepared by spray drying. *Food Technol. Biotechnol.* 55(2): 173–186.

He, Y., Hu, Z., Ren, M., et al. 2014. Evaluation of PHBHHx and PHBV/PLA fibers used as medical sutures. *J. Mater. Sci: Mater. Med.* 25: 561–571.

Hu, H., Li, Y., Zhou, Q., et al. 2016. Redox-sensitive hydroxyethyl starch–doxorubicin conjugate for tumor targeted drug delivery. *ACS Appl. Mater. Interfaces.* 8(45): 30833–30844.

Hu, Z., Zhang, D.Y., Lu, S.T., Li, P.W., and Li, S.D. 2018. Chitosan-based composite materials for prospective hemostatic applications. *Mar. Drugs.* 16: 273.

Huang, J., Huang, K., You, X., et al. 2018. Evaluation of tofu as a potential tissue engineering scaffold. *J. Mater. Chem. B* 6: 1328–1334.

Huang, M., Khor, E., and Lim, L.-Y. 2004. Uptake and cytotoxicity of chitosan molecules and nanoparticles: Effects of molecular weight and degree of deacetylation. *Pharm. Res.* 21(2): 344–353.

Huang, T.W., Young Y.H., Cheng, P.W., Chan, Y.H., and Young, T.H. 2009. Culture of nasal epithelial cells using chitosan-based membranes. *Laryngoscope* 119(10): 2066–2070.

ISO 10993-1:2018. 2018. *Biological Evaluation of Medical Devices – Part 1: Evaluation and Testing Within a Risk Management Process.* Geneva: International Organization for Standardization. Technical Committee: ISO/TC 194 Biological and Clinical Evaluation of Medical Devices.

Jameela, S.R., and Jayakrishnan, A. 1995. Glutaraldehyde cross-linked chitosan microspheres as a long acting biodegradable drug delivery vehicle: Studies on the *in vitro* release of mitoxantrone and *in vivo* degradation of microspheres in rat muscle. *Biomaterials* 16: 769–775.

Kalirajan, C., Dukle, A., Nathanael, A.J., Oh, T.H., and Manivasagam, G.A. 2021. Critical review on polymeric biomaterials for biomedical applications. *Polymers* 13: 3015.

Kang, S., He, Y., Yu, D.G., Li, W., and Wang, K. 2021. Drug-zein@lipid hybrid nanoparticles: Electrospraying preparation and drug extended release application. *Colloid. Surf. B: Biointerf.* 201: 111629.

Kim, H.S., Chen, J., Wu, L.P., et al. 2020. Prevention of excessive scar formation using nanofibrous meshes made of biodegradable elastomer poly(3-hydroxybutyrate-*co*-3-hydroxyvalerate). *J. Tissue Eng.* 23(11): 2041731420949332.

Kim, Y., Hu, Y., Jeong, J.-P., and Jung, S. 2022. Injectable, self-healable and adhesive hydrogels using oxidized succinoglycan/chitosan for pH-responsive drug delivery. *Carbohydr. Polym.* 284: 119195.

Ko, Y.G., Kawazoe, N., Tateishi, T., and Chen, G. 2010. Preparation of chitosan scaffolds with a hierarchical porous structure. *J. Biomed. Mater. Res.—B Appl. Biomater.* 93(2): 341–350.

Krogars, K. 2003. Academic dissertation. Aqueous-based amylose-rich maize starch solution and dispersion: A study on free films and coatings. *Mater. Sci.* 4.

Lau, E.T.L., Johnson, S.K., Mikkelsen, D., Halley, P.J., and Steadman, K.J. 2012. Preparation and *in vitro* release of zein microparticles loaded with prednisolone for oral delivery. *J Microencapsul.* 29(7): 706–712.

Lemos, P.V.F., Marcelino, H.R., Cardoso, L.G., de Souza, C.O., and Janice, J.I. 2021. Starch chemical modifications applied to drug delivery systems: From fundamentals to FDA-approved raw materials. *Int. J. Biol. Macromol.* 184: 218–234.

Li, D., Wei, Z., Sun, J., and Xue, C. 2022. Tremella polysaccharides-coated zein nanoparticles for enhancing stability and bioaccessibility of curcumin. *Curr. Res. Food Sci.* 5: 611–618.

Li, F., Chen, Y., Liu, S.B., et al. 2017. Size-controlled fabrication of zein nano/microparticles by modified anti-solvent precipitation with/without sodium caseinate. *Int J Nanomedicine* 12: 8197–8209.

Li, X., Zeng, D., Ke, P., Wang G., and Zhang, D. 2020. Synthesis and characterization of magnetic chitosan microspheres for drug delivery. *RSC Adv.* 10: 7163–7169.

Lin, J., Li, Y., Li, Y., et al. 2015. Drug/dye-loaded, multifunctional PEG-chitosan-iron oxide nanocomposites for methotraxate synergistically self-targeted cancer therapy and dual model imaging. *ACS Appl. Mater. Interfaces* 7(22): 11908–11920.

Lu, W., Shen, Y., Xie, A., and Zhang, W. 2013. Preparation and protein immobilization of magnetic dialdehyde starch nanoparticles. *J. Phys. Chem. B.* 117(14): 3720–3725.

Lübbert, C., Stoyanov, E., and Sadowski, G. 2021. Phase behavior of ASDs based on hydroxypropyl cellulose. *Int. J. Pharm.* 3: 100070.

Luo, L., Gan, L., Liu, Y., et al. 2015. Construction of nerve guide conduits from cellulose/soy protein composite membranes combined with Schwann cells and pyrroloquinoline quinone for the repair of peripheral nerve defect. *Biochem. Biophys. Res. Commun.* 457(4): 507–513.

Luo, L.-H., Wang, X.-M., Zhang, Y.-F., et al. 2008. Physical properties and biocompatibility of cellulose/soy protein isolate membranes coagulated from acetic aqueous solution. *J. Biomater. Sci. Polym.* 19(4): 479–496.

Mahmoudian, M., and Ganji, F. 2017. Vancomycin-loaded HPMC microparticles embedded within injectable thermosensitive chitosan hydrogels. *Prog. Biomater.* 6: 49–56.

Martin-Pastor, M., and Stoyanov, E. 2021. New insights into the use of hydroxypropyl cellulose for drug solubility enhancement: An analytical study of sub-molecular interactions with fenofibrate in solid state and aqueous solutions. *J. Polym. Sci.* 59(16): 1855–1865.

Mehta, R.Y., Missaghi, S., Tiwari, S.B., and Rajabi-Siahboomi, A.R. 2014. Application of ethylcellulose coating to hydrophilic matrices: A strategy to modulate drug release profile and reduce drug release variability. *AAPS Pharm. Sci. Tech.* 15: 1049–1059.

Mendibil, X., González-Pérez, F., Bazan, X., et al. 2021. Bioresorbable and mechanically optimized nerve guidance conduit based on a naturally derived medium chain length polyhydroxyalkanoate and poly(ε-caprolactone) blend. *ACS Biomater. Sci. Eng.* 7(2): 672–689.

Mesdour, S., Lepine, A., Erazo-Majewicz, P., Ducept, F., and Michon, C. 2008. Oil/water surface rheological properties of hydroxypropyl cellulose (HPC) alone and mixed with lecithin: Contribution to emulsion stability. *Colloid. Surf. A: Physicochem. Eng. Asp.* 331: 76–83.

Mohd Nasir, N.F., Sucinda, A., Cheng, E.M., et al. 2018. The study of brown rice starch effect on hydroxyapatite composites. *Int. J. Eng. Technol.* 7(2.5): 69–72.

Nasatto, P.L., Pignon, F., Silveira, J.L.M., Duarte, M.E.R., Noseda, M.D., and Rinaudo, M. 2015. Methylcellulose, a cellulose derivative with original physical properties and extended applications. *Polymers* 7: 777–803.

Ninan, N., Muthiah, M., Park, I.K., Elain, A., Thomas, S., and Grohens, Y. 2013. Pectin/carboxymethyl cellulose/microfibrillated cellulose composite scaffolds for tissue engineering. *Carbohydr. Polym.* 98: 877–885.

Okonogi, S., Khongkhunthian, S., and Jaturasitha, S. 2014. Development of mucoadhesive buccal films from rice for pharmaceutical delivery systems. *Drug Discov. Ther.* 8: 262–267.

Pérez-Guzmán, C.J., and Castro-Muñoz, R. 2020. A review of zein as a potential biopolymer for tissue engineering and nanotechnological applications. *Processes* 8(11): 1376.

Petreus, T., Stoica, B.A., Petreus, O., et al. 2014. Preparation and cytocompatibility evaluation for hydrosoluble phosphorous acid-derivatized cellulose as tissue engineering scaffold material. *J. Mater. Sci: Mater. Med.* 25: 1115–1127.

Prakash, P., Lee, W.-H., Loo, C.-Y., Wong, H.S.J., and Parumasivam, T. 2022. Advances in polyhydroxyalkanoate nanocarriers for effective drug delivery: An overview and challenges. *Nanomaterials* 12(1): 175.

Puppi, D., Pirosa, A., Lupi, G., Erba, P.A., Giachi, G., and Chiellini, F. 2017. Design and fabrication of novel polymeric biodegradable stents for small caliber blood vessels by computer-aided wet-spinning. *Biomed. Mater.* 12(3): 035011.

Qiu, Y., and Lee, P.I. 2017. Rational design of oral modified-release drug delivery systems. In *Developing Solid Oral Dosage Forms*, eds. Qui, Y., Zhang, G.G.Z., Mantri, R.V., Chen, Y., and Yu, L., pp. 1127–1160. Cambridge: Academic Press.

Quadrado, R.F.N., and Fajardo, A.R. 2020. Microparticles based on carboxymethyl starch/chitosan polyelectrolyte complex as vehicles for drug delivery systems. *Arab. J. Chem.* 13(13): 2183–2194.

Rassu, G., Porcu, E.P., Fancello, S., et al. 2019. Intranasal delivery of genistein-loaded nanoparticles as a potential preventive system against neurodegenerative disorders. *Pharmaceutics* 11: 8.

Requicha, J.F., Moura, T., Leonor, I.B., et al. 2014. Evaluation of a starch-based double layer scaffold for bone regeneration in a rat model. *J. Orthop. Res.* 32: 904–909.

Rodríguez-Félix, F., Del-Toro-Sánchez, C.L., Javier Cinco-Moroyoqui, F., et al. 2019. Preparation and characterization of quercetin-loaded zein nanoparticles by electrospraying and study of *in vitro* bioavailability. *J. Food. Sci.* 84(10): 2883–2897.

Roslan, M.R., Nasir, N.F.M., Cheng, E.M., and Amin, N.A.M. 2016. Tissue engineering scaffold based on starch: A review. *International Conference on Electrical, Electronics, and Optimization Techniques (ICEEOT)*. 2016: 1857–1860.

Sachan, N.K., Pushkar, S., Jha, A., and Bhattcharya, A. 2015. Sodium alginate: The wonder polymer for controlled drug delivery. *J. Pharm. Res.* 2(8): 1191–1199.

Seol, Y.J., Lee, J.Y., Park, Y.J., et al. 2004. Chitosan sponges as tissue engineering scaffolds for bone formation. *Biotechnol. Lett.* 26: 1037–1041.

Sevostianova, V.V., Antonova, L.V., Mironov, A.V., et al. 2020. Biodegradable patches for arterial reconstruction modified with RGD peptides: Results of an experimental study. *ACS Omega.* 5(34): 21700–21711.

Shamsuria, O., Fadilah, A.S., Asiah, A.B., Rodiah, M.R., Suzina, A.H., and Samsudin, A.R. 2004. *In vitro* cytotoxicity evaluation of biomaterials on human osteoblast cells CRL-1543; hydroxyapatite, natural coral and polyhydroxybutarate. *Med. J. Malays.* 59: 174–175.

Shao, Y., Li, L., Gu, X., Wang, L., and Mao, S. 2015. Evaluation of chitosan-anionic polymers based tablets for extended-release of highly water-soluble drugs. *Asian J. Pharm. Sci.* 10(1): 24–30.

Silva, S.S., Oliveira, J.M., Mano, J.F., and Rui, L. 2006. Reis. Physicochemical characterization of novel chitosan-soy protein/TEOS Porous hybrids for tissue engineering applications. *Mater. Sci. Forum.* 514–516:1000–1004.

Singh, R., Shitiz, K., and Singh, A. 2017. Chitin and chitosan: biopolymers for wound management. *Int. Wound J.* 14: 1276–1289.

Solway, D.R., Clark, W.A., and Levinson, D.J. 2011. A parallel open-label trial to evaluate microbial cellulose wound dressing in the treatment of diabetic foot ulcers. *Int. Wound J.* 8: 69–73.

Su, S.H., Nguyen, K.T., Satasiya, P., Greilich, P.E., Tang, L., and Eberhart, R.C. 2005. Curcumin impregnation improves the mechanical properties and reduces the inflammatory response associated with poly(L-lactic acid) fiber. *J. Biomater. Sci. Polym. Ed.* 16(3): 353–370.

Su, T., Wu, Q.X., Chen, Y., Zhao, J., Cheng, X.D., and Chen, J. 2019. Fabrication of the polyphosphates patched cellulose sulfate-chitosan hydrochloride microcapsules and as vehicles for sustained drug release. *Int. J. Pharm.* 555: 291–302.

Swain, K., Pattnaik, S., Mallick, S., and Chowdary, K.A. 2009. Influence of hydroxypropyl methylcellulose on drug release pattern of a gastroretentive floating drug delivery system using a 32 full factorial design. *Pharm. Dev. Technol.* 14: 193–198.

Teng, Z., Luo, Y., and Wang, Q. 2012. Nanoparticles synthesized from soy protein: Preparation, characterization, and application for nutraceutical encapsulation. *J. Agric. Food Chem.* 60(10): 2712–2720.

Torabizadeh, H., and Mikani, M. 2017. Inulinase immobilization on functionalized magnetic nanoparticles prepared with soy protein isolate conjugated bovine serum albumin for high fructose syrup production. *World Acad. Sci. Eng. Technol. Int. J. Biol. Biomol. Agric. Food Biotechnol. Eng.* 11: 546–553.

Vatankhah, E., Prabhakaran, M.P., Jin, G., Mobarakeh, L.G., and Ramakrishna, S. 2014. Development of nanofibrous cellulose acetate/gelatin skin substitutes for variety wound treatment applications. *J. Biomater. Appl.* 28: 909–921.

Volova, T., Shishatskaya, E., Sevastianov, V., Efremov, S., and Mogilnaya, O. 2003. Results of biomedical investigations of PHB and PHB/PHV fibers. *Biochem. Eng. J.* 16(2): 125–133.

Volova, T.G., Shumilova, A.A., Nikolaeva, E.D., Kirichenko, A.K., and Shishatskaya, E.I. 2019. Biotechnological wound dressings based on bacterial cellulose and degradable copolymer P(3HB/4HB). *Int. J. Biol. Macromol.* 131: 230–240.

Waghmare, V.S., Wadke, P.R., Dyawanapelly, S., Deshpande, A., Jain, R., and Dandekar, P. 2017. Starch based nanofibrous scaffolds for wound healing applications. *Bioact. Mater.* 3(3): 255–266.

Wasilewska, K., and Winnicka, K. 2019. Ethylcellulose – A pharmaceutical excipient with multidirectional application in drug dosage forms development. *Materials* 12: 3386.

Weinstein-Oppenheimer, C.R., Aceituno, A.R., Brown, D.I., et al. 2010. The effect of an autologous cellular gel-matrix integrated implant system on wound healing. *J. Transl. Med.* 8: 59.

Wsoo, M.A., Shahir, S., Bohari, S.P.M., Nayan, N.H.M., and Abd Razak, S.I. 2020. A review on the properties of electrospun cellulose acetate and its application in drug delivery systems: A new perspective. *Carbohydr. Res.* 491: 107978.

Yadav, N., Francis, A.P., Priya, V.V., et al. 2022. Polysaccharide-drug conjugates: A tool for enhanced cancer therapy. *Polymers* 14(5): 950.

Yalcin, S., and Gündüz, U. 2021. Synthesis and biological activity of siRNA and Etoposide with magnetic nanoparticles on drug resistance model MCF-7 Cells: Molecular docking study with MRP1 enzyme. *Nanomed. J.* 8: 98–105.

Yan, D., Li, Y., Liu, Y., Li, N., Zhang, X., and Yan, C. 2021. Antimicrobial properties of chitosan and chitosan derivatives in the treatment of enteric infections. *Molecules* 26: 7136.

Yan, H., Chen, X., Feng, M., et al. 2019. Entrapment of bacterial cellulose nanocrystals stabilized Pickering emulsions droplets in alginate beads for hydrophobic drug delivery. *Colloid. Surf. B.* 177: 112–120.

Yu, D., Xiao, S., Tong, C., et al. 2007. Dialdehyde starch nanoparticles: Preparation and application in drug carrier. *Chinese Sci. Bull.* 52: 2913–2918.

Zennifer, A., Senthilvelan, P., Sethuraman, S., and Sundaramurthi, D. 2021. Key advances of carboxymethyl cellulose in tissue engineering & 3D bioprinting applications. *Carbohydr. Polym.* 256: 117561.

Zhang, C.Y., Salick, M.R., Cordie, T.M., Ellingham, T., Dan, Y., and Turng, L.S. 2015. Incorporation of poly(ethylene glycol) grafted cellulose nanocrystals in poly(lactic acid) electrospun nanocomposite fibers as potential scaffolds for bone tissue engineering. *Mater. Sci. Eng. C.* 49: 463–471.

Zhang, C.Y., Zhang, W., Li-Bo Mao, L.B., Zhao, Y., and Yu, S.H. 2014. Biomimetic miner-
    alization of zein/calcium phosphate nanocomposite nanofibrous mats for bone tissue
    scaffolds. *Cryst. Eng. Comm.* 16: 9513–9519.
Zhao, K., Deng, Y., Chun Chen, J., and Chen G.Q. 2003 Polyhydroxyalkanoate (PHA) scaffolds
    with good mechanical properties and biocompatibility. *Biomaterials* 24: 1041–1045.
Zhao, L., Chen, Y., Li, W., et al. 2015. Controlled uptake and release of lysozyme from glycerol
    diglycidyl ether crosslinked oxidized starch microgel, *Carbohydr. Polym.* 121: 276–283.
Zheng, L., Li, S., Luo, J., and Wang, X. 2020. Latest advances on bacterial cellulose-based
    antibacterial materials as wound dressings. *Front. Bioeng. Biotechnol.* 8: 1–15.

8 (Bio)degradable Polymers in Everyday Life

Barbara Zawidlak-Węgrzyńska

CONTENTS

8.1 (BIO)DEGRABLE POLYMERS IN AUTOMOTIVE AND TRANSPORTATION INDUSTRY

The main goal of a sustainable automotive industry is to reduce fuel consumption and carbon dioxide (CO_2) emissions. One way of achieving this is to reduce the mass of vehicles. An important goal is also to reduce the amount of plastics from non-renewable sources used in vehicle manufacturing. This opens up the possibility of using bio-based and biodegradable polymers to produce some car parts. Leading car manufacturers are already successfully using bio-based plastics to reduce the environmental impact of synthetic polymers. One of the biodegradable polymers used in car manufacturing is polyesters. Also, some polyamides (PA)s undergo enzymatic degradation (Rydz et al., 2015).

DOI: 10.1201/9780429352799-9

Bio-based polymers, such as thermoplastic elastomers, fibre-reinforced thermo-setting polymers, shape memory and self-healing polymers for interior and exterior coatings, piezoresistive components for interior trim prototypes and strain gauges for vibration monitoring, have all the properties needed to produce high-quality vehicle components. Durabio™ is one of the commercial bioplastics produced from plant iso-sorbide and manufactured by Mitsubishi Chemicals (Japan). It was used by Renault for the dashboard of the Clio model. This material shows much better impact and heat resistance. It is also more resistant to weather conditions than conventional engi-neering plastics (MCPP, 2016). Another bio-based material made from tropical cas-tor beans is DSM's (U.S.) EcoPaXX® Q-HGM24 polyamide (PA410), which is used by Mercedes Benz A-class for its engine cover (DSM). One of the most commonly used ecomaterials is composites. These materials are characterised by good mechan-ical properties and relatively easy bonding with other materials. In the automotive industry, natural fibres from both renewable and non-renewable resources are used to produce composites. These include sisal, flax and jute, ramie, hemp and kenaf, cotton, coconut, pineapple and many others (Faruk et al., 2012; Azwa and Yousif, 2013; Yuhazri et al., 2018;Norizan et al., 2020; Nurazzi et al., 2020). Natural fibre-based polymer composites can be used in various structural components of vehicle bodies such as door handles, dashboard covers, panels, floorboards, scooter frames, secondary skin or interior and chassis (Figure 8.1).

Studies of PLA-based kenaf fibre composites using non-woven kenaf fibre and hybrid non-woven and woven kenaf fibre have shown that such composites can be

FIGURE 8.1   Applications of natural fibre composites in automotive components.

Source: reprinted with permission from Mohammed et al., A review on natural fiber reinforced polymer composite and its applications, *Int. J. Polym. Sci.*, no 2015 (2015):243947. For more details see the CC BY 3.0

successfully used in car door map pockets. These composites are lightweight and have good tensile and flexural strength compared to polypropylene (PP) due to the use of woven kenaf fabric in them (Nurazzi et al., 2021). UPM Formi's cellulose-based biocomposite (UPM, Finland) was used to create the Biofore concept car. The material is used for the front end, side sills, dashboard, door panels and interior panels, among other things (BE-Sustainable, 2014; Nissin, 2014). Toyota, General Motors and Opel have used such composites in its cars for floor mats, spare tire covers, door panels and seat backs (Akampumuza et al., 2017). In the Toyota Raum, the spare tire cover was made from a bioplastic – polylactide (PLA) reinforced with kenaf fibre (Bledzki, 2010). Mercedes Benz conducts research on PLA-based composites. It used flax, sisal and wood fibre composites in its door panels. Cotton fibres were used to insulate the instrument panel, and coconut fibres and natural rubber were used for the outer surfaces of the seats and backrests. In the truck, these natural fibres are used in composites for engine insulation, sun visors, interior insulation and bumpers (Mohammed et al., 2015). About 50% of all interior components, including safety subsystems, door and seat assemblies in commercial vehicles, are made of plastics.

In recent years, there has been steady progress in the research and development of new materials, including (bio)degradable plastics with well-defined properties from both performance and environmental perspectives. One example is PLA mixed with kenaf fibres (Anuar and Zuraida, 2011). This composite can be used to develop new materials for car door panels and dashboards (Mohd Radzuan et al., 2019). Using kenaf fibres and bio-based polymers to produce biocomposites for automotive applications will reduce the industry's dependence on fossil fuels such as oil. In addition, it will reduce greenhouse gas emissions through a better carbon balance over the vehicle's lifetime. The study of jute composites for their use as car side panels has shown that these composites can potentially be used as automotive panel materials (Rahman et al., 2021). Röchling Automotive has used Corbion Purac's Plantura material for its air filter housing and car interior trim. The material is a high-temperature PLA that can withstand temperatures of up to 120°C, making it suitable for durable applications (Plastics, 2014). Three-dimensional (3D) printing technology allows concept cars to be built but also enables components to be quickly printed from experimental materials and compared to traditionally made parts. With the advent of 3D printing and its rapid development, the market for materials used in 3D printing is growing rapidly, which also means the market for (bio)degradable polymers is growing and polymers such as PLA, poly(ε-caprolactone) (PCL) and polyhydroxyalkanoates (PHAs) are being used. Designer Erik Melldahl, in collaboration with BMW, has created a concept car called Maasaica. The car was made by 3D printing. What's more, materials that are completely biodegradable have been used for the body (Melldahl).

8.2   (BIO)DEGRADABLE POLYMERS IN AERONAUTICS AND SPACE INDUSTRY

Just as in the automotive industry, natural fibres can be used in aeronautical engineering (Balakrishnan et al., 2016; Jawaid, 2018; Kopparthy et al., 2021). Fibre mesh can be used to manufacture springs or trim tabs, ailerons and rudders in aircraft. The aerospace industry designs its aircraft structures with great precision for safety reasons, and the materials used in aerospace applications must therefore be completely free of defects. Unfortunately, natural fibres are susceptible to biodegradation in the presence of water and are not resistant to ultraviolet (UV) radiation. Consequently, they are not always suitable for use in the aerospace industry.

The BioForS project consortium is researching new fire-resistant and lightweight components for aircraft interiors that were previously made from fossil raw materials. For these applications, the consortium wants to use composites made from 100% renewable raw materials. These composites are based on biodegradable polymers based on vegetable oils in the form of films using natural flax fibres and flame retardants as a matrix. BioForS not only develops the base materials for these applications but also designs the composite assemblies and develops the manufacturing and moulding processes (Tyrrell, 2021).

8.3   (BIO)DEGRADABLE POLYMERS IN AGRICULTURE AND HORTICULTURE SECTOR

A recent trend in agricultural application is to use biodegradable polymers to replace plastics, where such substitutes are possible. Of particular significance, at this time, in agri- and horticulture is to use natural and synthetic (bio)degradable polymers such as agar, pectins, starches, alginates, cellulose and its derivatives, along with PLA, PCL and poly(vinyl alcohol) (PVA). Sosoyil retention sheeting, agricultural film, seed tape or fertiliser tape and binding materials are prime candidates for the use of biodegradable polymers. In the horticultural filed, threads, clips, stales, bags containing fertiliser, envelopes of ensilage and trays of seeds are targeted for applications of bioplastics (Vroman et al., 2009). Clay and plastic plant pots are being replaced by biodegradable equivalents and disposable composting containers are becoming popular items.

The benefits of using biodegradable products are self-evident: no cover biofilm needs to be removed to conserve the moisture, increase soil temperature or reduce weed growth once it has served its purpose (Vroman et al., 2009). Biodegradable mulch made from plant starch is currently available on the market. Starch, due to its poor mechanical properties such as brittleness, must be blended with other polymers. Mulch Film BioBags are marketed under the name BioAgri and are manufactured from material Mater-Bi® (Novamont, Italy), a biodegradable polyester/starch blend (Tofanelli and Wortman, 2020). BioAgri can be compostable in soil depending on climate and temperatures, and it is biodegradable within 1 to 24 months. This film foil can be used as a mulch soil cover to inhibit weed growth and retain the soil's moisture. Biomax TPS (DuPont, USA), Biopar (Biop, Germany), Paragon (Avebe, Netherlands), Biosafe™ (Xinfu Pharmaceutical Co., China), Eastar Bio™ (Novamont, Italy), C-Flex (Flextex Ltd, UK) and Ingeo® (NatureWorks, USA) are other commercially available products used in agriculture and horticulture containing plant starch. Another important application of biodegradable polymers is to meet the nutritional needs of plants and provide them with micro- and macroelements (nitrogen, potassium, calcium, sulphur, iron, chlorine, manganese and others). As these and many other elements are not presented in the natural environment in sufficient quantities for the growth of plants, it is necessary to use nutrients and artificial fertilisers. In turn, it is also necessary to limit the growth of weeds, which entails the use of herbicides and pesticides. To reduce the environmental impact of these chemicals, i.e. the possibility of their bioaccumulation in the food chain and potential contamination of neighbouring ecosystems, a number of systems have been developed in combination with biodegradable polymeric materials. The active ingredient can be dissolved, dispersed or encapsulated in a polymer matrix or coating, or it can be part of a macromolecular backbone or side chain (Roy et al., 2014). Chitosan microspheres crosslinked with genipin were used for the counter-controlled release of urea. It was found that the release rate of fertiliser increased both with an increasing in the concentration in the microspheres and with an increasing temperature (Hussain et al., 2012). Research was also conducted to assess the suitability of different polymer blends for coating urea (Figure 8.2). Their effects on the release of kinetics in different environments and nitrogen uptake by spinach were studied. The blend of starch, PVA and paraffin showed the best urea release efficiency. It was concluded that this coating mixture can be used wherever slow release of fertiliser is required (Beig et al., 2020).

Parasites present in the soil, along with weeds, take away nourishment from the soil. Solarisation, an approach developed in the 70s, involves covering the soil to be reconditioned with polymeric film. The decontamination takes place in 4–6 weeks. The problem is that at the end of the process the film must be removed and disposed of. Prior to disposal the film must be treated as waste which increases the cost. Often,

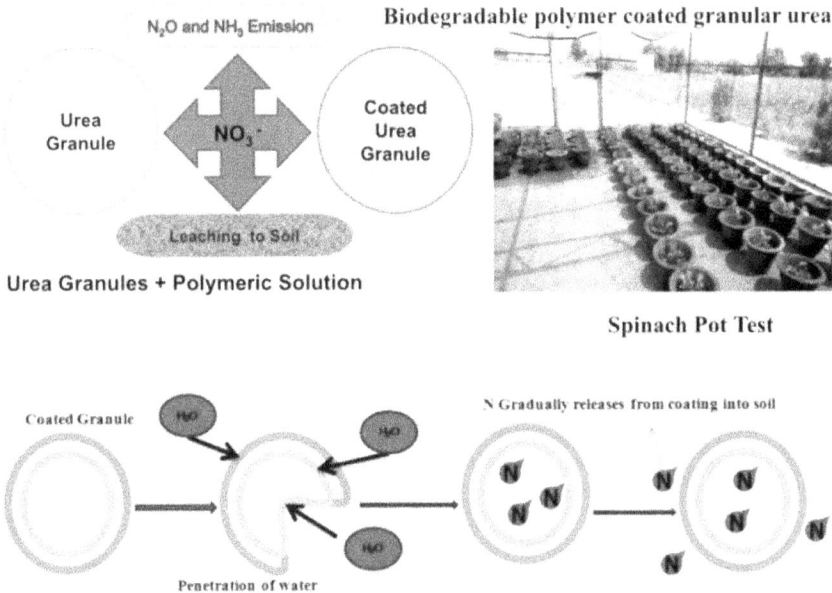

FIGURE 8.2 Use of a biodegradable polymer as a coating agent to slow the kinetics of nitrogen release from urea for improved spinach growth performance.

Source: reprinted with permission from Beig et al., Biodegradable polymer coated granular urea slows down n release kinetics and improves spinach productivity, *Polymers* no 12 (2020):2623. For more details see the CC BY 4.0

in violation of law, contaminated film is disposed of by burning which introduces the element of pollution. Polymer films for solarisation contain alginates, PVA and glycerol. They are biodegradable, made of natural polymer and do not have to be removed from soil after they are used (Di Mola et al., 2021).

8.4 (BIO)DEGRADABLE POLYMERS IN PACKAGING AND FOOD-SERVICES SECTORS

Delivery of food products to the point of sale or to the consumer requires a large and varied amount of packaging product, each designed to serve its unique purpose. The function of packaging is threefold: the material from which it is made must be approved for direct contact with food, the design of packaging must protect its contents from external conditions and also must serve to protect it from loss of flavour. Additional requirements address transportation and storage requirements. Apart from these functions, the packaging must also have an informative function and contain the required marketing information e.g. on allergen content (Shaikh et al., 2021) Currently, the most common packaging material are made of petrochemicals such as polystyrene (PS), PP, PA and poly(ethylene terephthalate) (PET). Those containers meet all regulatory requirements but the problem is that they resist degradation in the environment. Food packaging companies and the food industry are looking for new alternative sources of cheap, renewable and biodegradable materials to replace non-renewable petroleum-based ones. Biodegradable polymers and conventional bio-based polymers will drive rapid research and development (R&D) growth over the next decade due to strong demand from the packaging industry. Nowadays, natural and synthetic biodegradable alternatives to conventional polymers are increasingly being used as they are much more environmentally friendly, green and degradable and some of them can be used after the recycling process (Chaudhary et al., 2022). Made from renewable sources, natural polymers may be a good alternative to plastics due to the absence of waste-related problems – a lower environmental impact. Natural (biogenic) polymers can be divided into three classes based on their origin. Figure 8.3 shows the division of polymers of natural origin according to their extraction method.

8.4.1 STARCH

Starch can be one of the alternative polymers due to its biodegradability and non-toxicity. It can be used to produce biodegradable films or starch coatings for use in packaging. Starch film can be produced by various methods, which include injection moulding, extrusion, blow moulding and foaming (Jiang and Duan et al., 2020). However, the most common way to produce starch films and coatings is the solution casting method. Starch is readily available in large quantities and at a relatively low cost. It also has attractive physical properties (low gas permeability, barrier properties due to hydrophobic components, transparency/ability to impart desired colour – desirable optical and rheological properties) apart from

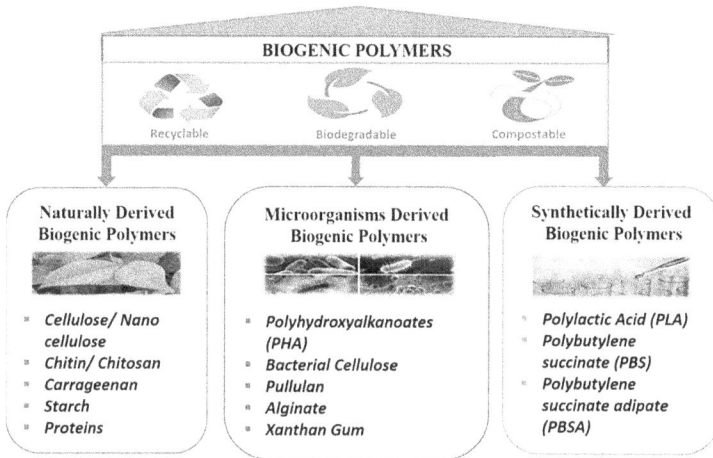

FIGURE 8.3 Division of biogenic polymers of natural origin by method of extraction.

Source: reprinted with permission from Chaudhary et al., Recent advancements in smart biogenic packaging: reshaping the future of the food packaging industry, *Polymers* 14 (2022): 829. For more details see the CC BY 4.0

mechanical properties that make it one of the top candidates for biodegradable food packaging. As starch can be used with practically no restrictions, the world market is using it more and more.

Thermoplastic starch (TPS) is used for food wrappings and cups, plates and other food containers. The mixture of starch and glycerol after the gelatinisation process is not suitable for packaging of high-moisture and liquid food products due to the hydroscopic properties of this material. Due to the poor mechanical properties and low moisture resistance of plasticised starch, which somewhat limits its application, this polymer can be blended with polymers such as PLA, PCL or PVA. This polymer/starch blend can be used to produce flexible and rigid films, allowing thermoforming into rigid trays for injection moulded packaging and coating paper and board (a requirement for direct food contact). TPS or plasticised starch is a good alternative to synthetic polymers. Research is also being conducted on fully biodegradable composites or so-called biocomposites. These are composites in which biodegradable polymers are mixed with natural fibres such as: cotton, ramie, jute fibre, kenaf or rice husk fibre (Girijappa et al. 2019). One example of commercially available starch and its blends is a starch-based biodegradable plastic resin called Solanyl BP produced by Solanyl Biopolymers Inc (Solanyl Biopolymers Inc). This is a resin made from recycled potato starch from which new and innovative short-life products, such as packaging, can then be moulded. BIOPLAST, produced by BIOTEC Biologische Naturverpackungen GmbH & Co. KG is another example of a material that can be used for packaging applications (BIOTEC). This is a plant-based resin containing large amounts of potato starch and PLA. What is important from the point of view of environmental impact is that all products made from this material are 100% biodegradable and compostable according to EN 13432 (CEN EN 13432:2000).

8.4.2 CELLULOSE

Another natural and biodegradable material used for packaging is cellulose and its derivatives such as cellulose acetate, cellulose sulphate, carboxymethyl cellulose (CMC), ethyl cellulose or methyl cellulose (Liu et al., 2021). This polymer can form transparent and rigid films and has a tensile strength similar to polypropylene. Due to its properties it is commonly used in packaging for bakery products. It has been shown that cellulose multilayer packaging has better water and fat barrier properties and also improved tensile strength (Wang and Deng et al., 2021). The feasibility study of using a CMC-based biodegradable material for packaging carried out on several gelatine/CMC/agar blends showed that the blend with 2.0% of glycerol was the best and optimal for potential use in food packaging, due to the fact that, in comparison with other samples, it had excellent properties such as the lowest water vapour permeability and the highest biodegradability (Yaradoddi et al., 2020). Cellulose-based films can also be used in so-called active packaging technology. It was demonstrated that, when in contact with food, cellulose-based packaging is able to effectively release antimicrobial agents which can be used for food preservation purposes (Cooksey, 2005). Substances such as nisin, natamycin, olive leaf extract, propolis and organic acids are widely studied antimicrobial agents that are used in cellulose-based packaging for food preservation (Kamarudin et al., 2022). Active packaging made of cellulose and its derivatives was used as packaging for fruit and vegetables. Cellulose/silver (Ag) nanoparticle film is used to extend shelf life and impregnate Ag particles to achieve an antibacterial effect and is used in packaging cabbage, tomato and fresh cut melon (Singh and Sahareen, 2017; Fernández et al., 2010). Another example is nanopaper made of bacterial cellulose and postbiotics, used to package meat to extend its shelf life under storage conditions (Yordshahi et al., 2020). Examination of cellulose acetate films containing bacteriophage solution showed that such films exhibited antimicrobial activity against *Salmonella typhimurium*. This means that these materials can be used to produce films with antimicrobial properties in food packaging (Gouvêa et al., 2015). Modified wood cellulose fibres were used to produce environmentally friendly film-forming materials. These films show a higher water vapour barrier than cellophane and have excellent mechanical properties (Sirviö et al., 2013). Modified cellulose fibres reinforced with PVA have been used to produce biodegradable composite films that can be used as packaging material. Studies have shown that modified cellulose has the effect of increasing the tensile strength of the composites (Tan et al., 2015).

8.4.3 CHITOSAN

Chitosan is another biodegradable and non-toxic biopolymer that can be used in packaging applications. It is obtained by the chemical *N*-deacetylation of chitin, which is a component of the carapace of marine crustaceans such as shrimp, oysters, crabs and lobsters. The industrial use of chitosan depends on the degree of deacetylation and the molar mass of the polymer (Hahn et al., 2020). However, its properties such as biodegradability depend on the relative proportions of N-acetyl-D-glucosamine and D-glucosamine residues. This polymer is very soluble in water, resistant to high temperatures, has good permeability to oxygen and carbon dioxide and has excellent mechanical properties. Due to its antimicrobial and antifungal properties, chitosan can be used in extending the shelf life of food and as an ingredient in the production of biodegradable edible films for food packaging. Films derived from chitosan are used to coat fresh fruits and vegetables, especially strawberries, blueberries and grapes. The active films developed using chitosan are mainly systems for the active release of antioxidants and antibacterial agents (Souza et al., 2017). These include essential oils, phenolic compounds and nisin, lysozyme, garlic extract and many others. A number of studies have shown that these additives allow products packaged in them to be stored for longer periods of time without the growth of bacteria and fungi (Zhang et al., 2019). For example, chitosan coatings containing thyme essential oil have been used to produce coatings on strawberries stored under refrigeration. The use of these coatings allowed strawberries to be stored under these conditions without deterioration for 15 days. The research also did not show any mould or fungal growth on the fruit stored in this way (Martínez et al., 2018). The effect of chitosan/clove oil packaging on the quality and shelf life of cooked pork sausages was also investigated. Supplementation of the film with essential oils of ginger and rosemary reduces oxidation processes in poultry meat (Pires et al., 2018). Cellulose sulphate films supplemented with rosemary were also shown to have good antimicrobial activity against *Bacillus cereus* and *Salmonella enterica* (Souza et al., 2019). On the other hand, the chitosan film containing the essential oil of *Perilla frutescens* (L.) Britton showed growth inhibition against *Escherichia coli*, *Staphylococcus aureus* and *Bacillus subtilis*. The obtained films may have applications as biodegradable, environmentally friendly packaging materials (Zhang et al., 2017). Research is also being conducted on the effect of active biocomponents, on mechanical properties such as tensile strength and fracture strain as well as barrier properties of edible chitosan films (Souza et al., 2018; Yuan et al., 2015). Studies of chitosan-based films have shown that, depending on the essential oil used, i.e. its composition, concentration and interactions with the polymer matrix, these additives can cause a decrease or an increase in the mechanical properties of the films obtained. A decrease in the mechanical properties of the films was observed after the addition of *Perilla frutescens* (L.) Britton oil to chitosan films and also after the addition of ginger oil (Sánchez-González and González-Martínez, 2010; Mahdavi et al., 2018; Zhang et al., 2018; Amor et al., 2021). Studies have shown that the decrease in the mechanical properties of the films may be influenced by the hydrophobicity of the oils and the heterogeneity of the dispersion, which disrupts the microstructure of the obtained films. On the other hand, an increase in mechanical properties was observed when anise (*Pimpinella anisum* L.) oil or rosemary essential oil was added to chitosan films. When these oils were added to the chitosan matrix, an increase was observed not only in tensile strength but also elongation at the break. The observed increase of mentioned properties may be related to the hydrophilicity and high molar mass of these compounds which has a beneficial effect on reducing film softening and increasing their mechanical strength. The barrier properties of the obtained chitosan films also depended on the type of biocomponent added. Water barrier properties decreased after the addition of essential oils such as *Perilla frutescens* (L.) Britton, bergamot, anise, tea tree and phenolic compounds including extract of propolis, apple, pine needle (*Cedrus deodara*), spirulina alga, purple and black rice and curcumin (Flórez et al., 2022). Various chitosan film additives are used to produce active and smart films. The purpose of such food packaging is to extend shelf life and maintain quality and sensory characteristics. These include packaging systems that release antioxidants and absorb oxygen and moisture and films containing indicators that react to changes in pH or temperature to monitor food quality (Flórez et al., 2022). Anthocyanins extracted from apple (*Plinia cauliflora*) fruits and purple sweet potato (*Ipomoea batatas* L.) skins were used as raw material for the production of colorimetric indicator films based on chitosan/PVA polymers (Capello et al., 2021). Studies on the influence of applied extracts on morphological, thermal and physicochemical properties have shown that apple fruit extracts can be used to produce colorimetric indicator films in food packaging. Purple tomato anthocyanin was used as an additive in the production of a chitosan matrix. The aim of the study was to produce a smart colorimetric indicator film for monitoring the freshness of milk and fish. It was shown that as the quality of food products decreased, the films changed colour, allowing freshness of the products to be monitored (Li et al., 2021b).

8.4.4   PROTEIN

Protein-based polymers (zein, gelatine, wheat gluten, soy and whey protein isolates) are further materials used to develop smart packaging films. These biopolymers can be used to produce coatings and films with good barrier properties against food-borne pathogens and gas transport.

8.4.4.1   Whey Protein

Whey protein is an animal protein, by-product of cheese production, used in the production of flexible films. Films obtained from whey protein concentrates and isolates are characterised by good water solubility, transparency, high water vapour permeability, good oxygen barrier and good biodegradability, suitable mechanical and optical properties (transparency) and the possibility of introducing functional compounds and active ingredients to improve their properties and potential applications in food packaging. The addition of antimicrobial agents such as acids (citric, lactic, malic and tartaric), nisin or essential oils (oregano, rosemary and garlic) to the obtained whey protein films has the effect of reducing the growth of pathogenic microorganisms and deteriorating food quality (Kandasamy et al., 2021). Physical and chemical methods such as UV, alkalisation and ultrasound can also be used to improve film strength. The use of UV radiation in the manufacture of films improves their mechanical properties. On the other hand, the addition of plasticisers makes these films flexible, transparent and colourless, and they show very good barrier properties against oxygen, oil and aromas. The addition of lipid components (fatty acids, waxes and vegetable oils) improves the barrier properties of the resulting films against moisture (Chen et al., 2019). This biodegradable polymer can find applications in the production of edible films and biodegradable packaging materials

(Kandasamy et al., 2021). Whey protein-based edible films can be used as films and coatings in several types of food such as dairy products, eggs, meat, fruits, vegetables, seafood, fish and nuts (Feng et al., 2018; Brink et al., 2019; Galus et al., 2021; Rossi-Márquez et al., 2021). Milk protein films (casein and whey proteins) obtained by the casting method were used as a potential application for packaging cheddar cheese. These films were plasticised with glycerol or sorbitol. The properties of the films were affected not only by the plasticiser itself but also by the type of polymer used. The plasticiser concentration affected the thickness of the resulting film as well as its mechanical properties (Wagh et al., 2014). Liquid whey protein concentrate with Chinese cinnamon bark extract (*Cinnamomum cassia*) was used to produce an antimicrobial coating for fresh, unripened cottage cheese. The effect of the coating for two different groups was evaluated for 31 days of storage. The first group consisted of the product packed with edible coating only and the second group consisted of the product packed with edible coating and vacuum packed. The results showed that the edible coating did not affect the appearance, flavour or aroma of the cheese but had an inhibitory effect on mould and yeast growth. This coating can be used to extend the shelf life of fresh curd (Mileriene et al., 2021). Films based on furcellaran-whey protein isolate and furcellaran-whey protein isolate incorporated with yerba mate extract or with white tea extract were used as packaging for fresh, soft rennet-type cottage cheese. Tests showed that coliform bacteria counts decreased in all samples tested. On the other hand, the amount of yeast and mould increased in all samples tested, except the film with yerba mate extract. However, cheese packed in these films showed better organoleptic properties than cheese packed in films made of low-density polyethylene (PE) (Pluta-Kubica et al., 2020). Edible films of trans-glutaminase cross-linked whey protein/pectin were used to coat freshly cut fruits and vegetables. The study showed that, after 10 days of storage, the coated fruits and vegetables did not lose firmness and retained their nutritional value (Rossi-Márquez et al., 2017). *Fucus vesiculosus* extract with antimicrobial and antioxidant properties was also used as an additive to produce an active coating from whey protein concentrate. The effect of packaging on the degree of lipid oxidation in raw chicken breasts was assessed by measuring malondialdehyde content. The study showed that, after 25 days of food storage in such packaging, there was no increase in malondialdehyde content (Andrade et al., 2021). Whey protein concentrate films containing a mixture of the essential oils *Cinnamomum cassia*, *Cinnamomum zeylanicum* and *Rosmarinus officinalis* were developed and applied to active food packaging. Salami was chosen as a food model. The oxidation status of the salami was evaluated and thus the suitability of the film for 180 days of product storage. The addition of a mixture of essential oils to the polymer film can retard lipid oxidation but does not improve the mechanical and barrier properties of the films obtained (Ribeiro-Santos et al., 2018).

8.4.4.2 Wheat Gluten

Wheat gluten is a natural protein, insoluble in water, which is a by-product obtained from wheat flour. This natural polymer consists of two proteins: gliadin – which contains low-molar-mass proteins – and glutenin, which consists of high-molar-mass proteins. Commercially used wheat gluten consists of 75–80% protein, 15–17%

carbohydrate, 5–8% fat and 0.6–1.2% ash (Nadathur et al., 2015). The study of antibacterial films from wheat gluten containing potassium sorbate obtained by the compression moulding method have shown that potassium sorbate acts as a plasticiser. These films show antifungal properties against *Aspergillus niger* and *Fusarium incarnatum* bacteria (Türe et al., 2012). The next study evaluated the antimicrobial properties of films obtained in wheat gluten containing 0.5, 1, 2 and 4 wt% vanillin. These films showed antimicrobial activity against *Escherichia coli* and *Staphylococcus aureus*, when the vanillin concentration was greater than 1%. Moreover, the addition of vanillin increased the stretchability and elongation of the wheat gluten films (Aarabi et al., 2015). Bionanocomposites containing wheat gluten, cellulose nanocrystals and additive titanium dioxide (TiO_2) nanoparticles were prepared by the casting/evaporation method. The resulting nanocomposite was used to coat sheets of unbleached paper. The effect of the percentage of additives in the nanocomposite on antimicrobial activity, mechanical properties and water sensitivity was investigated (El-Wakil et al., 2015). The paper coated with nanocomposite consisting of 7.5% cellulose nanocrystals and 0.6% TiO_2 had better antimicrobial properties against *Saccharomyces cervisiae*, Gram-negative bacteria *Escherichia coli* and Gram-positive bacteria *Staphylococcus aureus* than the paper coated with wheat gluten/cellulose nanocrystal nanocomposite. Furthermore, the paper showed significantly better mechanical properties. Hybrid coatings were prepared for cardboard substrates containing wheat gluten and silica. Application of these coatings to cardboard substrates resulted in a 4-fold reduction in the water vapour permeability coefficient (Rovera et al., 2020). Also, films from nisin/zirconium dioxide (ZrO_2)/PVA/wheat gluten were prepared and their properties were characterised. The films containing 40 wt% wheat gluten showed excellent antimicrobial activity against *Staphylococcus aureus* (Pang et al., 2019). Furthermore, the gas and water vapour permeability decreased with increasing wheat gluten content, and the tensile strength increased with increasing zinc peroxide (ZnO_2) content and increasing wheat gluten content (up to 60 wt%). The production of innovative, naturally derived bionanocomposites for food preservation was also described. These nanocomposites were made from wheat gluten and lignocellulosic nanofibres. Both coatings and films made from these polymers showed excellent preservation properties, extending the shelf life of fruits. Furthermore, the bionanocomposites showed antimicrobial properties and good water resistance. The coatings and packaging tested also exhibited excellent barrier properties against oxygen and water vapour (Chen et al., 2022).

8.4.4.3 Soy Protein

Soy protein is obtained from soybeans, which contain about 40–45% of this protein. It contains such amino acids as cysteine, arginine, lysine and histidine. Soy protein is found in three forms; in the form of defatted soy flour (contains about 56% protein), soy protein concentrate (contains 65–75% protein) and soy protein isolate (over 90% protein). This biopolymer has very good film-forming properties and the films obtained from this biodegradable polymer have many functional properties (viscosity, cohesiveness, ability to emulsify, absorb water and fat as well as ability to form fibres). The films obtained from this biodegradable polymer are mostly based on soy protein isolate. Composite films were prepared based on soy protein containing hyperbranched polysiloxane and dendritic tannic acid amines (Li et al., 2021a). Testing of these films showed an increase in tensile strength compared to films containing pure soy protein. In addition, these films exhibit UV-blocking properties, making them useful for packaging technology applications. Other films containing soy protein were obtained and enriched with a galactomannan fraction extracted from *Gleditsia triacanthos* L. The study investigated the effect of the addition of galactomannan on improving the physicochemical and mechanical properties of the films obtained. It was shown that films produced from the combination of these components are more resistant to water and have higher mechanical strength. The mixture of the two ingredients allowed obtaining films with improved properties that can be used in food packaging (González et al., 2019). Soy protein-based films with the addition of cellulose nanocrystals and *Cedrus deodara* pine needle extract were produced to obtain packaging materials with good mechanical properties, antioxidant capacity and water vapour barrier. The addition of cellulose nanocrystal increased the tensile strength and the addition of pine needle extract increased the water vapour barrier of soy-protein-based films. Furthermore, pine needle extract-added films showed good antioxidant properties (Yu et al., 2018). Soy protein films were obtained having a dual network structure composed of covalent and hydrogen bonds. The covalent bonds in the obtained films were formed by crosslinking the soy protein with triglycidylamine, and the hydrogen bond network was formed by the interaction of soy protein isolate with PVA. The resulting network structure improves the water resistance and strength of the film. In addition, the film exhibits anti-fungal properties, protecting products from mould (Xu et al., 2022).

8.4.4.4 Zein

Zein is a protein found in approximately 45–50% of maize proteins. Due to its non-polar amino acid content, this natural polymer is insoluble in water but readily soluble in alcohol. Zein is a biopolymer with the ability to form films, which has found applications in the packaging industry. The films obtained have advantages (tastelessness, good water vapour barrier, good gas barrier, abrasion and fat resistance; they are also biodegradable and biocompatible) but also disadvantages such as brittleness of films obtained without the addition of a plasticiser. The selection of a suitable plasticiser is important due to the miscibility of the components of a given film and its homogeneity. This is influenced by the molar mass of the plasticiser used, its chemical structure, the number and ratio of polar and non-polar groups and the linearity of the molecule. The possibility of using poly(ethylene glycol) (PEG) as a potential plasticiser was investigated (Zhou and Wang, 2021). Composite films with zein and methyl cellulose were prepared and oleic acid and PEG were used as plasticiser. The increase in the strength of the composite films compared to the zein film was demonstrated. Furthermore, the addition of zein reduced the water vapour permeability. The antimicrobial properties of the films when thymol was added to them were also investigated. The addition of thymol to the composite films affected their antimicrobial properties against *Escherichia coli* bacteria. Zein films with the addition of oleic acid, nanocarbonate and plasticiser (glycerol) were also obtained. The concentration of nanocarbonate improves the solubility of the obtained films in water and improves their mechanical properties, such as elasticity and resilience of the film (Ribeiro et al., 2015). Zein-based biodegradable films often containing antimicrobial additives are used in food packaging to coat products such as fruits and vegetables, nuts and meat products and can also be used to separate food layers such as cheese slices. An example is the use of nisin additive in the development of antibacterial edible zein coatings to reduce *Listeria monocytogenes* populations on nectarines and apples. The results show that coating fruit with zein films with nisin additive reduces *Listeria monocytogenes* contamination on fresh produce (Mendes-Oliveira et al., 2022). Another study evaluated the effect of the addition of cinnamon essential oil and chitosan nanoparticles on the physical, mechanical, structural and antimicrobial properties of the resulting zein films. The addition of cinnamon essential oil inhibited the growth of *Escherichia coli* and *Staphylococcus aureus* bacteria. On the other hand, the addition of chitosan nanoparticles improved the tensile strength and reduced the elongation of the obtained composites (Vahedikia et al., 2019). The aim of another study was to obtain and test active antibacterial packaging materials based on biodegradable and renewable materials (zeins) containing fatty acid esters, lauric acid monoglycerides and essential oils of oregano and thyme. The combination of eugenol and amphiphilic monoglyceride resulted in stretchable films with enhanced barrier properties and good antimicrobial efficacy against yeasts and moulds as well as Gram-positive and Gram-negative bacteria (Sedlarikova et al., 2022). Films produced based on chitosan, zein and PVA can potentially be used in food packaging (Bueno et al., 2021). Depending on the content of a given polymer in the blend, these films can have different properties. Tests have shown that a high PVA content can produce films which are easy to process, transparent, crack-free and odourless. Nanofibrillar films based on konjac glucomannan and zeins were produced by electrospinning technology. These films showed excellent antimicrobial properties against food-borne pathogens when curcumin was added to them. The films could potentially be used in food packaging (Wang et al., 2019).

8.4.5 POLYHYDROXYALKANOATES (PHA)

The next class of biodegradable polymers as a potential candidate for the packaging sector is the PHA. PHAs belong to a group of polyesters naturally produced by microorganisms as a storage carbon source. Materials for use as food packaging must ensure material quality, i.e., adequate mechanical and barrier properties during storage. The properties of PHA-based materials are significantly influenced by the structure of the monomeric copolymer composition. The extension of the use of materials for food packaging applications can be achieved through modifications of PHA, such as chemical or enzymatic post-synthetic modification, obtaining blends and composites as well as functional multilayer films (Koller, 2014). Poly(3-hydroxybutyrate) (PHB) and its copolymers poly(3-hydroxybutyrate-*co*-3-hydroxyoctanoate) (PHBO), poly(3-hydroxybutyrate-*co*-3-hydroxyoctadecanoate) (PHBOD) poly(3-hydroxybutyrate-*co*-3-hydroxyvalerate) (PHBV), poly(3-hydroxybutyrate-*co*-3-hydroxyhexanoate) (PHBHHx) and poly(3-hydroxybutyrate-*co*-4-hydroxybutyrate) (P3HB4HB) can be used in deep-drawn food packaging such as bottles, disposable cups and laminated films (Tripathi and Srivastava, 2015). Poly(3-hydroxybutyrate-*co*-3-hydroxyvalerate-*co*-3-hydroxyhexanoate) terpolyester made from fruit pulp using mixed microbial culture technology containing ester monomers in the following amounts: about 68 mol% 3-hydroxybutyrate (3HB), 17 mol% 3-hydroxyvalerate (3HV) and 15 mol% 3-hydroxyhexanoate (3HHx) units when blended with commercial PHBV and thermocompressed can be used to produce recyclable food packaging (Meléndez-Rodríguez et al., 2021). Furthermore, terpolyesters derived from industrial biowaste can be used as a plasticising additive for PHA-based packaging materials. Biodegradable materials with antimicrobial properties can be an interesting alternative in active food packaging (Castro-Mayorga and Fabra, 2016). The study of Ag-based nanocomposites with PHBV showed that, compared to the pure polymer, the oxygen permeability was reduced by about 56%, and therefore this packaging can be applied where we want to protect our product from the external environment. PHBV films obtained by electrospinning containing oregano oil and zinc flax nanoparticles showed antimicrobial properties against *Escherichia coli* and *Staphylococcus aureus*. The best results were obtained with films that contained a mixture of both agents (Figueroa-Lopez and Torres-Giner, 2020).

8.4.6 POLYLACTIDE (PLA)

PLA is one of the generally used (bio)degradable polymers. It is becoming an alternative as a green food packaging because of its high molar mass, resistance to water solubility, easy processing by thermoforming and (bio)degradability. PLA has the tensile strength modulus, flavour and odour barrier of PE and PET or flexible poly(vinyl chloride) (PVC), the temperature stability and process ability of polystyrene and the printability and grease resistance of PE. Processed PLA comes in the form of film, containers and as coating for paper and paperboards. However, PLA displays undesirable properties; it is more brittle and degrades easily as temperature rises. PLA demonstrated practical application in the production of beverage bottles, transparent rigid containers, bags, jars, food containers, disposable cups, coating for all types of packaging, packaging bags, household refuse bags and film packaging foams (Shaikh et al., 2021). This polymer can be blended with other polymers; for example: poly(ethylene oxide), PEG, PCL, poly(vinyl acetate), PHB, cellulose acetate, poly(butylene succinate) and poly(hexamethylene succinate) or with plasticisers such as oliglactide, glycerol, triacetine and low-molar-mass citrates (Ayse et al., 2017). The choice of plasticisers or polymers depends on the intended application. Selected additives cannot be toxic or prohibited by regulation when selected for direct contact with food or when accidental (unintended) contact is possible. The study of (bio)degradable food packaging materials with PLA and starch have shown that combination of these two polymers in appropriate proportions with the addition of suitable compatibilisers (epoxidised soybean oil and maleic anhydride or castor oil and hexamethylene diisocyanate) resulted in high-performance films that can meet a wide range of packaging requirements (Muller et al., 2017). PLA/starch multilayer films showed very good water vapour and gas barrier properties and higher mechanical resistance compared to films made from starch alone. Research on the PLA composite with cocoa bean shell fine powder has shown that the addition of cocoa bean shell results in an improvement of the physical properties of the composites (Yung modulus increases by 80% with the addition of 75% by mass of cocoa bean shell). The addition of cocoa bean shell gives the composites antioxidant activity and affects the rate of (bio)degradation of the composite in an aqueous environment (Papadopoulou et al., 2019). Material based on PLA and PHBV has shown effective application for production of the thermoformed containers for food packaging. Another strategy is to blend PLA with inorganic nanoparticles such as metal oxides and minerals (Ag, magnesium oxide (MgO), zinc oxide (ZnO), TiO_2, silver-copper (Ag-Cu), halloysite nanotubes, hydroxyapatite, silica, alumina, magnetite, zirconium oxide (ZrO) and calcium carbonate ($CaCO_3$)). These additives significantly improve the mechanical properties of PLA films (Mulla et al., 2021).

8.5 (BIO)DEGRABLE POLYMERS IN CONSUMER GOODS SECTOR

Biodegradable polymers are widely used to replace traditional materials such as paper, wood, plastics and other materials that do not degrade easily. Polymers used in these applications are derived from renewable sources, which slows down the depletion of limited fossil fuel resources. The consumer goods category includes a very large number of items, from household goods to leisure and sporting goods, toys and consumer goods. A new generation of adhesives, paints, motor lubricants and building materials (Fomin and Guzeev, 2001) contain biodegradable polymer materials. Biodegradable bamboo golf shirts and biodegradable fishing hooks have reached the hands of consumers. Toothbrush handles and adhesive tapes contain chemically modified plant cellulose (e.g. cellulose acetate) in their composition. Traditional materials used for coffee capsules, which are consumed in large quantities, are being replaced by starch-based materials or biodegradable polyesters in combination with PLA. Willowflex, a biodegradable material used in 3D printing, is made from non-genetically modified maize corn starch (Scott C, 2017). Its unique features include the fact that it is made from compostable material and is flexible. Fashion designer Babette Sperling uses Willowflex to produce clothes using the 3D printing method. Additionally, testing is under way to use it in the manufacture of toys. Cups, plates, bowls and shoes can also be made using this material. Another material used in the consumer goods sector is Seppa produced by Chieng Roung Industry Co., Ltd. This biodegradable material from marine shell and its applications includes Lego-like blocks, plastic balls, model toys, pens and art knives. PLA can also be used successfully as a filament in 3D printing to produce all kinds of toys because PLA is non-toxic. It can be used to print board games, customised toys for children, children's jewellery (neck ornaments, rings, earrings, bracelets and anklets) and educational toys (Tesseract, 2021).

8.6 (BIO)DEGRABLE POLYMERS IN TEXTILE SECTOR

The textiles industry is considered the second-largest economic undertaking in the world. However, textiles themselves are considered the fourth category of products that have an impact on the environment. In the countries of the European Union, the waste generated after the production process and at the end of the product's life cycle represents between 2 and 10% of the environmental impact. Depending on the type of fibres used in production (natural such as cotton, wool, linen, silk, kenaf, ramie as well as artificial fibres or synthetic polymers) and the type of production process (dry or wet), the production process of a garment can have a greater or lesser impact on the environment and the problems associated with it. The textile industry, instead of using toxic chemicals in the pre-treatment of natural fibres, has started to use enzymes that are able to impart specific functional characteristics to the treated textiles (Eid, and Ibrahim, 2021). Furthermore, typical chemical dyes are being replaced by natural dyes extracted from different parts of plants, such as bark, leaves, roots and flowers (tannin, flavonoids and quinonoids; Kasiri and Safapour, 2014). Research and experiments are also being carried out to replace traditional man-made polymer fibres with new, environmentally friendly materials. One such material could be PLA (Grancarić and Jerković, 2013). The use of recycled materials and 3D printing in garment design and production has the potential to reduce textile waste and CO_2 emissions. A systematic biomimetic approach to the development of high-performance, environmentally friendly fibre composites and technical textiles helps to understand the biological structures, processes and functions that can be used to produce new biomimetic microcomposites for the textile sector, including biogeotextiles and nonwovens (Grancarić and Jerković, 2013). One example is the use of bioplastics derived from renewable biomass, in this case potatoes, to produce disposable mackintoshes (Quilicuá). Soybean fibre, biodegradable and non-allergenic, can be blended with PVA. Resulting clothing is less durable but has a soft, elastic "feel". Zein was used to coat cotton fibres to impart hydrophobic properties and antimicrobial activity to cotton textiles. To obtain good antimicrobial properties, ellagic acid was encapsulated in zein particles. The textiles possessed antimicrobial activity against *Escherichia coli* and *Staphylococcus aureus* (Gonçalves et al., 2020).

8.7 (BIO)DEGRABLE POLYMERS IN ENERGY AND ENGINEERING SECTOR

One of the applications of biodegradable polymers is to improve energy sources. One of the natural and biodegradable materials used to produce components applicable for energy production can be cellulose. Cellulose itself, as well as cellulose nanocrystals, cellulose nanofibres and bacterial nanocellulose can be used in fuel cells (Dai and Ottesen, 2019). Cellulose nanofibres can also be used as proton fuel cell components such as ion exchange membrane and catalyst layer. The first fuel cell containing nanostructured cellulose membranes that could operate at 80°C was obtained in 2016 (Bayer and Cunning, 2016). A tetrafluoroethylene-perfluoro-3,6-dioxa-4-methyl-7-octenesulfonic acid copolymer (Nafion) and bacterial nanocellulose were used to produce the membrane in polymer electrolyte fuel cells (Jiang and Zhang, 2015). Chitosan is also a natural polymer that can be used as a material in fuel cells (polymer electrolyte fuel cell, alkaline fuel cell, biofuel cell and direct methanol fuel cell) or as membrane electrolytes and electrodes in low- and medium-temperature hydrogen cells. Chitosan with graphene oxide has been used to produce fuel cell membranes. The addition of graphene oxide to chitosan improved mechanical properties such as tensile strength from 40.1 MPa for chitosan membrane to 89.2 MPa for chitosan/ graphene oxide membranes and Young's modulus from 1.32 to 2.17 GPa for chitosan and chitosan/oxide membranes, respectively (Shao et al., 2013, Liu et al., 2014).

Another study focused on membranes made of chitosan/sulfonated graphene oxide. The addition of such a filler improved their mechanical properties and increased the proton conductivity 0.0117 S/cm for the pure chitosan membrane to 0.0267 S/cm for the chitosan/sulfonated graphene oxide membrane. The addition of phosphorylated graphene oxide to chitosan membranes, on the other hand, increases the strength properties and, moreover, has the effect of increasing the proton conductivity. The highest proton conductivity was 63.4 mS/cm at 95°C for a content of 2.5 wt% of phosphorylated graphene oxide (Weber Kreuer et al., 2008). Pectin is a natural polymer also used in the construction of electricity carriers. Pectin-based membranes can be used to build direct methanol fuel cells. Such membranes can be formed from nanocomposites containing pectin, PVA and sulphonated TiO_2 nanoparticles (Mohanapriya et al., 2020). Biodegradable polymers and ecomaterials also have found applications in the construction of various components of electronic parts. Composites based on keratin fibres and chemically modified soybean oil are used in electronics as new materials with low dielectric constant. Other electronic devices that use polymer-based ecomaterials include capacitors, communication devices, loud speakers with advanced functionalities and paper-based displays made from wood lignocellulose, microfibre-based polymers or batteries for microelectromechanical systems. Electroconductive plastics were developed to manufacture electronics from sustainable sources. This is accomplished by the spinning of different nanofibres individually or by hybridising with other materials and incorporating them with suitable bio-based polymers which can produce nanobiocomposites as possible superior structural components (lighter than their microcounterparts). These materials can find applications in a wide range of electrical, optical and biomedical devices, e.g., as components of various functional devices. Eco-friendly materials can be used in vacuum cleaners, as housing for power tools and for portable electronic equipment including cellular phones. The electronics industry is also using PLA-based (bio) degradable materials that can be functionalised with various additives. PLA-based composites such as PLA/cordenka composites or PLA and kenaf composite, which is used for compact discs or computer cases, are also used in this industry (Frackowiak et al., 2018). Besides, using special technology, PLA can be refined to obtain properties such as conductivity and improved mechanical and thermal properties. PLA is also suitable for 3D printing technology, which offers unlimited possibilities for the production of ready-to-use electronic components (Molitch-Hou, 2018).

Just like the automotive industry, the construction industry is also striving to reduce CO_2 emissions. Buildings made of environmentally friendly materials are supposed to provide a healthier place to live and work. Today, eco-friendly building materials include plastic laminated wood elements, polymerised waterproof masses and polymer compositions (polyesters with jute fibres or fibre-reinforced polymers). Bio-based and/or biodegradable polymers can be used as moulds for truss joints, rigid foams with high thermal insulation, new mortars or concretes containing waste bio-based polyurethane foams and piezoresistive strain gauges for monitoring mechanical deformation of structures. PLA-based composites can be used as internal window profiles. PLA-based materials with the addition of aluminium hypophosphite, bamboo charcoal and core-shell nanofibrous can provide flame retardant properties to some building materials (Feng and Sun, 2017, Wang and Zhang,

2020). The aim is also to reduce energy consumption in buildings. In this case, it is necessary to design materials that allow for better control of energy consumption. Composites of phase change materials based on bio-based materials with nanoplatelets of exfoliated graphite and carbon nanotubes seem to be suitable materials for such applications. These composites are characterised by high thermal conductivity and are suitable candidates for energy-saving materials (Yu et al., 2014). Another example, thermal insulation using composites and nanoporous materials based on biopolymers and silica such as cellulose/silica, pectin/silica and polysaccharide/silica aerogel, has also received much attention recently (Pierre, 2011; Zhao et al., 2015). Hybrid aerogels based on polysaccharides such as alginate, pectin, xanthan and guar and tetramethylorthosilicate have been produced and tested. The thermal properties of the obtained materials improved and the lowest thermal conductivity of 19 $mWm^{-1} K^{-1}$ was observed in the pectin-silica hybrid aerogel (Horvat et al., 2019).

8.8 CONCLUSION

The use of biodegradable polymers helps to reduce the environmental impact of the production and processing of petroleum-based plastics. However, biodegradable polymers are not free from disadvantages, which pose a number of problems in their applications. These problems are related to the degradation time of these polymers and the decrease in their durability, especially during their use. Products made from these plastics, when used in industry, are expected to retain their properties throughout their lifetime and only degrade at the end of that lifetime. This is why research is constantly being conducted to develop polymers and their composites with specific properties for a given type of application. Furthermore, some biodegradable polymers are produced from renewable raw materials such as agricultural waste, so it is possible to recycle this waste and use it to produce biodegradable polymers for everyday use and thus reduce the environmental impact of this waste as well.

REFERENCES

Aarabi, A., Ebad Idehaghani, H., and Saiedi, S. 2015. Antimicrobial effects of edible gluten films incorporated with vanillin. *J. Food Hyg.* 5(1–17): 1–12.

Akampumuza, O., Wambua, P.M., Ahmed, A., Li, W., and Qin, X.H. 2017. Review of the applications of biocomposites in the automotive industry. *Polym. Compos.* 38: 2553–2569.

Amor, G., Sabbah, M., Caputo, L., et al. 2021. Basil essential oil: Composition, antimicrobial properties, and microencapsulation to produce active chitosan films for food packaging. *Foods* 10(121): 1–16.

Andrade, M.A., Barbosa, C.H., Souza, V.G.L., et al. 2021. Novel active food packaging films based on whey protein incorporated with seaweed extract: Development, characterization, and application in fresh poultry meat. *Coatings* 11(2): 229.

Anuar, H., and Zuraida, A. 2011. Thermal properties of injection moulded polylactic acid – Kenaf fibre biocomposite. *Malays. Polym. J.* 6(1): 51–57.

Ayse, O., Özge, S., and Yasemin, Ç.S. 2017. Poly(lactic acid) films in food packaging systems. *Food Sci. Nutr. Technol.* 2: 000131.

Azwa, Z.N., and Yousif, B.F. 2013. Thermal degradation study of kenaf fibre/epoxy composites using thermogravimetric analysis. In *3rd Malaysian Postgraduate Conference (MPC2013)*, eds. Noor, M.M., Rahman, M.M., and Ismail, J., pp. 256–264. Toowoomba: University of Southern Queensland.

Balakrishnan, P., John, M.J., Pothen, L., Sreekala, M.S., and Thomas, S. 2016. Natural fiber and polymer matrix composites and their applications in aerospace engineering. In *Advanced Composite Materials for Aerospace Engineering*, eds. Rana, S., and Fangueiro, R., pp. 365–383. Amsterdam: Elsevier Science LTD.

Bayer, T., Cunning, B.V., Selyanchyn, R., Nishihara, M., Fujikawa, S., Sasaki, K., et al. 2016. High temperature proton conduction in nanocellulose membranes: Paper fuel cells. *Chem. Mater.* 28: 4805–4814.

Beig, B., Niazi, M.B.K., Jahan, Z., et al. 2020. Biodegradable polymer coated granular urea slows down n release kinetics and improves spinach productivity. *Polymers* 12: 2623.

BE-Sustainable. 2014. *Biofore Concept Car Show Cases the Potential of Biomaterials.* www.besustainablemagazine.com/cms2/upms-biofore-concept-car-drives-sustaina-ble-change-through-innovative-use-of-biomaterials (accessed February 29, 2022).

BIOTEC. *Bioplast.* https://en.biotec.de/bioplast (accessed March 25, 2022).

Bledzki, A.K., Faruk, O., and Jaszkiewicz, A. 2010. Cars from renewable materials. *Kompozyty.* 10(3): 282–288.

Brink, I., Šipailienė, A., and Leskauskaitė, D. 2019. Antimicrobial properties of chitosan and whey protein films applied on fresh cut turkey pieces. *Int. J. Biol. Macromol.* 130: 810–817.

Bueno, N.N.J., Corradini, E., de Souza, P.R., de S. Marques, V., Radovanovic, E., and Muniz, E.C. 2021. Films based on mixtures of zein, chitosan, and PVA: Development with perspectives for food packaging application. *Polym. Test.* 101: 107279.

Capello, C., Trevisol, T.C., Pelicioli, J., et al. 2021. Preparation and characterization of colorimetric indicator films based on chitosan/polyvinyl alcohol and anthocyanins from agri-food wastes. *J. Polym. Environ.* 29: 1616–1629.

Castro-Mayorga, J.L., Fabra, M.J., and Lagaron, J.M., 2016. Stabilized nanosilver based antimicrobial poly(3-hydroxybutyrate-*co*-3-hydroxyvalerate) nanocomposites of interest in active food packaging. Innov. *Food Sci. Emerg. Technol.* 33: 524–533.

CEN EN 13432:2000. *Requirements for Packaging Recoverable Through Composting and Biodegradation – Test Scheme and Evaluation Criteria for the Final Acceptance of Packaging.* Brussels: European Committee for Standardization. Technical Committee: CEN/TC 261 – Packaging.

Chaudhary, V., Punia Bangar, S., Thakur, N., and Trif, M. 2022. Recent advancements in smart biogenic packaging: reshaping the future of the food packaging industry. *Polymers* 14(4): 829.

Chen, H., Wang, J., Cheng, Y., Wang, C., Liu, H., Bian, H., Pan, Y., Sun, J., and Han, W. 2019. Application of protein-based films and coatings for food packaging: A review. *Polymers* 11(12): 2039.

Chen, Y., Li, Y., Qin, S., Han, S., and Qi, H. 2022. Antimicrobial, UV blocking, water-resistant and degradable coatings and packaging films based on wheat gluten and lignocellulose for food preservation. *Compos. B. Eng.* 238: 109868.

Cooksey, K. 2005. Effectiveness of antimicrobial food packaging materials. *Food Addit. Contam.* 22(10): 980–987.

Dai, Z., Ottesen, V., Deng, J., Helberg, R.M.L., and Deng, L. 2019. A brief review of nanocellulose based hybrid membranes for CO2 separation settings. *Fibers* 7(5): 40–58.

Di Mola, I., Ventorino, V., Cozzolino, E., Ottaiano, L., Romano, I., Duri, L.G., Pepe, O., and Mori, M. 2021. Biodegradable mulching vs traditional polyethylene film for sustainable solarization: Chemical properties and microbial community response to soil management. *Appl. Soil Ecol.* 163: 103921.

DSM. *EcoPaXX® Q-HGM24.* https://plasticsfinder.com/en/datasheet/EcoPaXX%C2%AE%20Q-HGM24/K570l (accessed May 16, 2022).

Eid, B.M., and Ibrahim, N.A. 2021. Recent developments in sustainable finishing of cellulosic textiles employing biotechnology. *J. Clean. Prod.* 284: 124701.

El-Wakil, N.A., Hassan, E.A., Abou-Zeid, R.E., Dufresne, A. 2015. Development of wheat gluten/nanocellulose/titanium dioxide nanocomposites for active food packaging. *Carbohydr. Polym.* 124: 337–346.

Faruk, O., Bledzki, A.K., Fink, H.P., and Sain, M. 2012. Biocomposites reinforced with natural fibers: 2000–2010. *Prog. Polym. Sci.* 37(11): 1552–1596.

Feng, J., Sun, Y., Song, P., et al. 2017. Fire-resistant, strong, and green polymer nanocomposites based on poly(lactic acid) and core-shell nanofibrous flame retardants. *ACS Sustain. Chem. Eng.* 5: 7894–7904.

Feng, Z., Wu, G., Liu, C., Li, D., Jiang, B., and Zhang, X. 2018. Edible coating based on whey protein isolate nanofibrils for antioxidation and inhibition of product browning. *Food Hydrocoll.* 79: 179–188.

Fernández, A., Picouet, P., and Lloret E. 2010. Cellulose-silver nanoparticle hybrid materials to control spoilage-related microflora in absorbent pads located in trays of fresh-cut melon. *Int. J. Food Microbiol.* 142: 222–228.

Figueroa-Lopez, K.J., Torres-Giner, S., Enescu, D., Cabedo, L., Cerqueira, M.A., Pastrana, L.M., and Lagaron, J.M. 2020. Electrospun active biopapers of food waste derived poly(3-hydroxybutyrate-co-3-hydroxyvalerate) with short-term and long-term antimicrobial performance. *Nanomaterials* 10: 506.

Flórez, M., Guerra-Rodríguez, E., Cazón, P., Vázquez, M. 2022. Chitosan for food packaging: Recent advances in active and intelligent films. *Food Hydrocoll.* 124: 107328.

Fomin, V.A., and Guzeev V.V. 2001. Biodegradable polymers, their present state and future prospects. *Prog. Rubber Plast. Recycl. Technol.* 17(3): 186–204.

Frackowiak, S., Ludwiczak, J., and Leluk, K. 2018. Man-made and natural fibres as a reinforcement in fully biodegradable polymer composites: A concise study. *J. Polym. Environ.* 26: 4360–4368.

Galus, S., Mikus, M., Ciurzyńska, A., et al. 2021. The effect of whey protein-based edible coatings incorporated with lemon and lemongrass essential oils on the quality attributes of fresh-cut pears during storage. *Coatings* 11: 745.

Girijappa, Y.G.T., Rangappa, S.M., Parameswaranpillai, J., and Siengchin, S. 2019. Natural fibers as sustainable and renewable resource for development of eco-friendly composites: A comprehensive review. *Front. Mater.* 6: 1–14.

Gonçalves, J., Torres, N., Silva, S., Gonçalves, F., Noro, J., Cavaco-Paulo, A., Ribeiro, A., and Silva, C. 2020. Zein impart hydrophobic and antimicrobial properties to cotton textiles. *React. Funct. Polym.* 154: 104664.

González, A., Barrera, G.N., Galimberti, P.I., Ribotta, P.D., and Igarzabal, C.I.A. 2019. Development of edible films prepared by soy protein and the galactomannan fraction extracted from Gleditsia triacanthos (Fabaceae) seed. *Food Hydrocoll.* 97: 105227.

Gouvêa, D.M., Mendonça, R.C.S., Soto M.L., and Cruz, R.S. 2015. Acetate cellulose film with bacteriophages for potential antimicrobial use in food packaging. *LWT – Food Sci. Technol.* 63(1): 85–91.

Grancarić, A.M., Jerković, I., and Tarbuk, A. 2013. Bioplastics in Textiles. *Polymers* 34(1): 9–14.

Hahn, T., Tafi, E., Paul, A., Salvia, R., Falabella, P., and Zibek, S. 2020. Current state of chitin purification and chitosan production from insects. *J. Chem. Technol. Biotechnol.* 95: 2775–2795.

Horvat, G., Pantić, M., Knez, Ž., and Novak, Z. 2019. Preparation and characterization of polysaccharide – silica hybrid aerogels. *Sci. Rep.* 9(1): 16492.

Hussain, M.R., Devi, R.R., and Maji, T.K. 2012. Controlled release of urea from chitosan microspheres prepared by emulsification and cross-linking method. *Iran Polym. J.* 21: 473–479.

Jawaid, M., and Thariq, M. 2018. Sustainable composites for aerospace applications. In *Woodhead in Composites Science and Engineering*, eds. Jawaid, M., and Thariq, M., pp. 171–209. Cambridge: Woodhead Publishing.

Jiang, G.P., Zhang, J., Qiao, J.L., Jiang, Y.M., Zarrin, H., Chen Z., et al. 2015. Bacterial nanocellulose/Nafion composite membranes for low temperature polymer electrolyte fuel cells. *J. Power Sources* 273: 697–706.

Jiang, T., Duan, Q., Zhu, J., Liu, H., and Yu, L. 2020. Starch-based biodegradable materials: Challenges and opportunities. *Adv. Ind. Eng. Polym. Res.* 3(1): 8–18.

Kamarudin, S.H., Rayung, M., Abu, F., et al. 2022. A review on antimicrobial packaging from biodegradable polymer composites. *Polymers* 14(1): 174.

Kandasamy, S., Yoo, J., Yun, J., Kang, H.B., Seol, K.H., Kim, H.W., and Ham, J.S. 2021. Application of whey protein-based edible films and coatings in food industries: An updated overview. *Coatings* 11: 1056.

Kasiri, M.B., and Safapour, S. 2014. Natural dyes and antimicrobials for green treatment of textiles. *Environ. Chem. Lett.* 12: 1–13.

Koller, M. 2014. Poly(hydroxyalkanoates) for food packaging: Application and attempts towards implementation. *Appl. Food Biotechnol.* 1: 3–15.

Kopparthy, S.D.S., and Netravali, A.N. 2021. Review: Green composites for structural applications. *Composites Part C: Open Access* 6: 100169.

Li, J., Jiang, S., Wei, Y., et al. 2021a. Facile fabrication of tough, strong, and biodegradable soy protein-based composite films with excellent UV-blocking performance. *Compos. B. Eng.* 211: 108645.

Li, Y., Wu, K., Wang, B., and Li, X. 2021b. Colorimetric indicator based on purple tomato anthocyanins and chitosan for application in intelligent packaging. *Int. J. Biol. Macromol.* 174: 370–376.

Liu, Y., Ahmed, S., Sameen D.E., Wang, Y., Lu, R., Dai, J., Li, S., and Qin, W. 2021. A review of cellulose and its derivatives in biopolymer-based for food packaging application. *Trends Food Sci. Technol.* 112: 532–546.

Liu, Y., Wang, J., Zhang, H., Ma, C., Liu, J., Cao, S., et al. 2014. Enhancement of proton conductivity of chitosan membrane enabled by sulfonated graphene oxide under both hydrated and anhydrous conditions. *J. Power Sources* 269: 898–911.

Mahdavi, V., Hosseini, S.E., and Sharifan A. 2018. Effect of edible chitosan film enriched with anise (*Pimpinella anisum* L.) essential oil on shelf life and quality of the chicken burger. *Food Sci. Nutr.* 6(2): 269–279.

Martínez, K., Ortiz, M., Albis, A., Gilma Gutiérrez Castañeda, C., Valencia, M.E., and Tovar, G.C.D. 2018. The effect of edible chitosan coatings incorporated with thymus capitatus essential oil on the shelf-life of strawberry (*Fragaria x ananassa*) during cold storage. *Biomolecules* 8(4): 155.

MCPP. 2016. *Durabio™ New Bio-Based Engineering Plastic*. www.mcpp-global.com/en/mcpp-europe/products/brand/durabio™ (accessed May 16, 2022).

Meléndez-Rodríguez, B., Torres-Giner, S., Reis, M.A.M., et al. 2021. Blends of poly(3-hydroxybutyrate-*co*-3-hydroxyvalerate) with fruit pulp biowaste derived poly(3-hydroxybutyrate-*co*-3-hydroxyvalerate-*co*-3-hydroxyhexanoate) for organic recycling food packaging. *Polymers* 13: 1155.

Melldahl, E. *Biodegradable and 3D Printable Concept Car from BMW*. https://trends.directindustry.com/project-15990.html (accessed February 29, 2022).

Mendes-Oliveira, G., Gu, G., Luo, Y., Zografos, A., Minas, I., and Nou, X. 2022. Edible and water-soluble corn zein coating impregnated with nisin for *Listeria monocytogenes* reduction on nectarines and apples. *Postharvest Biol. Technol.* 185: 111811.

Milleriene, J., Serniene, L., Henriques, M., et al. 2021. Effect of liquid whey protein concentrate-based edible coating enriched with cinnamon carbon dioxide extract on the quality and shelf life of Eastern European curd cheese. *J. Dairy. Sci.* 104(2): 1504–1517.

Mohammed, L., Ansari, M.N.M., Pua, G., Jawaid, M., and Islam, M.S. 2015. A review on natural fiber reinforced polymer composite and its applications. *Int. J. Polym. Sci.* 2015: 243947.

Mohanapriya, S., Rambabu, G., Bhat, S.D., Raj, V. 2020. Pectin based nanocomposite membranes as green electrolytes for direct methanol fuel cells. *Arab. J. Chem.* 13(1): 2024–2040.

Mohd Radzuan, N.A., Ismail, N.F., Fadzly Md Radzi, M.K., et al. 2019. Kenaf composites for automotive components: enhancement in machinability and moldability. *Polymers* 11: 1707.

Molitch-Hou, M. 2018. *The significance of completely biodegradable 3D-printed plastic.* www.engineering.com/story/the-significance-of-completely-biodegradable-3d-printed-plastic (accessed May 27, 2022).

Mulla, M.Z., Rahman, M.R.T., Marcos, B., Tiwari, B., and Pathania, S. 2021. Poly lactic acid (PLA) nanocomposites: Effect of inorganic nanoparticles reinforcement on its performance and food packaging applications. *Molecules* 26: 1967.

Muller, J., González-Martínez, C., and Chiralt, A. 2017. Combination of poly(lactic) acid and starch for biodegradable food packaging. *Materials* 10: 952.

Nadathur, S.R., Wanasundara, J.P.D., and Scanlin, L. 2015. Proteins in the diet: Challenges in feeding the global population. In *Sustainable protein sources*, eds. Nadathur, S.N., Wanasundara, J.P.D., and Scanlin, L., pp. 1–19. Amsterdam: Elsevier Science LTD.

Nissin, O. 2014. *The Biofore Concept Car.* https://kipdf.com/the-biofore-concept-car-oscar-nissin-project-engineer-young-researchers_5ab81d531723dd349c81fd60.html (accessed March 22, 2022).

Norizan, M.N., Abdan, K., Ilyas, R.A., and Biofibers, S.P. 2020. Effect of fiber orientation and fiber loading on the mechanical and thermal properties of sugar palm yarn fiber reinforced unsaturated polyester resin composites. *Polimery* 65: 34–43.

Nurazzi, N.M., Asyraf, M.R.M., Fatimah Athiyah, S., et al. 2021. A review on mechanical performance of hybrid natural fiber polymer composites for structural applications. *Polymers* 13(13): 2170.

Nurazzi, N.M., Khalina, A., Sapuan, S.M., Ilyas, R.A., Rafiqah, S.A., and Hanafee, Z.M. 2020. Thermal properties of treated sugar palm yarn/glass fiber reinforced unsaturated polyester hybrid composites. *J. Mater. Res. Technol.* 9: 1606–1618.

Pang, M., Cao, L., Lili, C., Yi, S., and Hualin, W. 2019. Properties of nisin incorporated ZrO_2/poly(vinyl alcohol)-wheat gluten antimicrobial barrier films. *CyTA J. Food* 17(1): 400–407.

Papadopoulou, E.L., Paul, U.C., Tran, T.N., et al. 2019. Sustainable active food packaging from poly(lactic acid) and cocoa bean shells. *ACS Appl. Mater. Interfaces.* 11(34): 31317–31327.

Pierre, A.C. 2011. History of aerogels. In *Aerogels Handbook, Advances in Sol-gel Derived Materials and Technologies*, eds. Aegerter, M.A., Leventis, N., and Koebel, M.M., pp. 3–18. New York: Springer.

Pires, J.R.A., de Souza, V.G.L., and Fernando, A.L. 2018. Chitosan/montmorillonite bionanocomposites incorporated with rosemary and ginger essential oil as packaging for fresh poultry meat. *Food Packag. Shelf Life* 17: 142–149.

Plastics. 2014. *Latest Generation of PLA.* www.plastics.gl/exhibit/latest-generation-of-pla/ (accessed June 28, 2022).

Pluta-Kubica, A., Jamróz, E., Kawecka, A., Juszczak, L., and Krzyściak, P. 2020. Active edible furcellaran/whey protein films with yerba mate and white tea extracts: Preparation, characterization and its application to fresh soft rennet-curd cheese. *Int. J. Biol. Macromol.* 155(15): 1307–1316.

Quilicuá, *Fantastic Bioplastic.* www.equilicua.com/513-potatoes (accessed May 27, 2022).

Rahman, M., Viduran V., Islam, K.S., and Khan., A.M. 2021. Development of jute hybrid composites for use in the car panels. *Glob. J Eng Sci.* 7(3).

Ribeiro, W.X., Filho, J.F.L., and Cortes, M.S. 2015. Characterization of biodegradable film based on zein and oleic acid added with nanocarbonate. *Food Technol.* 45: 1890–1894.

Ribeiro-Santos, R., de Melo, N.R., Andrade, M., et al. 2018. Whey protein active films incorporated with a blend of essential oils: Characterization and effectiveness. *Packag. Technol. Sci.* 31: 27–40.

Rossi-Márquez, G., Di Pierro, P., Mariniello, L., Esposito, M., Giosafatto, C.V.L., and Porta, R. 2017. Fresh-cut fruit and vegetable coatings by transglutaminase-crosslinked whey protein/pectin edible films. *LWT* 75: 124–130.

Rossi-Márquez, G., Helguera, M., Briones, M., Dávalos-Saucedo, C.A., and Di Pierro, P. 2021. Edible coating from enzymatically reticulated whey protein-pectin to improve shelf life of roasted peanuts. *Coatings* 11: 329.

Rovera, C., Türe, H., Hedenqvist, M.S., and Farris, S. 2020. Water vapor barrier properties of wheat gluten/silica hybrid coatings on paperboard for food packaging applications. *Food Packag. Shelf Life* 26: 100561.

Roy, A., Singh, S., Bajpai J., and Bajpai A. 2014. Controlled pesticide release from biodegradable polymers. *Open Chem.* 12(4): 453–469.

Rydz, J., Sikorska, W., Kyulavska, M., and Christova, D. 2015. Polyester-based (bio)degradable polymers as environmentally friendly materials for sustainable development. *Int. J. Mol. Sci.* 16(1): 564–596.

Sánchez-González, L., González-Martínez, C., Chiralt, A., and Cháfer, M. 2010. Physical and antimicrobial properties of chitosan-tea tree essential oil composite films. *J. Food Eng.* 98(4): 443–452.

Scott, C. 2017. *Fashion Designer Babette Sperling Uses WillowFlex Filament to 3D Print Secret Messages in Natural Materials.* https://3dprint.com/161341/babette-sterling-fashion-design (accessed March 20, 2022).

Sedlarikova, J., Janalikova, M., Peer, P., Pavlatkova, L., Minarik, A., and Pleva, P. 2022. Zein-based films containing monolaurin/eugenol or essential oils with potential for bioactive packaging application. *Int. J. Mol. Sci.* 23(1): 384.

Shaikh, S., Yaqoob, M., and Aggarwal, P. 2021. An overview of biodegradable packaging in food industry. *Curr. Res. Food. Sci.* 30: 503–520.

Shao, L., Chang, X., Zhang, Y., Huang, Y., Yao, Y., and Guo, Z. 2013. Graphene oxide cross-linked chitosan nanocomposite membrane. *Appl. Surf. Sci.* 280: 989–992.

Singh, M., and Sahareen T. 2017. Investigation of cellulosic packets impregnated with silver nanoparticles for enhancing shelf-life of vegetables. *Lebensm. Wiss. Technol.* 86: 116–122.

Sirviö, J.A., Liimatainen, H., Niinimaki, J., and Hormi, O. 2013. Sustainable packaging materials based on wood cellulose. *RSC Adv.* 3: 16590–16596.

Solanyl Biopolymers Inc. *A New Generation of Smart Materials.* https://solanylbiopolymers.com (accessed March 25, 2022).

Souza, V.G.L., Fernando A.L., Pires J.R.A., Rodrigues P.F., Lopes A.A.S., and Fernandes F.M.B. 2017. Physical properties of chitosan films incorporated with natural antioxidants. *Ind. Crops Prod.* 107: 565–572.

Souza, V.G.L., Pires, J.R.A., Vieira, É.T., Coelhoso, I.M., Duarte, M.P., and Fernando, A.L. 2019. Activity of chitosan-montmorillonite bionanocomposites incorporated with rosemary essential oil: From in vitro assays to application in fresh poultry meat. *Food Hydrocoll.* 89: 241–252.

Souza, V.G.L., Rodrigues, P.F., Duarte, M.P., and Fernando, A.L. 2018. Antioxidant migration studies in chitosan films incorporated with plant extracts. *J. Renew. Mater.* 6: 548–558.

Tan, B.K., Ching, Y.C., Poh, S.C., Abdullah, L.C., and Gan, S.N. 2015. A review of natural fiber reinforced poly(vinyl alcohol) based composites: Application and opportunity. *Polymers* 7: 2205–2222.

Tesseract. 2021. *Here's What PLA 3D Printing Filament Can be put to Use: Awesomely Innovative Toys for Kids and Adults.* www.tesseract3d.com/pla-3d-printing-filament-uses-innovative-toys-kids-adults/?doing_wp_cron=1653661125.4148879051208496093750 (accessed May 27, 2022).

Tofanelli, M.B.D., and Wortman, S.E. 2020. Benchmarking the agronomic performance of biodegradable mulches against polyethylene mulch film: A meta-analysis. *Agronomy* 10: 1618.

Tripathi, A.D., Srivastava, S.K., and Yadav, A. 2015. Biopolymers: Potential biodegradable packaging material for food industry. In *Polymers for Packaging Applications*, eds. Alavi, S., Thomas, S., Sandeep, K.P., Kalarikkal, N., Varghese, J., and Yaragalla, S., pp. 115–153. Palm Bay: Apple Academic Press.

Türe, H., Gällstedt, M., and Hedenqvist, M.S. 2012. Antimicrobial compression-moulded wheat gluten films containing potassium sorbate. *Food Res. Int.* 45(1): 109–115.

Tyrrell, M. 2021. Sustainable materials arrive in aviation industry. *News*, 17 February. www.aeromag.com/sustainable-materials-arrive-in-aviation-industry (accessed March 28, 2022).

Vahedikia, N., Garavand, F., Tajeddin, B., Cacciotti, I., Jafari, S.M., Omidi, T., and Zahedi, Z. 2019. Biodegradable zein film composites reinforced with chitosan nanoparticles and cinnamon essential oil: Physical, mechanical, structural and antimicrobial attributes. *Colloids Surf. B: Biointerfaces* 1(177): 25–32.

Vroman, I., and Tighzert, L. 2009. Biodegradable polymers. *Materials* 2(2): 307–344.

Wagh, Y.R., Pushpadass, H.A., Emerald, F.M., and Nath, B.S. 2014. Preparation and characterization of milk protein films and their application for packaging of Cheddar cheese. *J. Food Sci. Technol.* 51(12): 3767–3775.

Wang, L., Mu, R.J., Li, Y., Lin, L., Lin, Z., and Pang, J. 2019. Characterization and antibacterial activity evaluation of curcumin loaded konjac glucomannan and zein nanofibril films. *LWT* 113: 108293.

Wang, S., Zhang, L., Semple, K., Zhang, M., Zhang, W., and Dai, C. 2020. Development of biodegradable flame-retardant bamboo charcoal composites, Part I: Thermal and elemental analyses. *Polymers* 12: 2217.

Wang, W., Gu, F., Deng, Z., Zhu, Y., Zhu, J., Guo, T., Song, J., and Xiao, H. 2021. Multilayer surface construction for enhancing barrier properties of cellulose-based packaging. *Carbohydr. Polym.* 255: 117431.

Weber, J., Kreuer, K.D., Maier, J., and Thomas, A. 2008. Proton conductivity enhancement by nanostructural control of poly(benzimidazole)-phosphoric acid adducts. *Adv. Mater.* 20: 2595–2598.

Xu, Y., Han, Y., Chen, M., et al. 2022. A soy protein-based film by mixed covalent cross-linking and flexibilizing networks. *Ind. Crops Prod.* 183: 114952.

Yaradoddi, J.S., Banapurmath, N.R., Ganachari, S.V., et al. 2020. Biodegradable carboxymethyl cellulose based material for sustainable packaging application. *Sci. Rep.* 10: 21960.

Yordshahi, A.S., Moradi, M., Tajik, H., and Molaei R. 2020. Design and preparation of antimicrobial meat wrapping nanopaper with bacterial cellulose and postbiotics of lactic acid bacteria. *Int. J. Food Microbiol.* 321: 108561.

Yu, J.S., Chung, S.G., and Kim, O.S. 2014. Bio-based PCM/carbon nanomaterials composites with enhanced thermal conductivity. *Sol. Energy Mater. Sol. Cell* 120: 549–554.

Yu, Z., Sun, L., Wang, W., Zeng, W., Mustapha, A., and Lin, M. 2018. Soy protein-based films incorporated with cellulose nanocrystals and pine needle extract for active packaging. *Ind. Crops Prod.* 112: 412–419.

Yuan, G., Lv, H., Yang, B., Chen, X., and Sun, H. 2015. Physical properties, antioxidant and antimicrobial activity of chitosan films containing carvacrol and pomegranate peel extract. *Molecules* 20: 11034–11045.

Yuhazri, M.Y., Amirhafizan, M.H., Abdullah, A., Yahaya, S.H., Lau, S.T.W., and Kamarul, A.M. 2018. Kenaf fibre composites as promising green-composites for automotive car door map pocket application. *Int. J. Mech. Mechatron. Eng.* 18(2): 15–21.

Zhang, J., Zou, X., Zhai, X., Huang, X.W., Jiang, C., and Holmes, M. 2019. Preparation of an intelligent pH film based on biodegradable polymers and roselle anthocyanins for monitoring pork freshness. *Food Chem.* 272: 306–312.

Zhang, X., Xiao, G., Wang, Y., Zhao, Y., Su, H., and Tan, T. 2017. Preparation of chitosan-TiO₂ composite film with efficient antimicrobial activities under visible light for food packaging applications. *Carbohydr. Polym.* 169: 101–107.

Zhang, Z.J., Li, N., Li, H.Z., Li, X.J., Cao, J.M., and Zhang, G.P. 2018. Preparation and characterization of biocomposite chitosan film containing *Perilla frutescens* (L.) Britt. essential oil. *Ind. Crops Prod.* 112: 660–667.

Zhao, S., Malfait, W.J., Demilecamps, A., et al. 2015. Strong, thermally superinsulating biopolymer-silica aerogel hybrids by cogelation of silicic acid with pectin. *Angew. Chem. Int. Ed.* 54(48): 14282–14286.

Zhou, L., and Wang, Y. 2021. Physical and antimicrobial properties of zein and methyl cellulose composite films with plasticizers of oleic acid and polyethylene glycol. *LWT* 140: 110811.

9 Principles Concerning Sustainability

Present and Future of (Bio)degradable Polymers – Final Conclusions

Joanna Rydz

Bioeconomy covers all sectors and systems that rely on biological resources (animals, plants, microorganisms and derived biomass, including organic waste), their functions and principles. It includes and links: terrestrial and marine ecosystems and the services they provide; all primary production sectors that use and produce biological resources (agriculture, forestry, fisheries and aquaculture) and all economic and industrial sectors that use biological resources and processes to produce food, feed, bio-based products, energy and services (COM(2018)673 final).

The value chain of (bio)degradable polymers describes the full range of activities to take a product from conception to end use and, in fact, to its end-of-life (EoL). The value chain refers to the full life cycle of a product, including material sourcing, production, consumption and disposal and/or recycling (DanuBioValNet project, 2019).

Both of these issues come together in the context of sustainable development. Sustainable development is development that meets the current needs without compromising the future of generations to come and ensures a balance among economic growth, care for the environment and social well-being. Sustainability goals are based on the belief that eradicating poverty and deprivation must go hand-in-hand with strategies that improve health and education, reduce inequality and stimulate economic growth, while combating climate change and preserving oceans and forests (A/RES/70/1, 2015). Science plays an important role and provides the knowledge needed to achieve sustainable development, particularly in the areas of health, agriculture, food security, education, water, energy and biodiversity (Independent Group of Scientists, 2019).

Rapid economic growth, urbanisation and a growing population are driving increased consumption, which in turn leads to a significantly increase in global annual waste generation. Solid waste management is not only an individual but also a global issue. Waste affects human health and living conditions, the environment and economic

DOI: 10.1201/9780429352799-10

development. Poorly managed waste pollutes the oceans, creates barriers, transmits diseases, harms animals and increases the problems associated with incinerating waste. Greenhouse gasses from waste management are a major contributor to climate change. Plastic pollutions of all solid waste are particularly problematic (The World Bank, 2018). If not properly collected and managed, they will pollute and affect ecosystems, especially marine ecosystems, for thousands of years. In urban agglomerations, it is very important to improve waste management systems and adapt them to society's demand for a "closed loop" system. Some specific waste streams, such as plastics, can be reintegrated into the economy at the end-of-life stage (Nichols and Smith, 2019).

The way forward for the use of sustainable plastics is sustainable chemistry, i.e. the design, production and use of environmentally friendly chemical products and processes that prevent pollution, reduce or eliminate the use and generation of hazardous waste and reduce the risk to human health and the environment (OECD, 1999). Green chemistry is similarly defined, but sustainable chemistry is more of a scientific concept that aims to improve the efficiency of the use of natural resources to meet human needs for chemical products and services (The Organisation for Economic Co-operation and Development definition; Hogue, 2019). Sustainable chemistry is a broader view of the environment, i.e. the end-of-life of the products, not only from the synthesis side. Chemistry is a science but also an important industrial sector. It cannot change the behaviour and properties of chemicals under given conditions, but our decisions lead to specific environmental consequences, good or unwanted. Green chemistry is an approach requiring chemical synthesis that generates less waste, with less energy consumption and greater safety for workers and the environment (Zuin, 2021). The practice of sustainable chemistry goes beyond the environmental and human health roots of green chemistry towards a more systemic, life-cycle-based and interdisciplinary perspective for the design of new chemical and related processes. The end-of-life phase takes into account recycling or reuse and impacts related to waste management in order to develop a circular economy. A full life cycle assessment takes into account all impacts e.g. greenhouse gas equivalents to be quantified and assessed in terms of their cumulative effects on the material or the life cycle of the product (Constable, 2021). Only a small part of the exploited resources is transformed into desired products, leaving large amounts of waste and hazardous substances. So the problem is to provide humanity with enough food, energy, chemicals and materials in a sustainable way, without harming the environment (Song and Han, 2015).

Currently, bio-based and/or biodegradable plastics (bioplastics) account for around 1.5% of the total plastics production (PlasticsEurope, 2021). However, the demand for products, especially packaging made of sustainable plastics, is starting to increase due to greater consumer awareness, relevant legislation, risks concerning climate changes, decreasing biodiversity and the increasing economic and environmental costs of dependence on fossil (non-renewable) resources. Bio-based and/or (bio)degradable plastics as well as biocomposites can offer solutions that help companies remain competitive and meet growing consumer expectations by delivering more sustainable valuable products. Bioplastics contribute to more sustainable commercial plastic life cycles, as part of a circular economy, where polymers are made from renewable or recycled raw materials using carbon-neutral energy and, at the end-of-life, products are reused or recycled (Rosenboom et al., 2022).

A closed-loop economy is a concept that aims to rationally use resources and reduce the negative environmental impact of manufactured products, which should remain on the market as long as possible, and waste generation should be minimised as much as possible. Plastic end-of-life options include both recycling and energy recovery (including incineration). Recycling is a conscious effort to reduce the amount of polymer waste deposited in landfills by industrial use of it to create raw materials and energy again. The mechanical (reuse of waste as a full-valuable raw material for further processing), chemical (recycling of raw materials) or organic (composting and anaerobic digestion) recycling method is chosen depending on the polymer material involved, the origin of the waste, its possible toxicity and its flammability. Although (bio)degradable polymers are suitable for recycling by all methods, organic recycling is considered the most appropriate technology for the disposal of compostable waste as it is economically justified. It is designed for industrially compostable plastics, such as cellulose films, starch blends or polyesters. Biological processing of organic waste also leads to a reduction in landfills and thus methane emissions from them. If the absence of toxicity is added to their biodegradability, a biotechnology product of great industrial importance is obtained. Natural polymers and polymers from renewable raw materials, such as plant fibres, starch, cellulose, PLA and PHA, as well as biodegradable aliphatic-aromatic polyesters, fit well into the concept of a closed-loop economy and appear to be a good alternative to conventional plastics. The use of (bio)degradable polymers results in reduced environmental pollution and offers the possibility of closing the life cycle of products obtained from them (Khemani and Scholz, 2012; Rydz et al., 2015a; Rydz et al., 2015b; Achenie and Pavurala, 2018; Sikorska et al., 2021).

Blue growth, i.e. supporting the sustainable development of all marine and maritime sectors, has recently gained importance as an option aimed at improving the quality of marine ecosystems. It is a long-term strategy to support the sustainable development of the marine and maritime sectors, implemented by the European Commission to fully exploit the potential of Europe's oceans, seas and coasts as a driver of Europe's green economy with a high innovation potential and to improve competitiveness and the quality of jobs to achieve the goals of the Europe 2020 strategy for smart, sustainable and inclusive growth (Smart Specialisation Platform). Ocean policy and the principles of the blue economy are essential to achieving the transformation set out in the European Green Deal and to close the circularity of the European economy. The contribution of the oceans to the production of green energy, the greening of transportation and the sustainable production of food, i.e. the contribution of a healthy ocean, is therefore essential (COM(2021) 240 final).

In summary, the whole issue of sustainable development revolves around inter- and intragenerational justice based on the common good, complementarity and compromise of the environment, economy and society which should be realised through responsible human behaviour and actions in the international, national, community and individual arenas to uphold and promote the principles of this paradigm (Mensah, 2019).

REFERENCES

A/RES/70/1. 2015. *Transforming our world: The 2030 Agenda for Sustainable Development.* New York: General Assembly of United Nations.

Achenie, L.E.K., and Pavurala, N. 2018. On the modeling of oral drug delivery. In *Quantitative Systems Pharmacology: Models and Model-Based Systems with Applications*, *Computer-Aided Chemical Engineering*, vol. 42, ed. Manca, D., pp. 305–324. Cambridge: Elsevier.

COM(2018)673 final. *Communication from the Commission to the European Parliament, The Council, The European Economic and Social Committee and the Committee of the Regions. A sustainable Bioeconomy for Europe: Strengthening the Connection Between Economy, Society and the Environment.* Brussels: European Commission.

COM(2021)240 final. *Communication from the Commission to the European Parliament, The Council, The European Economic and Social Committee and the Committee of the Regions on a New Approach for a Sustainable Blue Economy in the EU Transforming the EU's Blue Economy for a Sustainable Future.* Brussels: European Commission.

Constable, D.J.C. 2021. Green and sustainable chemistry – The case for a systems-based, interdisciplinary approach. *iScience* 24: 103489.

DanuBioValNet Project. 2019. *The Bio-based Packaging Value Chain in the Danube Region.* Project co-funded by European Union funds (ERDF, IPA). www.interreg-danube.eu/danubiovalnet (accessed June 30, 2022).

Hogue, C. 2019. Differentiating between green chemistry and sustainable chemistry in Congress. In *Chemical & Engineering News*. Washington, DC: American Chemical Society. https://cen.acs.org/environment/green-chemistry/Differentiating-between-green-chemistry-sustainable/97/web/2019/07 (accessed June 23, 2022).

Independent Group of Scientists appointed by the Secretary-General, Global sustainable development report. 2019. *The future is now – Science for achieving sustainable development.* New York: United Nations.

Khemani, K., and Scholz, C. 2012. Introduction and overview of degradable and renewable polymers and materials. In *Degradable polymers and materials: principles and practice.* 1114, eds. Khemani, K., and Scholz, C. Washington, DC: American Chemical Society.

Mensah, J. 2019. Sustainable development: Meaning, history, principles, pillars, and implications for human action: Literature review. *Cogent. Soc. Sci.* 5(1): 1653531.

Nichols, W., and Smith, N. 2019. *Waste generation and recycling indices 2019 overview and findings.* Bath: Verisk Maplecroft.

OECD. 1999. Report of the survey on sustainable chemistry activities. *Venice: Proceedings of the OECD Workshop on Sustainable Chemistry.* www.oecd.org/officialdocuments/publicdisplaydocumentpdf/?doclanguage=en&cote=env/jm/mono(99)19/PART3 (accessed June 23, 2022).

PlasticsEurope 2021. Plastics – the facts 2021. *An Analysis of European Plastics Production, Demand and Waste Data.* https://plasticseurope.org/pl/knowledge-hub/plastics-the-facts-2021 (accessed April 19, 2022).

Rosenboom, J.-G., Langer, R., and Traverso, G. 2022. Bioplastics for a circular economy. *Nat. Rev. Mater.* 7: 117–137.

Rydz, J., Sikorska, W., Kyulavska, M., and Christova, D. 2015a. Polyester-based (bio)degradable polymers as environmentally friendly materials for sustainable development. *Int. J. Mol. Sci.* 16 (1): 564–596.

Rydz, J., Zawidlak-Węgrzyńska, B., and Christova, D. 2015b. Degradable polymers. In *Encyclopedia of biomedical polymers and polymeric biomaterials*, ed. Mishra, M.K., pp. 2327–2349. Boca Raton: CRC Press.

Sikorska, W., Musioł, M., Zawidlak-Węgrzyńska, B., and Rydz, J. 2021. End-of-life options for (bio)degradable polymers in the circular economy. *Adv. Polym. Technol.* 2021: 6695140.

Smart Specialisation Platform. *Blue growth.* https://s3platform.jrc.ec.europa.eu/blue-growth (accessed June 30, 2022).

Song, J., and Han, B. 2015. Green chemistry: A tool for the sustainable development of the chemical industry, *Natl. Sci. Rev.* 2(3): 255–256.

The World Bank. 2018. *What a waste: An updated look into the future of solid waste management.* www.worldbank.org/en/news/immersive-story/2018/09/20/what-a-waste-an-updated-look-into-the-future-of-solid-waste-management (accessed January 27, 2020)

Zuin, V.G., Eilks, I., Elschami M., and Kümmerer K. 2021. Education in green chemistry and in sustainable chemistry: Perspectives towards sustainability. *Green Chem.* 23: 1594–1608.

Index

For Product Safety Concerns and Information please contact our EU
representative GPSR@taylorandfrancis.com
Taylor & Francis Verlag GmbH, Kaufingerstraße 24, 80331 München, Germany